always up to date

The law changes, but Nolo is always on top of it! We offer several ways to make sure you and your Nolo products are always up to date:

1 **Nolo's Legal Updater**

We'll send you an email whenever a new edition of your book is published! Sign up at **www.nolo.com/legalupdater**.

2 **Updates @ Nolo.com**

Check **www.nolo.com/update** to find recent changes in the law that affect the current edition of your book.

3 **Nolo Customer Service**

To make sure that this edition of the book is the most recent one, call us at **800-728-3555** and ask one of our friendly customer service representatives. Or find out at **www.nolo.com**.

5th edition

Consultant & Independent Contractor Agreements

By attorney Stephen Fishman

Fifth Edition	OCTOBER 2005
Editors	JANET PORTMAN, LISA GUERIN, AND AMY DELPO
Cover Design	TONI IHARA
Illustrations	MARI STEIN
Book Design	TERRI HEARSH
Production	SARAH HINMAN
Proofreading	MARTIN ARONSON
CD-ROM Preparation	DOUG VARN
Index	ELLEN SHERRON
Printing	DELTA PRINTING SOLUTIONS, INC.

Fishman, Stephen.
    Consultant & independent contractor agreements / by Stephen Fishman.--5th ed.
      p. cm.
    ISBN 1-4133-0373-0 (alk. paper)
     1. Contracts for work and labor--United States. 2. Consulting contracts--United States.
3. Independent contractors--Legal status, laws, etc.--United States. I. Title: Consultant
and independent contractor agreements. II. Title.

KF898.F567 2005
346.7302'4--dc22

2005050861

For information on bulk purchases or corporate premium sales, please contact the Special
Sales Department. For academic sales or textbook adoptions, ask for Academic Sales. Call
800-955-4775 or write to Nolo, 950 Parker Street, Berkeley, CA 94710.

Acknowledgments

Many thanks to:

Janet Portman, Amy DelPo, and Lisa Guerin for their outstanding editing.

Jake Warner for his help.

Terri Hearsh and Sarah Hinman for the outstanding book design and production.

Ely Newman and André Zivkovich for disk preparation.

Ellen Sherron for the thorough index.

Martin Aronson for proofreading.

About the Author

Stephen Fishman is an attorney in San Francisco who has been writing about the law for more than 20 years. His many publications include *The Copyright Handbook: How to Protect & Use Written Works; Working for Yourself: Law & Taxes for Independent Contractors, Freelancers & Consultants;* and *Working With Independent Contractors,* all published by Nolo.

Table of Contents

A1   Appendix I: How to Use the CD-ROM

A2   Appendix II: Sample Agreements for Use by Hiring Firm

A3   Appendix III: Sample Agreements for Use by IC

1

How to Use This Book

This book contains agreements for people who hire consultants and independent contractors (ICs), and for people who work as such. This chapter explains how to use this book to put together agreements that will serve your business needs, legally and efficiently.

If you need some legal background on hiring independent contractors, refer to Chapter 2. If you would like an introduction to the legalities of working as a consultant or IC, refer to Chapter 3.

 This book is intended to serve as a supplement to two books that provide detailed guidance on all the legal issues involved in either hiring ICs or working as one. These are:

- *Working With Independent Contractors*, and
- *Working for Yourself: Law & Taxes for Independent Contractors, Freelancers & Consultants.*

Both books were written by Stephen Fishman and are published by Nolo.

At the back of this book, you will find a CD-ROM that contains all of the forms discussed in this book. In Appendix I of this book, you will find directions for using the CD-ROM.

A. Why Use Written Agreements

Using written independent contractor or consulting agreements benefits the independent contractors and consultants, as well as the clients who hire them. For the sake of simplicity, the terms "consultant" and "IC" are often used interchangeably in this book.

1. Oral Agreements Are Legal But Dangerous

Most contracts need never be written down to be legally valid. (For exceptions to this general rule, see "Some Agreements Must Be in Writing," below.) For example, a client and an IC can enter into a contract over the phone or during a lunch meeting at a restaurant. No magic words need be spoken, and nothing has to be written on a piece of paper. They just have to agree that the IC will perform services for the client in exchange for something of value—usually money. Theoretically, most oral agreements are as valid as a 50-page contract drafted by a high-powered law firm.

Example: Gary, a freelance translator, receives a phone call from a vice president of Acme Oil Co. He asks Gary to translate some Russian oil industry documents for $2,000. Gary says he'll do the work for the price. Gary and Acme have a valid oral contract.

In the real world, however, using oral agreements is like driving without a seatbelt—there's no problem as long as you don't have an accident, but if you do have an accident, you'll wish you had buckled up. An oral IC agreement can work just fine provided that you and the other party remember the contract's terms in the same way and fulfill them as expected by the other party.

Unfortunately, things don't always work so perfectly. Courts are crowded with lawsuits filed by people who entered into oral agreements with each other. Costly misunderstandings can develop if an IC performs services without a clear written description of what he or she is supposed to do and what will happen if it isn't done. Such misunderstandings may be innocent—you and the other party may have misinterpreted each other or failed to listen carefully. Or they may be purposeful—without a written document to prove otherwise, the other side can claim that you orally agreed to anything.

A good written IC agreement is your legal lifeline. If properly drafted, it will help prevent disputes by making clear exactly what's been agreed to. If problems develop, it will provide ways to solve them. If you and the other party end up in court, it will establish your legal duties to each other.

SOME AGREEMENTS MUST BE IN WRITING

Some types of agreements must be in writing to be legally enforceable. Each state has a law, usually called the "statute of frauds," that lists the types of contracts that must be written to be valid. A typical list includes:

- any contract that cannot possibly be performed in less than one year

    Example: John is hired to perform consulting services for the next two years for $2,000 per month. Since the agreement cannot be performed in less than one year, it must be in writing to be legally enforceable.

- contracts for the sales of goods—that is, tangible personal property, such as a computer or car—worth $500 or more

- a promise to pay someone else's debt

    Example: John is hired to perform consulting services for Acme Corporation. John is worried he won't be paid on time, so Sheila, Acme's president, personally guarantees John's payment—that is, she promises to pay John out of her own pocket if Acme Corporation doesn't. The guarantee must be in writing to be legally enforceable.

- contracts involving the sale of real estate
- real estate leases lasting more than one year, and
- any transfer of copyright ownership. (Copyright issues are covered in Chapter 11.)

2. Advantages of Written Agreements

Written agreements do more for you than help to avoid misunderstandings or dishonest dealings. There are important practical reasons why you should always sign a written IC agreement before work begins.

a. Defining projects

The process of deciding what to include in an agreement forces both the IC and the client to think carefully, perhaps for the first time, about exactly what the IC is supposed to do. Hazy or ill-defined ideas or expectations stand out when they're reduced to writing, spurring further discussion and negotiation until they are reduced to a concrete contract specification of the work that the IC will perform. This gives both sides a yardstick by which to measure the IC's performance and is the best way to avoid later disputes about whether the IC has performed adequately.

b. Establishing IC status

A well-drafted IC agreement will also help establish that the worker is an IC, not the client's employee. This is vital both for the client and the IC. (See Chapter 2, Section B, for a detailed discussion of the importance of distinguishing employees from independent workers.) However, a written IC agreement is not a magic legal bullet. It will *never* by itself turn an employee into an IC. What really counts is how the worker is actually treated, not a formality like signing an agreement. (See Chapter 2 for information on how to treat a worker as an IC.) Still, a written agreement can weigh in favor of IC status.

c. Assuring payment terms

If you're an IC, a written agreement clearly setting out your fees will help ward off disputes about how much the client agreed to pay you. If a client fails to pay and you have to sue for your fee, the written agreement will be proof of how much you're owed. Relying on an oral agreement with a client can make it very difficult for you to get paid in full or at all. Conversely, clients will want a clear fee understanding to avoid claims of underpayment.

B. Agreements Contained in This Book

This book contains a number of consultant and independent contractor agreements you can adapt to fit your needs. It includes:

- two general-purpose IC agreements that can be used by almost anyone who hires an IC or works as one (see Chapter 5), and

- several different agreements tailored to specific kinds of work ICs commonly perform, including:
  - ✓ independent consulting (see Chapter 6)
  - ✓ household work (see Chapter 7)
  - ✓ sales (see Chapter 8)
  - ✓ accounting and bookkeeping (see Chapter 9)
  - ✓ computer software consulting (see Chapter 10)
  - ✓ creative work, such as writing, illustrating, graphic art, and music (see Chapter 11)
  - ✓ construction (see Chapter 12), and
  - ✓ messenger, courier, and delivery services (see Chapter 13).

You'll find two different agreements for most of the above categories—one for use by the hiring firm and another for use by the IC. This was done so that each agreement could best protect the interests of the person or firm using it. However, none of the agreements is unduly one-sided.

Each chapter contains a detailed description of all the provisions in the agreement it covers. Many of the provisions contain alternatives you may choose from, such as whether the IC will be paid by the hour or a fixed fee. Or they may require you to provide additional information, such

as an address. Be sure to read the instructions carefully as you complete your agreement.

All of the agreements are included on the CD-ROM in the back of the book. (See Appendix I for a discussion of how to use the CD-ROM.) If you do not have access to a computer, you can use the tear-out sample forms in Appendix II (forms for use by the hiring firm) or Appendix III (forms for use by the independent contractor). You may wish to photocopy the original and save it for next time.

C. Putting Your Agreement Together

Make sure your agreement is properly signed and put together. If it isn't, it might not be legally valid. This is not difficult if you know what to do. This section provides all the instructions you need to get it right.

1. Signatures

It's best for both parties to sign the agreement and to do it in ink. The two of you need not be together when you sign, and it isn't necessary to sign at the same time. There's no legal requirement that the signatures be located in any specific place in a business contract, but they are customarily placed at the end of the agreement—that helps signify that both parties have read and agreed to the entire document.

It's very important that both parties sign the agreement properly. Failure to do so can have drastic consequences. How to sign depends on the legal form of the business of the person signing.

In this section, we take a look at the possible legal forms and their effects on signature requirements.

NO INITIALS NEEDED

It's not necessary that both parties initial every page of an agreement. This is sometimes done to make it more difficult for one party to remove a page and add a new one with different terms without the other side knowing about it. If you're afraid the other side might do something like this, you can insist on initialing the pages. Otherwise, save your ink for the signature page.

a.   Sole proprietorships

A person is a sole proprietor if he or she is running a one-person business and hasn't incorporated or formed a limited liability company. The vast majority of ICs and consultants are sole proprietors, as are many clients.

If you or the other party is a sole proprietor, you can each simply sign your own name and nothing more. That's because a sole proprietorship, unlike a corporation or partnership, is not a separate legal entity from the person who owns it. Therefore, if Susie Davis runs custom shopping tours for wealthy tourists, for example, her agreements with her clients can be signed "Susie Davis."

However, if you use a fictitious business name, it's best for you to sign on behalf of your business.

Example: Chris Kraft is an IC sole proprietor who runs a marketing research business. Instead of using his own name for the business, he calls it AAA Marketing Research. He should sign his contracts like this:

AAA Marketing Research
By: _____
    Chris Kraft

b.   Partnerships

If either you or the other party is a partnership, you must sign the agreement on behalf of the partnership, which means that the partnership must be identified in the signature block. Identifying the partnership is very important: If a partner signs only his or her name without mentioning the partnership, the partnership is not bound by the agreement—only the individual partner will be bound. This means that you couldn't go after the partnership's money or assets if the signing partner breaches the agreement. Instead, you could obtain only the signing partner's assets, which will likely be less than the partnership's.

Conversely, if you're a partner in a partnership and mistakenly sign an agreement as an individual, you're setting yourself up as the legal target if something goes wrong and the other side decides to sue. Since the other side won't be able to sue the partnership, it will look solely to you for legal recourse.

Which partner should sign? If the partnership is a general partnership (every partner invests and participates in managing the business), any partner can sign. But some partnerships are limited partnerships, which means that there is at least one general partner, but also some partners who invest in but don't participate in the business. Limited partners should never sign agreements—by law, they have no authority to bind the partnership. The agreement should always be signed by a general partner.

Only one partner needs to sign. The signature block for the partnership should state the partnership's name and the name and title of the person signing on the partnership's behalf.

Example: The Argus Partnership contracts with Sam for marketing research. Randy Argus is one of the general partners. He signs the contract on the partnership's behalf like this:

The Argus Partnership
A Michigan Partnership
By:

    Randy Argus, a General Partner

It's possible for a person who is not a partner to legally sign on behalf of a partnership. In this circumstance, the signature should be accompanied by a partnership resolution stating that the person signing the agreement is authorized to do so. The partnership resolution is a document signed by one or more of the general partners stating that the person named has the authority to sign contracts on the partnership's behalf. Attach the resolution to the end of the agreement.

c.   Corporations

If either you or the other party is a corporation, the agreement must be signed by someone who has authority to sign contracts on the corporation's behalf. The corporation's president or chief executive officer (CEO) is presumed to have this authority.

If someone other than the president or CEO of a corporation signs—for example, the vice president, treasurer, or other corporate officer—ask to see a board of directors' resolution or corporate bylaws authorizing that individual to sign. If the person signing doesn't have authority, the corporation won't be legally bound by the contract. Attach the resolution to the end of the agreement.

Keep in mind that if you own a corporation and you sign personally instead of on your corporation's behalf, you'll be personally liable for the contract. It's likely that the main reason you've gone to the trouble to form a corporation is to avoid such liability. So signing improperly is self-defeating.

The signature block for a corporation should state the name of the corporation and the name and title of the person signing on the corporation's behalf.

Example: Susan Ericson is the president of Kiddie Krafts, Inc. Since she is the president, any contracts signed by her need not be accompanied by a corporate resolution showing she has authority to bind the corporation. The signature block for contracts she signs should look like this:

Kiddie Krafts, Inc.
A California Corporation
By: _____

    Susan Ericson, President

d.   Limited liability companies

The owners of limited liability companies are called members. Members may hire others, called managers, to run their company for them. An agreement with a limited liability company should be signed by a member or manager.

Example: AcmeSoft LLC, a limited liability company, hires Sally to perform freelance programming services. The contract is signed on AcmeSoft's behalf by Edward Smith, the company's manager. The signature block should appear in the contract like this:

AcmeSoft LLC
A California Limited Liability Company
By: _____

    Edward Smith, Manager

2.   Dates

When you sign an agreement, include the date and make sure the other party does, too. You can simply put a date line next to the place where each person signs—for example:

_____ Date:_____
Jamie Alvarez

You and the other party don't have to sign on the same day. Indeed, you can sign weeks apart. However, unless the agreement provides a specific starting date, it goes into effect on the date it's signed. If the parties sign on different dates, the agreement begins on the date the last person signed. Until the agreement is signed by both parties, neither is bound by it.

3.   Attachments or Exhibits

An easy way to keep an agreement focused on the essential aspects of your arrangement is to use attachments, also called exhibits, to list lengthy details such as performance specifications. (You can also use attachments when someone who normally wouldn't have authority to sign has been given that power by the partnership or corporation, as explained above.) Putting practical details in a separate but attached document makes the main body of the contract shorter and easier to follow.

If you have more than one attachment or exhibit, they should be numbered or lettered—for example, Attachment 1 or Exhibit A. The attachments don't have to be dated or signed. Be sure that the main body of the agreement mentions that the attachments or exhibits are included as part of the contract.

4.   Altering the Agreement

Sometimes it's necessary to make last-minute changes to a contract just before it's signed. If you use a computer to prepare the agreement, it's best to make the changes on the computer and print out a new agreement.

However, it's not legally necessary to prepare a new contract. Instead, the changes may be handwritten or typed onto all existing copies of the agreement. If you use this approach, be sure that all those signing the agreement also sign their initials as close as possible to the place where the change is made. If both people who sign the entire document don't also initial each change, questions might arise as to whether both parties really agreed to the change.

5.   Copies of the Agreement

Prepare at least two copies of your agreement. Make sure that each copy contains all the needed exhibits and attachments. Both you and the other party should sign both copies—and each should keep one signed original.

Be sure to keep your copy of the agreement in a safe place. If you're hiring an IC, set up a separate vendor file for the IC. In this file, keep the IC agreement, the IC's invoices, copies of IRS Form 1099 and any other information that shows the worker is operating an independent business. Don't keep independent contractor records with your employee personnel records.

6.   Faxing and Emailing Agreements

It has become very common for people doing business with each other to communicate by fax machine or email. Especially when you are hashing out the details of contract clauses, it is very convenient to fax or email drafts back and forth. However, there are some potential problems with using faxes and email.

a.   Preserving confidentiality

One problem is preserving confidentiality. You never know who might end up receiving or reading a fax or email message. For this reason, it's wise to place a confidentiality legend such as the following on your faxes or emails:

The messages and documents transmitted with this notice contain confidential information belonging to the sender.

If you are not the intended recipient of this information, you are hereby notified that any disclosure, copying, distribution, or use of the information is strictly prohibited. If you have received this transmission in error, please notify the sender immediately.

This legend can be placed on a fax cover sheet or at the beginning of an email message.

If your negotiations are particularly sensitive, you may wish to encrypt your email messages so others can't read them. Inexpensive encryption programs such as PGP (Pretty Good Privacy) are readily available. To learn how to obtain both free and commercial versions of PGP, go to http://cryptography.org/getpgp.htm.

b.   Signing the finished agreement

When a final agreement is reached, you'll both need to sign the contract. One approach is for one party to sign the contract and fax it to the other, who signs it, makes a photocopy, and faxes it back. This way, you each have a copy of the agreement signed by both sides.

A faxed signature is probably legally sufficient if neither party disputes that it is a fax of an original signature. However, if the other party claims that a faxed signature was forged, it could be difficult or impossible to prove it's genuine, since it is very easy to forge a faxed signature with modern computer technology. Forgery claims are rare, however, so this is usually not a problem. Even so, it's a good practice for you and the other party to follow up the fax by exchanging signed originals via mail or air express.

If both you and the other party are advanced computer users, you can "sign" the contract digitally and send it by email. This involves using encryption technology to create a digital signature. This is a technically complex subject beyond the scope of this book. You can find a good deal of material on digital signatures on the Internet. A good place to search is www.perkinscoie.com. Put "digital signature" into the search field, and

then double click on "Digital Signature Law and Policy Resource Center." There is also a comprehensive book on digital signature law called *The Law of Electronic Commerce*, by Benjamin Wright (Aspen Law and Business).

7.   Cover Letter

Although not necessary, it can be helpful to include a cover letter when you send the agreement to the independent contractor to sign. You can see a sample of such a letter below. It reminds the independent contractor that he or she is not entitled to unemployment or other employee benefits and should never refer to himself or herself as your employee. Ask the independent contractor to sign the letter along with the independent contractor agreement and return both to you.

 The cover letter for use by hiring firms is on the forms disk under the file name LETTER. A tear-out copy is included in Appendix II.

D.   Changing the Agreement After It's Signed

No contract is engraved in stone. You and the other party can always modify or amend your contract if you both agree to the changes. You can even agree to call the whole thing off and cancel your agreement.

The key to changing a contract is cooperation. Neither party is ever obligated to accept a proposed modification to a contract. Either of you can say no to the proposed change and accept the consequences—for example, the other side may go ahead with a unilateral change or stop performing altogether. Either way, you may end up with a court battle over breaking the original contract. You're usually better off reaching some sort of accommodation with the other side, unless the person is totally unreasonable.

Sample Cover Letter to Independent Contractor

August 1, 20xx

Joe Contractor
123 Main St
Anytown, TX 12345

Dear Mr. Contractor:

Enclosed, please find the independent contractor agreement for your services. The agreement makes clear that you are an independent contractor (self-employed), and not an employee of Widget Company. Please read it carefully. If you have any questions about your work status, please do not hesitate to contact me.

Because you are an independent contractor, Widget Company will not withhold any taxes from your pay. You must pay all your state and federal taxes yourself. Usually, you'll have to pay estimated taxes four times a year.

In addition, Widget Company will not provide you with unemployment insurance coverage. When your services end, you will not be legally entitled to apply for unemployment benefits based on your term of service with Widget Company.

Failure to preserve your independent contractor status could prove costly not only to Widget Company but to you personally because it could result in your loss of valuable tax deductions. To help preserve your status, please do not identify yourself as a Widget Company employee, either orally or in writing on tax or other government forms, your business cards, letterhead, resume, marketing literature, or any other document. If you are asked what your status was while working with Widget Company, please state that you were a self-employed independent contractor.

Please inform Widget Company immediately if the IRS or other government agency contacts you regarding your work status while performing services for us.

Please sign the acknowledgement of this letter, below, and return it with a copy of the signed independent contractor agreement.

Very truly yours,
Widget Company

by: _____
John Q. Widget

I have read and agree to be bound by the terms of this letter and the Independent Contractor Agreement.

Name of Contractor_____

By: _____

Date:_____

Unless your contract is one that must be in writing to be legally valid—for example, an agreement that can't be performed in less than one year—it can usually be modified by an oral agreement. In other words, you need not write down the changes.

Example: Art signs a contract with Zeno, who will build an addition to his house. Halfway through the project, Art decides that he wants Zeno to do some extra work not covered by their original agreement. Art and Zeno have a telephone conversation in which Zeno agrees to do the extra work for extra money. Although nothing is put in writing, their change to their original agreement is legally enforceable.

ICs and their clients change their contracts all the time and never write down the changes. The flexibility afforded by such an informal approach to contract amendments might be just what you want. However, misunderstandings and disputes often arise from this approach. It's always best to have some sort of writing showing what you've agreed to do. You can do this informally. For example, you can simply send a confirming letter following a telephone call with the other party summarizing the changes you both agreed to make. If the other side doesn't correct your letter in writing, the existence of the letter creates a legal presumption that your version of the contract change is correct. This will be very helpful if the other side later claims he or she never agreed to the change or that the change is set forth incorrectly in the letter. Be sure to keep a copy of the letter for your files.

Example: Janet, a much sought-after editor, agrees to perform editing services for Steve. Her written agreement with Steve provides that her work will be completed by April 1. How-

ever, Janet finds the project more time-consuming than she anticipated. She calls Steve and asks him for an extension. Steve agrees to give Janet until May 1 to complete the work. But in return, Janet agrees to a 5% reduction in her fee for the work. Janet sends Steve the following confirming letter setting forth this contract change:

March 15, 20XX

Steve Blair
100 Main Street
Marred Vista, CA 90000

Dear Steve:

This letter serves to confirm our phone conversation of March 14, 20XX. You agree to extend the deadline for completion of my editing work on your book "The History of Sparta" until May 1, 20XX. In return, I shall reduce my fee for the work by 5%. Instead of charging you $1,000, I will charge $950.

Thank you for your cooperation in this matter.

Like the Spartans, I remain with it or on it,

Janet Swift

If the change involves a contract provision that is very important, it's wise to insist on a written amendment signed by both parties. The amendment should set forth all the changes and state that the amendment takes precedence over the original contract provision. For example, an amendment specifying that payment will be by the piece instead of by the hour should state that the new arrangement replaces the original understanding.

 A tear-out form for a contract amendment is contained in Appendixes II and III, and on the CD-ROM. ■

Hiring Consultants and Independent Contractors

Many businesses routinely use the services of nonemployees, whether those nonemployees are called consultants, independent contractors, vendors, or nothing at all. For convenience, these people will be referred to in this chapter as independent contractors or ICs.

Businesses can benefit significantly by treating a worker as an independent contractor rather than as an employee. But there are risks as well. This chapter provides an overview of these risks and rewards, and discusses how government agencies determine whether workers qualify as ICs or employees.

For a detailed analysis of all the legal issues involved in working with independent contractors, see *Working With Independent Contractors,* by Stephen Fishman (Nolo).

A. Benefits and Drawbacks of Using Independent Contractors

A hiring firm that classifies workers as independent contractors reaps many financial and other benefits. But misclassifying an employee as an IC can also be costly.

1. Benefits of Using Independent Contractors

The main reason most businesses use independent contractors is probably to save money. Consider the following expenses (in addition to a salary) and paperwork that a business has to deal with when it hires an employee instead of an independent contractor:

- **Federal tax withholding.** The employer must withhold federal income tax from the wages paid to an employee. Each year, the employer must send the employee IRS Form W-2, which shows how much he or she earned and how much was withheld.

- **Social Security and Medicare taxes.** Both the employer and the employee have to pay a share of Social Security and Medicare taxes. These taxes must be paid with the withheld federal income tax.

- **Federal unemployment taxes.** Employers must pay federal unemployment taxes.

- **State taxes.** Employers must also pay state unemployment taxes and, in most states, withhold state income taxes from employees' paychecks.

- **Workers' compensation insurance.** Employers must usually provide workers' compensation insurance coverage for employees in case they are injured on the job.

- **Employment benefits.** Although they are not legally required to do so, most employers give their employees health insurance, sick leave, paid holidays, and vacations. More generous employers also provide pension benefits for their employees.

- **Office space and equipment.** An employer normally provides an employee with office space and whatever equipment necessary for the job.

All of these items add enormously to the cost of hiring and keeping an employee. Typically, more than one-third of all employee payroll costs go towards Social Security, unemployment insurance, health benefits, and vacation.

A business incurs none of these obligations when it hires an independent contractor instead of an employee. It need not withhold or pay any taxes. Perhaps most importantly, an independent contractor need not be provided with health insurance, a pension plan, or any other employee benefits.

A business that hires an IC need only report all payments to the IC by filing Form 1099-MISC with the IRS. Even this form need not be filed if an IC is incorporated or is paid less than $600 in a calendar year. (For a complete discussion of IRS tax reporting requirements for firms that work with ICs, refer to Nolo's *Working With Independent Contractors,* by Stephen Fishman.)

There is another important reason businesses often prefer to use independent contractors: to avoid making a long-term commitment to the worker. An independent contractor can be hired solely to accomplish a specific task, which allows a business to obtain specialized expertise only for a short time. The hiring firm need not go through the trauma, severance costs, and potential lawsuits brought on by laying off or firing an employee.

2. Drawbacks and Risks of Using Independent Contractors

You might now be thinking, "I'll never hire an employee again; I'll just use independent contractors." Before doing this, you should know that there are some substantial drawbacks and risks involved in using independent contractors.

a. Advantages of hiring employees

Even though there are many financial benefits to using ICs rather than employees, companies of course continue to hire employees. There are many good reasons for this, including the following:

- When you hire employees, you can provide them with detailed training on how to do the job just as you want it done. You can then closely supervise and otherwise control the way the employees perform on the job. This is something you can't do with an IC. Instead, ICs must be left alone to perform the agreed upon services without substantial help or interference from you. (See Section B, below.)

- Ordinarily, employees will not be working for your competitors while they're working for you. In contrast, an IC you hire today may go to work for a competitor tomorrow.

- When you hire employees, you can depend on having the same workers available day after day. Although it may not be as true today as it once was, generally, employees who are well treated can be relied upon to stick around for a while. In contrast, ICs have no loyalty to anyone but themselves and their own bottom line. Having workers constantly coming and going can be inconvenient and disruptive. And the quality of work you get from different ICs may be uneven.

b. Government scrutiny

Another reason businesses continue to hire employees is fear of the IRS and other government agencies. The IRS and most states want to see as many workers as possible classified as employees, not as independent contractors. This way, the IRS and states can immediately collect taxes based on automatic and involuntary payroll withholding, rather than waiting for ICs to pay estimated taxes voluntarily four times a year.

Because no taxes are deducted from their pay, ICs have many more opportunities to evade taxes than do employees. Moreover, substantial numbers of ICs habitually underreport their income to the IRS. This is something an employee can't do, because employers must report all employee payments to the IRS on IRS Form W-2. In short, the government gets more tax money faster when workers are treated as employees rather than as ICs.

If the IRS concludes that an employer has misclassified an employee as an independent contractor, it may impose substantial assessments, penalties, and interest against the employer. Being ordered to pay massive amounts of back taxes and penalties can easily put a small company out of business. (For a detailed discussion of IRS assessments for worker misclassification see Nolo's *Working With Independent Contractors,* by Stephen Fishman.)

An employer's woes do not necessarily end with the IRS. If the state version of the IRS or the state unemployment or workers' compensation agencies suspect that workers have been misclassified as ICs, they may also audit the employer and order that it pay back taxes or unemployment or workers' compensation insurance. (For a detailed discussion of state worker classification audits, see Nolo's *Working With Independent Contractors,* Chapters 9 and 10, by Stephen Fishman.)

B. Workers Who Qualify as Independent Contractors

When you first hire a worker, it's up to you to decide whether that person is an employee or an independent contractor. If you classify the worker as an IC and consistently treat the worker as such, you'll obtain the benefits outlined in Section A, above. However, your decision is subject to review by many government agencies, including:

- the IRS
- your state unemployment compensation agency
- your state workers' compensation agency
- your state tax department (if your state has income taxes), and
- the U.S. Department of Labor and the National Labor Relations Board.

These agencies work independently of each other. Your classification decisions may be questioned by one, many, or all of them. You want to be able to satisfy all agencies that any worker you classify as an IC really is one.

So when is a worker an IC? Quite simply, whenever the worker is running an independent business and you treat that person accordingly. Good examples of ICs are professionals with their own practices such as doctors, lawyers, dentists, and accountants. However, a person doesn't have to be a professional with an advanced college degree to be an IC. So long as a worker is running a

business and you treat the worker as an independent business, then the worker can be an IC.

To decide whether a worker is an independent businessperson or an employee, most government agencies assess the degree of control that the hiring firm has over the worker. Independent contractors maintain personal control over the way they do the work contracted for, including the details of when, where, and how the work is done. The hiring firm's control is limited to accepting or rejecting the final result of the work. An independent contractor is just that—independent.

Example: Ray, the owner of a new fast food joint called BetterBurger, hires Sal, a graphic designer, to design a logo for its advertisements, menus, and napkins. Even if he wanted to, Ray couldn't control the details of how Sal performed this service because Sal works at his own office miles away from BetterBurger. Sal provides his own tools, sets his own work schedule, and decides how to design the logo. Ray's control is limited to accepting or rejecting Sal's final result—that is, he can refuse to pay Sal if he doesn't like the design. Sal is clearly an IC, not Ray's employee.

On the other hand, if you have the right to control how the worker does the job, that worker is an employee. This is so whether or not you actually exercise that right—that is, it doesn't matter if you in fact decline to control the details of how the worker does the job.

And it makes no difference whether a person works only part time. Even a part-time worker will be considered an employee if the employer has the right to exercise control.

Example 1: Mary takes a job as a hamburger cook at BestBurger. BestBurger personnel train her carefully in how to make a Best-Burger hamburger, including the type and amount of ingredients to use, the temperature at which the hamburger should be cooked, and so forth. Once Mary starts work, BestBurger managers closely supervise how she does her job.

Mary is BestBurger's employee. Virtually every aspect of Mary's behavior on the job is under BestBurger control—including what time she arrives at and leaves work, when she takes her lunch break, what she wears, and the sequence of the tasks she must perform. If Mary proves to be an able and conscientious worker, her supervisors may not look over her shoulder very often, but they have the right to do so at any time.

Example 2: BestBurger has been discovered by the gourmet burger crowd, who are demanding a fitting end to their elegant meal. Best-Burger looks for a pastry chef and decides on Antoine. Antoine makes delectable pastries in his commercial kitchen for several restaurants in town. He agrees to deliver to BestBurger one dozen fruit tarts and 15 cakes every morning, but the ingredients and presentation are up to Antoine. Since he controls the production in every important respect, Antoine is an independent contractor.

IF YOU WANT TO BE A SUPERVISOR, YOU MUST BE AN EMPLOYER

When you classify workers as ICs, you relinquish your right to closely supervise them. ICs must be left more or less alone to perform the agreed-upon services without substantial guidance or interference.

Not all hiring firms are comfortable about granting such a high degree of autonomy to people who perform services for them. If you want to control how a worker performs, you should classify the worker as an employee.

Once you've decided that you are not going to have the right to control a worker, how do you make it clear that the person is a bona fide independent contractor? The government agencies you have to deal with can't look into your mind to see whether the right to control exists, nor will they merely take your word for it. They will rely primarily on indirect or circumstantial evidence indicating control or lack of it—for example, whether you provide a worker with tools and equipment, pay by the hour, or have the right to fire the worker. This is what government auditors will be asking you about if you're audited.

Government auditors examine a number of different factors to determine whether a hiring firm has the right to control a worker. Different agencies use different sets of factors. The IRS looks at 14 main factors. Other agencies examine 11 factors, others only three. (See "Worker Classification Factors," at the end of this section, for specific information about who requires what.) This section includes virtually every factor any auditor might consider.

As you read through this section, don't be overwhelmed—you don't need to memorize it, and it's not necessary to satisfy every factor for a worker to be considered an IC. How much is enough? The best answer we can give you is that the factors considered by the agency involved—be it the IRS, state unemployment compensation agency, or other agency—must weigh in favor of IC status. Obviously, the more factors that indicate IC status, the better off you'll be if you're audited.

FORMS TO HELP YOU CLASSIFY WORKERS

To supplement the information in this chapter, we provide two forms that will help you ensure that the workers you hire are independent contractors. These forms are not legally required, but they can be a big help—both for your own purposes and for defending your case with a government agency, should it come to that. If you use these forms, be sure to keep them in your independent contractor files.

- **Independent contractor questionnaire:** Ask prospective independent contractors to complete this questionnaire. It elicits the information that you will need to classify the worker, and it can serve as excellent evidence for the correctness of your classification. You can find the questionnaire on the CD-ROM under the file name QUESTION. A tear-out version is in Appendix II.
- **Documentation checklist:** This asks the independent contractor to provide you with copies of written records that will help show the worker is an independent contractor, not an employee. Few contractors will be able to provide all of the documents on the checklist, but the more you can get, the better. You can find the documentation checklist on the CD-ROM under the file name DOCCHECK. A tear-out version is in Appendix II.

For a detailed discussion of how various agencies determine whether someone is an IC or an employee, see *Working With Independent Contractors*, by Stephen Fishman (Nolo).

1. Making a Profit or Loss

Employees. Typically, employees are paid for their time and labor and have no liability for business expenses.

Independent Contractors. ICs can earn a profit or suffer a loss as a result of the services they provide. ICs are entrepreneurs. They make money if their businesses succeed but risk going broke if they fail. Whether ICs make money depends on how well they use their ingenuity, initiative, and judgment in conducting their business.

> **Example:** Jack retires from his job as an actuary for a large insurance company and decides to become a professional golfer. He purchases $2,000 worth of golf clubs and qualifies for the Senior PGA Golf Tour. He pays his own traveling expenses and entrance fees for tournaments. If Jack plays well, he will win prize money and may earn a profit from his golfing. But if he plays poorly, he may earn nothing and lose money. How much money Jack makes or loses is entirely up to him. Jack is an IC and proud of it!

2. Working on Specific Premises

Employees. Employees must work where their employers tell them, usually on the employer's premises.

Independent Contractors. Usually, ICs are able to choose where to perform their services.

Work at a location specified by a hiring firm implies control by the firm, especially where the work could be done elsewhere. A person working at a hiring firm's place of business is physically within the firm's direction and supervision. If the person can choose to work off the premises, obviously the firm has less control.

However, many ICs perform tasks that can be done only at the hiring firm's premises—for example, an IC painter or rug layer must perform the services on the client's premises. In this event, this factor is not considered in determining the worker's status.

3. Offering Services to the General Public

Employees. Employees offer their services solely to their employers.

Independent Contractors. ICs offer services to the general public.

Since they are independent businesspeople, normally ICs make their services available to the public. ICs often advertise and will work for anyone who agrees to their terms.

4. Right to Fire

Employees. Typically, an employee can be discharged by the employer at any time and for almost any reason.

Independent Contractors. An IC's relationship with a hiring firm can be terminated only according to the terms of their agreement.

If you have a right to fire a worker at any time for any reason or for no reason at all, government auditors may conclude that you have the right to control that worker. The ever-present threat of dismissal must inevitably cause a worker to follow your instructions and otherwise do your bidding. This type of control is not present when both you and the IC know you can't arbitrarily terminate the IC's services without risking a lawsuit for breach of contract.

5. Furnishing Tools and Materials

Employees. Typically, employees receive all the tools and materials necessary to do their jobs from their employers.

Independent Contractors. Typically, ICs furnish their own tools and materials.

The fact that a hiring firm furnishes tools and materials, such as computers and construction equipment, tends to show control because the firm can determine which tools the worker is to use and, at least to some extent, in what order and how they will be used.

Sometimes, however, ICs have to use a hiring firm's tools or materials. For example, a computer consultant may have to perform work on the hiring firm's computers. The fact that the tools are provided in such a situation should be irrelevant.

6. Method of Payment

Employees. Usually, employees are paid by unit of time.

Independent Contractors. Typically, ICs are paid a flat rate for a project.

Workers who are paid by unit of time—for example, by the hour, week, or month—are apt to be classed as employees. This is because the hiring firm assumes the risk that the services provided will be worth what the worker is paid. To protect its investment, the hiring firm demands the right to direct and control the worker's performance. In this way, the hiring firm makes sure it gets a day's work for a day's pay.

Generally, payment by the job or on a straight commission indicates that the worker is an IC. However, in many professions and trades, payment is customarily made by unit of time. For example, lawyers, accountants, and psychiatrists typically charge by the hour. Where this is the general practice, the method of payment factor will not be given great weight.

7. Working for More Than One Firm

Employees. Although employees can have more than one job at a time, employers can require loyalty and prevent employees from taking some alternative jobs.

Independent Contractors. Usually, ICs have multiple clients or customers.

Many employees have more than one job at a time. However, employees owe a duty of loyalty toward their employers—that is, employees cannot engage in activities that harm or disrupt the employer's business. This restricts employees' outside activities. For example, ordinarily an employee wouldn't be permitted to take a second job with a competitor of the first employer. An employee who did so would be subject to dismissal.

Generally, ICs are subject to no such restrictions. They can work for as many clients or customers as they want. Having more than one client or customer at a time is very strong evidence of IC status. People who work for several firms at the same time are generally ICs because they're usually free from control by any one of the firms.

8. Continuing Relationship

Employees. Employees have a continuing relationship with their employers.

Independent Contractors. Generally, ICs work on one project and then move on.

Although employees can be hired for short-term projects, this type of relationship is more typical of ICs. Typically, an employee works for the same employer month after month, year after year, sometimes decade after decade. Such a continuing relationship is one of the hallmarks of employment. Indeed, one of the main reasons businesses hire employees is to have workers available on a long-term basis.

9. Investment in Equipment or Facilities

Employees. Generally, employees have no investment in equipment or facilities, both of which are owned by their employers.

Independent Contractors. ICs have an investment in the equipment and facilities appropriate for their businesses.

This factor includes equipment and premises necessary for the work, such as office space, furniture, and machinery. It does not include tools, instruments, and clothing commonly provided by employees in their trade—for example, uniforms that are commonly provided by the employees themselves. Nor does it include education, experience, or training.

When a worker has made a significant investment in the equipment and facilities he needs to perform his services, he has a good argument that he should be considered an IC. By making such a financial investment, the worker risks losing it if the business is not profitable. Also, the worker is not dependent upon a hiring firm for the tools and facilities needed to do the work. Owning the tools and facilities also implies that the worker has the right to control their use.

On the other hand, lack of investment indicates dependence on the hiring firm for tools and facilities; this is another hallmark of an employer-employee relationship.

Some types of employees typically provide their own inexpensive tools. For example, carpenters may use their own hammers, and accountants may provide their own calculators. Providing such inexpensive tools doesn't show that a worker who is otherwise an employee is an IC instead. But a worker who provides a $3,000 computer or $10,000 lathe is more likely to be considered an IC.

10. Business or Traveling Expenses

Employees. Employees' job-related business and traveling expenses are paid by the employer.

Independent Contractors. Typically, ICs pay their own business and traveling expenses.

If the hiring firm pays a worker's business and traveling expenses, the worker will often be viewed as an employee. To be able to control such expenses, the employer must retain the right to regulate and direct the worker's actions.

On the other hand, people who are paid per project and who pay expenses out of their own pockets are generally ICs. Workers who are accountable only to themselves for expenses are free to work according to individual methods and means—the hallmark of an IC.

Of course, some ICs typically bill their clients for certain expenses. For example, accountants tend to bill clients for travel, photocopying, and other incidental expenses. This does not in itself make them employees, since their clients do not control them.

11. Right to Quit

Employees. Normally, an employee may quit the job at any time without incurring any liability to the employer.

Independent Contractors. ICs are legally obligated to complete the work they agreed to do.

Normally, employees work "at will." They can quit whenever they want to without incurring liability, even if it costs the employer substantial money and inconvenience. And, conversely, they can be fired at will (see Section 4, above).

ICs are legally obligated to complete the job they signed on to do. If they don't, they are liable to the hiring firm for any losses caused by their stoppage.

Example: The Lazy Eight Motel hires John, a licensed building contractor, to construct a new wing. John agrees to complete the work by May 1. Halfway through construction, John receives a much more lucrative offer to work on a new office building project. John stops work and abandons the project. Lazy Eight is forced to hire a new contractor to complete the wing. As a result, completion of the wing is delayed by several months. John is liable to Lazy Eight for the revenue it lost due to his failure to complete the wing as agreed.

12. Instructions

Employees. Employers have the right to give their employees oral or written instructions that the employees must obey concerning when, where, and how they are to work.

Independent Contractors. ICs need not comply with instructions on how to perform their services; they decide on their own how to do their work. They are, however, obliged to deliver the final product as called for in the agreement or contract.

Watch out—this matter of instructions can be tricky. That's because there is no requirement that instructions actually be given to an employee. The IRS and other government agencies will focus instead on whether you have the right to give them. Even though you have not given a worker instructions, the IRS could conclude that you have set up the relationship so that you have the right to do so. The agency may view this power to instruct as an indication of employment status.

Fortunately, in many situations the issue of instructions won't be problematic for IC status. If a worker is running an independent business and you are just one client or customer among many, it's likely you don't have the right to give the worker instructions about how to perform the services. Your right is usually limited to accepting or rejecting the final results.

Example: The local AcmeBurger has a plumbing problem. The manager looks in the Yellow Pages and calls Jake the plumber to come and fix the problem. Jake shows up with his assistant Tony. The manager explains the problem, and Jake gives him an estimate of how much it will cost to fix it. The manager gives the go ahead, and Jake and Tony begin work. When the work is finished, the manager can refuse to pay if he thinks the plumbing has not been properly repaired. However, if he had presumed to tell Jake how to go about doing the plumbing repair, Jake would probably have told him to get lost and gone on to his next customer.

On the other hand, you probably will have the right to give instructions to workers who are not running an independent business and are largely or solely dependent upon you for their livelihood.

Example: Joe the tailor abandons his own tailor shop when he's hired to perform full-time tailoring services for Acme Suits, a large haberdashery chain. Joe is completely dependent upon Acme for his livelihood. Acme managers undoubtedly have the right to give Joe instructions, even if they don't feel the need to do so because Joe is such a good tailor.

Note that you may give an IC detailed guidelines as to the end results to achieve without destroying his status as an IC. For example, a software programmer may be given highly detailed specifications describing the software programs to be developed, or a building contractor may be given detailed blueprints showing precisely what the finished building should look like. Since these instructions relate only to the end results and not to how they must be accomplished, they do not turn the programmer and building contractor into employees.

13. Sequence of Work

Employees. Employees may be required to perform services in the order or sequence set for them by the employer.

Independent Contractors. ICs decide for themselves the order or sequence in which they work.

This factor is closely related to the right to give instructions. If a person must perform services in the order or sequence set by the hiring firm, it shows that the worker is not free to use discretion in working, but must follow established routines and schedules.

Often, because of the nature of the occupation, the hiring firm either does not set the order of the services or sets them infrequently. It is sufficient to show control, however, if the hiring firm retains the right to do so. For example, a salesperson who works on commission is usually permitted latitude in mapping out work activities. But normally the hiring firm has the discretion to require the salesperson to report to the office at specified times, follow up on leads, and perform certain tasks at certain times. Such requirements can interfere with and take precedence over the salesperson's own routines or plans. They indicate control by the hiring firm and employee status for the salesperson.

14. Training

Employees. Employees may receive training from their employers.

Independent Contractors. Ordinarily, ICs receive no training from those who purchase their services.

Training may be done by teaming a new worker with a more experienced one, by requiring attendance at meetings or seminars, or even by correspondence. Training shows control because it indicates that the employer wants the services performed a particular way. This is especially true if the training is given periodically or at frequent intervals.

Usually, ICs are hired precisely because they don't need any training. They possess special skills or proficiencies that the hiring firm's employees do not have.

15. Services Performed Personally

Employees. Employees are required to perform their services on their own—that is, they can't get someone else to do their jobs for them.

Independent Contractors. Generally, ICs are not required to render services personally; for example, they can have their own employees do all or part of the work.

Ordinarily, an IC doesn't have to do all the work personally. This is part and parcel of running a business. For example, if you hire an accountant to prepare your tax return, normally the accountant has the right to have employee assistants do all or part of the work under supervision—a process known as delegation. However, the law of contracts imposes significant restrictions on an IC's ability to delegate work that involves personal services—for example, painting a picture or creating another work of art.

Requiring someone whom you hire to personally perform all the work indicates that you want to control how the work is done, not just the end results. If you were just interested in end results, you wouldn't care who did the work; you'd just make sure the work was done right when it was finished.

16. Hiring Assistants

Employees. Although employees may supervise and oversee the work of assistants, these assistants are hired and paid by the employer—ultimately, they work for the employer, not the employee.

Independent Contractors. ICs hire, supervise, and pay their own assistants.

Usually, government auditors will be very impressed by the fact that a worker hires and pays assistants. This is something employees simply do not do and is strong evidence of IC status because it shows risk of loss if the IC's income cannot meet payroll expenses.

17. Set Working Hours

Employees. Ordinarily, employees have set hours of work.

Independent Contractors. ICs are masters of their own time; ordinarily they set their own work hours.

Obviously, telling a worker when to come to work and when to leave shows that you have control over that worker.

18. Working Full Time

Employees. An employee may be required to work full time at the employer's business.

Independent Contractors. ICs are free to work when and for whom they choose.

Requiring a worker to work full time indicates that you have control over how much time the worker spends on your job. In practice, it restricts that worker from working elsewhere.

19. Oral or Written Reports

Employees. Employees may be required to submit regular oral or written reports to the employer regarding the progress of their work.

Independent Contractors. Generally, ICs are not required to submit regular reports; they are responsible only for end results.

Submitting reports shows that the worker is compelled to account for individual actions. Reports are an important control device for an employer. They help determine whether directions are being followed or whether new instructions should be issued.

This factor focuses on regular reports detailing a worker's day-to-day performance. It does not include the common practice among ICs to submit infrequent interim reports to hiring firms when they are working on long or complex projects. These reports are typically tied to specific completion dates, timelines, or milestones written into the contract. For example, a building contractor may be required contractually to report to the hiring firm when each phase of a complex building project is completed. Submission of a report like this will not in itself make the contractor look like an employee in the eyes of the government.

20. Integration Into Business

Employees. Typically, employees provide services that are an integral part of the employer's day-to-day operations.

Independent Contractors. Typically, ICs' services are not part of the hiring firm's day-to-day business operation.

Integration in this context has nothing to do with race relations. It simply means that the workers are a regular part of the hiring firm's overall operations. According to most government auditors, the hiring firm would likely exercise control over such workers because they are so important to the success of the business.

> **Example:** Fry King is a fast-food outlet. It employs 15 workers per shift who prepare and sell the food. Jean is one of the workers on the night shift. Her job is to prepare all the french fries for the shift. Fry King would likely go out of business if it didn't have someone to prepare the french fries. Since french fry preparation is a regular or integral part of Fry King's daily business operations, Jean's work is that of an employee.

By contrast, ICs generally have special skills that the hiring firm calls upon only sporadically.

> **Example:** Over the course of a year, Fry King hires a painter to paint its business premises, a lawyer to handle a lawsuit by a customer who suffered from food poisoning, and an accountant to prepare a tax return. All of these things may be important or even essential to Fry King—otherwise it wouldn't have them done—but they are not a part of Fry King's normal overall daily operations of selling fast food.

21. Skill Required

Employees. Workers whose jobs require a low level of skill and experience are more likely to be employees.

Independent Contractors. Workers with jobs requiring complex skills are more likely to be ICs.

The skill required to do a job is a good indicator of whether the hiring firm has the right to control a worker. This is because you are far more likely to have control over the way low-skill workers do their jobs than you do over the way high-skill workers do their jobs. For example, if you hire a highly skilled repair person to maintain an expensive and complex photocopier, you probably don't know enough about photocopiers to supervise the work or even tell the repair person what to do. All you are able to figure out is whether the repair person's results meet your requirements—that is, whether the photocopier works or not once the repairs are complete.

But when you hire a person to do a job that does not require high skills or training, such as answering telephones or cleaning offices, you are normally well qualified to supervise the work and give step-by-step directions. Generally, workers in such occupations expect to be controlled by the person who pays them—that is, they expect to be given specific instructions as to how to work, to be required to work during set hours, and to be provided with tools and equipment.

For these reasons, highly skilled workers are far more likely to be ICs than low-skill workers. But having an advanced professional degree or possessing unique skills does not in itself make that person an IC. The question of control must always be addressed. Corporate officers, doctors, and lawyers can be employees just like janitors and other manual laborers if they are subject to a hiring firm's control.

> **Example:** Dr. Welby leaves his lucrative solo medical practice to take a salaried position teaching medicine at the local medical school. When Dr. Welby ran his own practice, he was an IC in business for himself. He paid all the expenses for his medical practice and collected all the fees. If the expenses exceeded the fees, he lost money.
>
> As soon as Dr. Welby took the teaching job, he became an employee of the medical school. The school pays him a regular salary and provides him with employee benefits, so he has no risk of loss as he did when he was in private practice. The school also has the right to exercise control over Welby's work activities—for example, requiring him to teach certain classes. It also supplies an office and all the equipment Welby needs. He is no longer in business for himself.

22.   Worker Benefits

Employees. Employees usually receive benefits such as health insurance, sick leave, pension benefits, and paid vacation.

Independent Contractors. ICs ordinarily receive no similar workplace benefits.

If you provide a worker with employee benefits such as health insurance, sick leave, pension benefits, and paid vacation, it's only logical for courts and government agencies to assume that you consider the worker to be your employee

subject to your control. To keep the benefits, it's likely that the worker would obey your orders. You'll have a very hard time convincing anyone that a person you provide with employee benefits is not your employee.

On the other hand, typically you are not required to provide an IC with any benefits other than payment for completing the work.

23.   Tax Treatment of the Worker

Employees. Usually, employees have federal and state payroll taxes withheld by their employers and remitted to the government.

Independent Contractors. Ordinarily, ICs pay their own taxes.

Ordinarily, employers must withhold and pay federal and state payroll taxes for their employees, including Social Security taxes and federal income taxes. A firm that hires an IC ordinarily need not remit or withhold any taxes for the worker.

Treating a worker as an employee for tax purposes—that is, remitting federal and state payroll taxes for the worker—is very strong evidence that you believe the worker to be your employee and you have the right to exercise control.

24.   Intent of the Hiring Firm and Worker

Employees. Normally, people who hire employees intend to create an employer-employee relationship.

Independent Contractors. Normally, people who hire ICs intend to create an IC–hiring firm relationship.

Some government agencies and courts also consider the intent of the hiring firm and worker. If it appears they honestly intended to create an IC relationship, it's likely that the hiring firm would neither believe it had control nor attempt to exercise control over the worker. One way to establish intent to create an IC relationship is for the hiring firm and IC to sign an independent contractor agreement like those in this book.

On the other hand, if it appears that you never intended to create a true IC relationship and merely classified the worker as an IC to avoid an employer's legal obligations, then you probably had the right to control the worker.

25. Custom in the Trade or Industry

Employees. Workers who are normally treated like employees in the trade or industry in which they work are likely to be employees.

Independent Contractors. Workers who are normally treated like ICs in the trade or industry in which they work are likely to be ICs.

The custom in the trade or industry involved is important as well. If the work is usually performed by employees, employee status is indicated. And workers treated customarily as ICs may get the benefit of that custom, as long as there is in fact no employee-like control exerted over them.

Example: The long-standing custom among logging companies in the Pacific Northwest is to treat tree fellers—people who cut down trees—as ICs. Customarily, they are paid by the tree, receive no employee benefits, and are free to work for many logging companies, not just one. None of the logging companies withholds or pays federal or state payroll taxes for tree fellers. The fellers pay their own self-employment taxes. This long-standing custom is strong evidence that the workers are ICs.

OTHER AGENCIES, OTHER TESTS

Not all government agencies use the right-of-control test described above to decide whether a worker is an IC or employee. Some workers' compensation and labor law agencies use an economic reality test. Under this test, workers are employees if they are economically dependent upon the businesses for which they render services. This difficult-to-apply test is designed to protect low-pay and low-skill workers—that is, to find that such workers are employees for purposes of workers' compensation and labor laws intended to help workers. (For a detailed discussion, see Nolo's *Working with Independent Contractors*, by Stephen Fishman.)

WORKER CLASSIFICATION FACTORS

The following chart lists the worker classification factors that various agencies consider. Some states use more than one test and so may be listed twice. Others use different sets of factors and so are not listed here.

	IRS	Unemployment Compensation Agencies in States Listed in Footnote 1	Unemployment Compensation Agencies in States Listed in Footnote 2	Workers' Compensation Agencies in States Listed in Footnote 3	Workers' Compensation Agencies in States Listed in Footnote 4	U.S. Labor Department	Copyright Ownership
No training	✔		✔		✔		
Assistants can do work			✔		✔		✔
Worker can realize profit or loss	✔		✔		✔	✔	
Work not done on hiring firm's premises			✔		✔		
No right to fire worker	✔		✔		✔	✔	
Worker offers services to public			✔		✔		
Worker furnishes tools and materials	✔		✔		✔		
Payment by the project	✔		✔		✔		✔
Worker has multiple clients or customers			✔		✔		✔
No continuing relationship with hiring firm			✔		✔	✔	✔
Worker has significant investment in equipment and facilities	✔		✔		✔	✔	
Worker pays own business expenses	✔		✔		✔		
Worker has no right to quit			✔		✔		
Worker sets order or sequence of work			✔		✔		
Worker provides own training			✔		✔		
Written agreement	✔						

	IRS	Unemployment Compensation Agencies in States Listed in Footnote 1	Unemployment Compensation Agencies in States Listed in Footnote 2	Workers' Compensation Agencies in States Listed in Footnote 3	Workers' Compensation Agencies in States Listed in Footnote 4	U.S. Labor Department	Copyright Ownership
Worker need not perform services personally			✔		✔		
Worker sets own hours			✔		✔		✔
Full-time work not required			✔		✔		
Reports not required			✔		✔		
Work outside hiring firm's usual business	✔	✔		✔	✔	✔	✔
Custom in the community			✔		✔		
Worker highly skilled			✔		✔		✔
Initiative, judgment needed to succeed			✔		✔	✔	
Parties' intent			✔		✔		
Worker not subject to hiring firm's control	✔	✔	✔		✔	✔	
Worker has independent trade or business		✔		✔	✔		✔
Worker's tax treatment			✔		✔		✔
Worker receives no employee benefits	✔		✔		✔		✔
Work typically not supervised			✔		✔		
No right to assign additional projects							✔

[1] Alaska, Arkansas, Connecticut, Delaware, Georgia, Hawaii, Illinois, Indiana, Louisiana, Maine, Maryland, Massachusetts, Nebraska, Nevada, New Hampshire, New Jersey, New Mexico, Rhode Island, Tennessee, Vermont, Washington, West Virginia.

[2] Alabama, Arizona, California, Florida, Iowa, Kentucky, Michigan, Minnesota, Mississippi, Missouri, New York, North Carolina, North Dakota, Ohio, South Carolina, Texas, Utah.

[3] Alaska, Arkansas, Hawaii, Iowa, Mississippi, Missouri, New Jersey, North Dakota.

[4] Alabama, Arizona, Arkansas, California, Colorado, Connecticut, Delaware, Florida, Georgia, Hawaii, Idaho, Illinois, Indiana, Iowa, Kansas, Kentucky, Louisiana, Maine, Maryland, Massachusetts, Minnesota, Mississippi, Missouri, Montana, Nebraska, Nevada, New Hampshire, New Jersey, New Mexico, New York, North Carolina, North Dakota, Ohio, Oklahoma, Pennsylvania, Rhode Island, South Carolina, South Dakota, Tennessee, Texas, Utah, Vermont, Virginia, West Virginia, Wyoming.

C. When Written IC Agreements Are Vital

You could view every worker in America as being somewhere on an IC-employee continuum. At one end are workers who are clearly ICs. You have little to fear from government auditors when you hire such workers.

Example: You start a new business and hire Malcolm, a local certified public accountant (CPA), to perform accounting and tax preparation services. Malcolm has a well-established CPA firm that represents dozens of clients. He has several employees and his own office, and he advertises in the local Yellow Pages. Malcolm is running an independent business. There is no way Malcolm will be viewed as your employee by the IRS or any other government agency.

At the other end of the continuum are workers who are clearly employees. No matter what you call them, such workers will be classified as employees by the IRS and other government auditors.

Example: You hire Shirley to perform bookkeeping services for your business. You pay Shirley a monthly salary, provide her with office space at your workplace, reimburse her for all her expenses, and provide her with health insurance and sick leave. Shirley works for you full time, and you closely supervise her work. Shirley is your employee. She is not running her own business and is obviously subject to your control on the job.

But in between these two extremes, there is a vast middle ground where work relationships have some elements of employment and others of independence.

Example: Instead of hiring Shirley to perform bookkeeping services for your business, you hire John. John has two other clients besides you. He works partly in your office and partly at home. You pay him by the hour and reimburse him for expenses. Is John an IC or your

part-time employee? It's hard to say for sure. This is something an IRS or other government auditor may question.

Obviously, if you want to hire an IC, you'd be better off hiring workers who clearly qualify as ICs. However, this is often not possible—perhaps the only person you can find to do the work doesn't have all the indicia that shout "IC" to a government auditor. When workers fall into this uncertain middle ground, it's most important to use a carefully drafted, written independent contractor agreement. While such an agreement can never make a worker qualify as an IC by itself, it can help tip the balance in favor of IC status in close cases.

However, don't forget that a written agreement saying the worker is an IC won't help you if it is abundantly clear that the worker is really an employee. A government auditor will disregard a written IC agreement that's a sham.

How will you know if your situation falls within the great middle ground? If you answer "yes" to any of the following questions, the worker may still qualify as an IC, but it's likely that the IRS or other government agency will question the worker's status if you're audited. In these situations, it's vital that you use a written independent contractor agreement.

- Does the worker work for you full time?
- Is the worker going to work for you for a substantial period of time?
- Do you pay or reimburse the worker's business or travel expenses?
- Do you pay the worker by the hour or other unit of time?
- Is the worker performing services you also pay employees to perform?
- Can you terminate the worker at any time?
- Do you provide the worker with materials, equipment, or facilities?
- Does the worker work at your premises even though the work does not require it?

- Do you require the worker to perform the services personally?
- Are similar workers customarily treated like employees in your trade or industry?
- Is the worker performing services that require little or no specialized skills?

You should also use a written agreement if you are hiring a real estate salesperson or commission a door-to-door salesperson who sells consumer products. A special federal law provides that using such agreements automatically establishes such workers as ICs for IRS purposes. (See Chapter 8 of this book for a door-to-door salesperson agreement.)

It is also highly advisable to use a written agreement if the worker is going to create a copyrightable work of authorship for you. This has nothing to do with the worker classification issue. Rather, a written agreement is necessary to clearly establish who will own the IC's work product. In the absence of such an agreement, ICs, not you, could own the work they create on your behalf. (See Chapter 11.) ■

Working as a Consultant or Independent Contractor

The term independent contractor is another way of describing people who are self-employed. Other words used to describe such people include consultant, freelancer, sole proprietor, and entrepreneur. Whatever they are called, they all have one thing in common: They are running their own independent businesses. For the sake of simplicity, we refer to all such people as ICs in this book.

This chapter provides an overview of the benefits and drawbacks of this status and explains why you should use written agreements with your clients.

For a complete discussion of all the legal issues involved in working as an independent contractor or consultant, see *Working for Yourself: Law & Taxes for Independent Contractors, Freelancers & Consultants*, by Stephen Fishman (Nolo).

A. Benefits of Working as an Independent Contractor

Being an IC can give you more freedom than employees have and can result in tax benefits.

1. You Are Your Own Boss

When you're an IC, you're the boss, with all the risks and rewards that entails. This freedom from control is very appealing to many ICs. In most situations, you set your own work schedules, hire your own employees or assistants, schedule vacations when you want them, and work for people ·or firms you choose.

However, you're not a completely free agent. (Who is?) Even though you're an IC, you are legally obligated to honor your promises to your clients. For example, if you promise to finish a project by a certain date, you are obligated to do so. If you fail to live up to your promises, you could be sued by your clients, and they could recover damages for any losses they incurred due to your nonperformance.

2. You May Earn More Than Employees

Often, you can earn more as an IC than as an employee in someone else's business. According to *The Wall Street Journal*, ICs are usually paid at least 20% to 40% more per hour than employees performing the same work. This is because hiring firms don't have to pay half of an ICs' Social Security taxes and unemployment compensation taxes, or provide workers' compensation coverage and employee benefits such as health insurance, paid vacation, and sick leave. Of course, how much you're paid is a matter for negotiation between you and your clients. ICs whose skills are in great demand may receive far more than employees doing similar work.

3. Tax Benefits

Being an IC also provides you with many tax benefits that employees don't have. For example, no federal or state taxes are withheld from your paychecks as they must be for employees. Instead, ICs normally pay estimated taxes directly to the IRS four times a year. This gives you more control over your income. Even more important, you can take advantage of many business-related tax deductions that are limited or not available at all for employees. This may include, for example, office expenses (including those for home offices), travel expenses, entertainment and meal expenses, equipment, insurance costs, and more. (For a detailed discussion of the tax benefits of being an IC, see Nolo's *Working for Yourself: Law & Taxes for Independent Contractors, Freelancers, & Consultants,* by Stephen Fishman.)

B. Drawbacks of IC Status

Despite its advantages, being an IC is no bed of roses. Here are some of the major drawbacks and pitfalls.

1. No Steady Income

As discussed above, one of the best things about being an IC is that you're on your own. But this can be one of the worst things about it as well. When you're an employee, you are entitled to be paid as long as you have your job, even if your employer's business is slow. This is not the case when you're an IC. If you don't have business, you don't make money.

2. No Employer-Provided Benefits

Although not required to by law, employers usually provide their employees with health insurance, paid vacations, and paid sick leave. More generous employers may also provide retirement benefits, bonuses, and even employee profit sharing. When you're an IC, you get no such benefits.

Also, ICs don't have the safety net provided by unemployment insurance, nor do hiring firms provide them with workers' compensation coverage. ICs can buy their own disability insurance to protect their income if they become sick or disabled, but such insurance is expensive and hard to obtain.

3. Risk of Not Being Paid

Some ICs have great difficulty getting their clients to pay them on time or at all. This is not a problem employees usually have because employers know that they can be fined by the state labor department if they don't pay their employees on time. In contrast, when you're an IC, no government agency will help you get paid. It's up to you to take whatever steps are necessary to get paid, including suing the client. When you're an IC, you bear the risk of loss from deadbeat clients.

4. Liability for Business Debts

If, like most ICs, you're a sole proprietor or partner in a partnership, you are personally liable for your business debts. An IC whose business or partnership fails could lose most of what he or she owns. You may be able to avoid personal liability for some business debts by incorporating or forming a limited liability company. (For a detailed discussion, see Nolo's *Working for Yourself: Law & Taxes for Independent Contractors, Freelancers & Consultants,* by Stephen Fishman.)

5. Tax Drawbacks

Finally, your tax returns will likely be far more complex when you work as an IC than they were when you worked as an employee. In addition, you'll have to pay all of your Social Security and Medicare taxes yourself. Hiring firms must pay half their employees' Social Security taxes but need make no contributions for ICs. This means you'll be paying a 15.3% Social Security and Medicare tax instead of the 7.65% that employees pay. You must pay these taxes directly to the IRS in the form of quarterly estimated taxes.

For many, keeping track of your quarterly earnings and paying the estimated tax is the single worst thing about being an IC. However, the business deductions you may qualify for because you're in business for yourself may offset this additional Social Security tax liability.

C. Who Decides Whether You're an IC?

Initially, it's up to you and each client or customer you contract with to decide whether you are going to work as an IC or an employee. However, as the old line goes, saying it doesn't make it so. Your decision is subject to review by numerous federal and state agencies, including the IRS, your state tax department, your state unemployment tax agency, and your state workers' compensation agency. These agencies will look at the details of your work relationship to determine whether your initial characterization of the relationship has been honored in reality.

1. How Misclassifications Are Discovered

The IRS or your state tax department might question your status in a routine audit of your tax returns. This is particularly likely if all or most of your income comes from just one source.

More commonly, however, you'll come to the government's attention indirectly, when it investigates the classification practices of a firm that hired you to do work. The IRS checks routinely to see whether workers have been misclassified as ICs whenever it audits a business. Worker status may also be questioned in audits by your state unemployment or workers' compensation agency.

Government auditors may question you and examine your and the hiring firm's records. Because the rules for determining whether you're an IC or employee are vague and subjective, it's often easy for the government to claim that you're an employee even though both you and the hiring firm sincerely believed you qualified as an IC. (See Chapter 2 for a discussion of worker classification rules.)

2. How Misclassification Can Hurt You

If the IRS or another government agency determines that you were misclassified as an IC, the company that hired you will have to pay assessments and penalties. No penalties or assessments will be imposed on you. However, this does not mean being reclassified as an employee won't adversely affect you. Unfortunate fall out might include:

- **Losing the job.** It's not uncommon for the hiring firm to dispense with the improperly classified IC's services because it doesn't want to pay the additional expenses involved in having to treat the IC as an employee. If the IC has a contract with the hiring firm calling for completion of a specific project, the firm would be legally required to honor the existing contract. But once the project is finished

and the contract ends, the firm would be free to refuse to hire the IC again. If the IC is performing services on an ongoing basis, the firm could terminate the relationship even more easily.

- **Losing the job as a casualty of settlement negotiations.** A common feature of an IRS settlement agreement with a hiring firm is to require the termination of contracts with ICs—with no input from the ICs. The IC may be able to sue the firm and obtain the amount it should have received under the contract, but he or she can forget about getting any additional work from the firm.

- **Getting paid less money.** The hiring firm may insist on reducing your compensation for new projects to make up for the extra expenses it will incur by treating you as an employee.

- **Adverse tax consequences.** An IRS determination that you should be classified as an employee can also have adverse tax consequences for you. For starters, you'll be told to file amended tax returns for the years involved. This extra work has one advantage: You'll be entitled to claim a refund for half the Social Security and Medicare taxes you paid on the compensation you received as a misclassified IC. (Since you were really an employee, you were supposed to pay only half of the Social Security tax due on your income.) The IRS will require the employer to pay the entire Social Security tax it should have paid all along, and you get a refund for the amount you overpaid.

Sounds like a windfall, right? Not so fast—you'll need to balance this found money against tax breaks you might have taken as an IC but must now forfeit. For example, if you took advantage of your ability as an IC to claim valuable business deductions for business expenses such as a home office, travel, and health insurance premiums, you'll find either that these are not deductible for employees at all or that the deductions are limited. You'll end up owing more taxes if the value of these lost deductions exceeds the amount of your Social Security and Medicare tax refund. Your employer will also have to start withholding Social Security and income taxes from your pay.

Even if being reclassified as an employee doesn't result in any financial losses, it's likely you'll be treated very differently on the job. For example, the hiring firm—now your employer—will probably expect you to follow its orders and may attempt to restrict you from working for other companies. Typically, employers don't like their employees to work for competitors and are free to fire those who do so. ■

Enforcing Consultant and Independent Contractor Agreements

One of the most important reasons to put your agreements in writing is to make sure that you can enforce the agreement if it isn't carried out. Of course, having a written agreement doesn't guarantee that the other party will perform as promised—but it does make it much easier to resolve contract disputes or get a court or arbitrator to enforce the agreement.

There is no single way to enforce a consultant or independent contractor agreement. This chapter explores all of your options, including:

- negotiation (see section B, below)
- mediation (see section C, below)
- arbitration (see section D, below), and
- litigation (see section E, below).

But first, we discuss a few issues you should consider before you start thinking about how to enforce your contract.

A. Preliminary Considerations

Before accusing the other party of failing to live up to its contract obligations (called "breaching the contract" in legal lingo), you need to make sure that you're on solid ground. After all, you don't want to embarrass yourself—and sour your relationship with the other party—by complaining of a breach that hasn't actually occurred. Before you decide what to do, ask yourself these preliminary questions, addressed in detail below:

- What does the agreement require? (See Section 1, below.)
- Has the other party failed to do what is required by the agreement (in other words, has the other party breached the agreement)? (See Section 2, below.)
- Have you taken steps to minimize the amount of damage caused by the other party's failure to live up to the agreement? (See Section 3, below.)

1. What Does the Agreement Require?

Before you do anything else, take a careful look at your agreement. Your contract defines your rights and responsibilities. Here are the key provisions to examine:

- *Contract specifications:* First, look at the contract specifications (called "specs" for short). These are located in Clause 1, "Services to Be Performed," of most of the agreements in this book. The specs define what the IC is supposed to do. The IC has no duty to do work that isn't mentioned in the specifications. If there is an argument about whether the IC has performed adequately, a mediator, arbitrator, judge, or jury will look at the specs and compare them to the work actually performed to try to determine who's right.

- *Deadline:* How long does the agreement give the IC to complete the work? Usually, this date is contained in the "Term of Agreement" clause of our form agreements. Has the deadline for completion already passed, or does the IC still have time to finish the job?

- *Payment terms:* The payment terms are contained in the "Terms of Payment" clause of our contracts. Is the IC entitled to be paid a set fee or by the hour? Are partial payments required? How long does the client have to pay the IC after the work is completed? Is the client late paying the IC?

- *Dispute resolution:* What does the contract say about how disputes are supposed to be resolved? Our sample contracts provide several alternatives—court litigation, mediation, or mediation coupled with arbitration. The alternative you've chosen will have a big impact on how you resolve your dispute. (See Sections C, D, and E, below, for more about these methods for resolving disputes.)

- *Attorney fees:* The agreement may also contain a clause requiring the losing side to pay the winner's attorney fees in any dispute. Again, this may have a big impact on how you decide to proceed. If the loser is required to pay the winner's attorney fees, you should be more reluctant to sue or be sued unless you have a very good case.

- *Limitations on the IC's liability:* In cases where the IC may have breached the agreement, check to see if the contract contains a provision limiting the IC's liability. For example, the contract may limit the IC's liability to a set dollar amount or provide that the IC is not liable for certain types of damages (known as special, incidental, or consequential damages—see Section E4, below, for an explanation of what these types of damages entail). An IC with this type of liability protection will be in a stronger negotiating position than one without it.

2. Has the Contract Been Breached?

A contract is breached when one (or both) of the parties fails to perform as promised. However, the law recognizes that some breaches are more serious than others.

A serious, or material, breach of the contract occurs when one party has substantially failed to live up to its end of the deal. The client or IC has failed to perform the bulk of its duties under the contract; as a result, the other side has not gotten what it bargained for.

In contrast, a minor contract breach occurs when the contract has been substantially fulfilled and only small, insignificant defects or problems remain.

Whether a breach is material or minor is very important, as a legal matter. If the breach is material, the innocent party (the one who hasn't breached the contract) has the legal right to terminate the contract immediately and sue the breaching party for damages—it doesn't have to continue to meet its obligations under the contract. In contrast, if the breach is minor, the innocent party doesn't have the legal right to terminate the contract. Instead, it must continue to perform as promised—but it can also sue the other party for any damages caused by the minor breach.

Whether a breach is material or minor is a judgment call, but usually you should be able to tell by using your common sense.

Example 1: Bill, a freelance writer, is writing a book about toxic waste disposal. He enters into a contract with Barbara, a photographer, for photos of various toxic waste dumps around the country. The contract calls for a minimum of 10 photos of 20 specified dump sites. Barbara turns in photographs of only five of the 20 sites. This is a material breach of the contract. Bill has the right to terminate the contract with Barbara—that is, he is not legally required to pay her. He can also bring suit against Barbara for the damages caused by her material breach of the contract.

Example 2: Assume instead that Barbara photographs all 20 of the dump sites, but delivers the photographs one week late. Although it is annoying that Barbara is late, this violation of the contract terms is minor, and Bill must still pay her. If he suffers any damage because of Barbara's tardiness (for example, his publisher fines him $100 for turning in his book late), he can sue Barbara for that economic loss.

3. Have You Taken Reasonable Steps to Minimize Your Losses?

When the other party fails to live up to a contract, you have a duty to take reasonable steps to minimize the economic impact of the breach. This is called mitigating your damages. You cannot recover money for damages that you could have avoided by taking reasonable efforts. For example, if an IC breaches a contract, the client cannot just sit back and let the resulting damages accrue without taking action to avoid them. In most cases, this means the client will have to find someone else to do the work.

Example 1: Recall that Bill in the previous examples contracted with Barbara to take photographs for his book on toxic waste. Assume that Barbara fails to deliver any photos by the deadline. Bill does nothing to try to obtain other photographs and therefore cannot deliver his book to his publisher. His publisher cancels Bill's publishing agreement and forces Bill to return his $50,000 advance. Bill sues Barbara for breach of contract, claiming $50,000 in damages. He cannot recover this amount—he took no reasonable steps to mitigate the damages caused by Barbara's breach.

Example 2: Assume instead that Bill hires another photographer to take the pictures for his book and gets it published. Hiring the second photographer ends up costing Bill an extra $5,000. Bill sues Barbara for breach of contract and may recover $5,000 in damages. Bill took reasonable steps to mitigate his damages, but Barbara's contract breach still cost him money—and he's entitled to collect it from her.

B. Negotiation

Rule #1 of contract enforcement is "negotiate, don't litigate." You'll almost always be better off reaching a resolution by talking to the other side rather than by going to court.

Most disputes over consultant and IC contracts concern the IC's performance under the agreement. In these situations, the client believes that the IC has failed to do the work properly—because the work is not done at all, the work is only partially completed, the work is late, or the work is performed poorly. As a result, the client either fails to pay the IC or pays only part of the amount due.

If the problems are not significant—in other words, if the contract breach is minor—you should be able to settle the dispute informally. Usually, minor breaches are not worth the time, cost, and trouble of filing a lawsuit. Even material breaches can be resolved if the parties are willing to work together.

Call or meet with the other side to talk out your differences. Don't rely on letters at this stage unless the other side refuses to talk to you. Before you start your negotiation discussions, make sure you're prepared. First, you should have a copy of your contract in front of you—and be ready to explain why you think the other party did not fulfill its obligations. If there is a dispute about whether the services were performed properly, be prepared with sensible arguments and evidence to back up your position. Keep careful notes of all your meetings and phone conversations.

How should you go about negotiating? First, even if you're angry at the other side, don't be overly aggressive or hostile. This gets you nowhere. Stay calm and businesslike. A good way to start the ball rolling is to make a reasonable settlement offer, but on the high end of what you want. See how the other side responds. If it makes a low counteroffer, don't get angry, but don't jump to accept it either. Make your own counteroffer. For example, if you've offered to settle for $10,000 and the other side comes back with a $5,000 counteroffer, make a counteroffer of $7,500. There's a good chance your counteroffer will be accepted; or at least the other side might improve its own counteroffer. Be patient. Remember, negotiation is a process of give and take. It might take a while to reach a settlement you can live with, but this is time well spent if you can avoid an expensive lawsuit.

 For additional guidance on negotiation, see:

- *Getting to Yes: Negotiating Agreement Without Giving In*, by Roger Fisher and William Ury (Houghton Mifflin)
- *Effective Approaches to Settlement: A Handbook for Lawyers and Judges*, by Wayne Brazel (Prentice Hall)
- *Effective Legal Negotiation and Settlement*, by Charles Craver (Michie)

What should you settle for? First of all, calculate how much you could collect in damages if you sued the other side in court and won (see the discussion of contract damages in Section E, below). Be realistic about this calculation. If there are problems with your case or the other side has good points, consider them in arriving at your figure. For example, a contractor whose work product doesn't completely live up to the contract specifications should be prepared to settle for less than the full contract price. Likewise, a client who has obtained at least part of the work contracted for must be prepared to pay for the fair value of that work.

Even if your contract allows you to collect attorney fees if you sue and win, don't consider such fees in arriving at your potential damages. You won't have any such fees if you settle the case yourself. Shave 10% to 20% off your damages amount to arrive at your initial settlement offer. The 10% to 20% reduction is in exchange for your not having to go to court.

The other side may accept your settlement offer; then again, it might not. You may have to do some hard bargaining. Keep in mind that any settlement offer you make is not legally binding on you unless the other side accepts it. For example, you could make an oral or written demand for $20,000, then offer to compromise for $15,000 and, if your compromise offer is turned down, still sue for $20,000.

Be sure to document your settlement efforts—write down what was said during phone calls or at meetings. Also, keep copies of all correspondence and email relating to the negotiations.

If you reach a settlement, promptly write down the terms in a confirming letter and have all involved sign it. This need not be a complex legal document; it should simply restate the agreement you've reached.

> **Example:** Gwen, a self-employed computer programmer, is hired by Acme Corp. to create a custom software program for $10,000. Gwen delivers the program on time. However, Acme refuses to pay Gwen because it claims the program doesn't fully live up to the contract specifications. Gwen meets with Acme's president and admits that the program doesn't do everything it's supposed to do. However, she demonstrates that the program fulfills at least 80% of Acme's requirements. Therefore, Gwen offers, to accept $8,000 as full payment instead of the $10,000 stated as full payment in the original contract. Acme's president agrees. Gwen drafts and sends to Acme the following confirming letter:

Sept. 15, 20xx

John Jackson
Acme Corporation
123 Main St.
Marred Vista, CA 90000

Re: Your contract #1234

Dear John:
As we agreed, I'm willing to accept $8,000 as a full and complete settlement of your account.

This sum must be paid to me by Oct. 1, 20xx, or this offer will become void.

Thank you for your cooperation, and I look forward to receiving full payment.

Very truly yours,

Gwen Green

C. Mediation

If informal negotiations fail, the next alternative you should consider is mediation. Look at the dispute resolution clause of your agreement. This clause may require mediation before you take any other steps, such as beginning an arbitration or filing a lawsuit. If the contract requires mediation, you have no choice—you must go to a mediator unless the other side agrees to dispense with this step. If your agreement does not require mediation, you may still mediate your dispute, but only if the other side agrees.

People often confuse mediation with arbitration (see Section D, below), but they are very different procedures. In mediation, a neutral third person (called a mediator) meets with the people involved in the dispute and suggests ways they could resolve their controversy. Typically, the mediator either sits both sides down together or shuttles between them.

Where the underlying problem is actually a personality conflict or simple lack of communication, a good mediator can often help those involved in the dispute find their own compromise settlement. Where the argument is more serious, a mediator may at least be able to lead them to a mutually satisfactory ending of both the dispute and their relationship that will obviate time-consuming and expensive litigation.

If you've ever had a dispute with a friend or relative that another friend or relative helped resolve by meeting with you both and helping you talk things over, you've already been through a process very like mediation. It's an informal and inexpensive way to try to resolve your dispute.

Mediation is nonbinding. That means that if either person involved in the dispute doesn't like the outcome of the mediation, he or she can ask for binding arbitration or go to court.

For detailed guidance on all aspects of mediation, see *Mediate, Don't Litigate,* by Peter Lovenheim and Lisa Guerin (Nolo). You can also find information on mediation in *The Lawsuit Survival Guide,* by Joseph Matthews (Nolo), a comprehensive guide to civil litigation.

D. Arbitration

If those involved in a dispute cannot resolve it by mediation, they often submit it to arbitration. This may be required by your agreement, or you and the other side may agree to arbitration even though it's not required.

Even if your agreement doesn't require arbitration, you should consider it instead of going to court. Usually, arbitration is much faster than court litigation. Most arbitrations are concluded in fewer

than six months. Court litigation often takes years. No one can ever tell how much a court case will cost, but it's usually a lot. Typically, business lawyers charge from $150 to $250 per hour. Unless the amount of money involved is small and the case can be tried in small claims court (see Section E6), arbitration is usually far cheaper than a lawsuit. Typically, a private dispute resolution company will charge about $500 to $1,000 for a half day of arbitration. The speed and low cost of arbitration may convince both you and the other side to arbitrate rather than litigate your contract dispute.

An arbitrator's role is very different from that of a mediator. Unlike a mediator who seeks to help the parties resolve their dispute themselves, an arbitrator personally imposes a solution, like a judge in a courtroom.

Normally, the arbitrator hears both sides at an informal hearing. You can be represented by a lawyer at the hearing, but it's not required. The arbitrator acts as both judge and jury: After the hearing, the arbitrator issues a decision called an award. The arbitrator follows the same legal rules a judge or jury would follow in deciding whether you or the other side has a valid legal claim and should be awarded money.

Arbitration can be either binding or nonbinding. If arbitration is nonbinding, either party can take the matter to court if he or she doesn't like the outcome. Binding arbitration is usually final. You can't go to court and try the dispute again if you don't like the arbitrator's decision—except in unusual cases where you can show the arbitrator was guilty of fraud, misconduct, or bias. In effect, binding arbitration takes the place of a court trial.

If the losing party to a binding arbitration doesn't pay the money required by an arbitration award, the winner can easily convert the award into a court judgment that can be enforced just like any other court judgment—in other words, a binding arbitration award is just as good as a judgment you could get from a court.

For more information on arbitration, see Chapter 9 of *Mediate, Don't Litigate,* by Peter Lovenheim and Lisa Guerin (Nolo). You can also find information on arbitration in *The Lawsuit Survival Guide,* by Joseph Matthews (Nolo), a comprehensive guide to civil litigation.

FINDING A MEDIATOR OR ARBITRATOR

It is usually up to you and the other side to decide who should serve as a mediator or arbitrator. Normally, you can choose anyone you want unless your contract restricts your choice. You can use a professional mediator or arbitrator, or just someone you both respect. A professional organization may be able to refer you to a good mediator or arbitrator. Businesses often use private dispute resolution services that maintain a roster of mediators and arbitrators—often retired judges, attorneys, or businesspeople with expertise in a particular field. Most private dispute resolution companies offer both mediation and arbitration. Two of the best known are:

- The American Arbitration Association. This is the oldest and largest private dispute resolution service, with offices in most major cities. It handles both mediations and arbitrations. The main office is in New York City; you can reach it by calling 212-778-7879. The American Arbitration Association also has a very informative website, www.adr.org.
- Judicate. This Philadelphia-based firm emphasizes mediation and handles disputes in all 50 states. You can reach it by calling 800-488-8805.
- The National Arbitration Forum is another nationwide arbitration organization. The Forum's website is www.arg-forum.com and its phone number is 800-474-2371.

E. Litigation

If your attempts to settle the dispute through informal negotiations, mediation, and/or arbitration fail, your remaining alternative is to sue in court. In a typical breach of contract case, the person who sues—called the plaintiff—asks the judge to issue a judgment against the person being sued—called the defendant. Usually, the plaintiff wants money, also known as damages.

For a comprehensive guide to every aspect of litigating a lawsuit with the help of a lawyer, see *The Lawsuit Survival Guide,* by Joseph Matthews (Nolo). If you don't want to hire a lawyer, get a copy of *Represent Yourself in Court,* by Paul Bergman and Sara Berman-Barrett (Nolo), a step-by-step guide to bringing or defending a lawsuit on your own.

1. Send a Final Demand Letter

Before you file a lawsuit, send the other side a final demand letter. This should be a concise businesslike letter in which you summarize the history of your dispute, make your final settlement offer, and make it clear that you'll go to court if the offer isn't accepted. Many potential defendants will settle after receiving such a letter because they don't want to get involved in litigation.

> **Example:** Natalie, a freelance translator, contracted with Jack Jinx to translate a website into French. She performed the work but was not paid. Jinx refused to discuss settlement with her—indeed, he wouldn't answer her phone calls, emails, or letters. She sends him the following final demand letter:

May 1, 20xx

Jack Jinx
456 First St.
Erehwon, NY 10011

Dear Mr. Jinx:

I contracted with you to translate the website jinx.com into French. A copy of my contract is attached. The work was to be completed by Feb. 1, 20xx, for a fee of $5,000. I submitted my translation by the deadline, but as of today have received no payment. My attempts to contact you have gone unanswered.

This is to notify you that if you do not make full payment of the $5,000 contract amount by May 15, 20xx, a lawsuit will be filed against you. A recorded judgment will be a lien against your property and have an adverse affect on your credit rating.

I hope to hear from you immediately so that this matter can be resolved without filing a lawsuit.

Very truly yours,

Natalie Kalmus

2. Is It Worth Going to Court?

Before you sue, ask yourself one important question: Is it really worth the trouble? If only a minor issue is involved, it probably isn't. Minor problems should be settled out of court or through mediation. If your damages are substantial, however, a lawsuit may be in order if you can't get satisfaction any other way. But always remember this: It's worthwhile to file a lawsuit only if you can collect any judgment you obtain.

If you sue and win, the court will order the defendant to pay you a specified sum of money. This is known as a court judgment. Unfortunately, if the defendant fails or refuses to pay the judgment, the court will not help you collect it. You've got to do it yourself or hire someone to help you. There are many legal tools you can use to collect your judgment. For example, you can file liens on the defendant's property that make it impossible for the defendant to sell the property without paying you, get hold of the defendant's bank accounts, or even have business or personal property (such as the defendant's car) seized by local law enforcement and sold. However, none of these tools will be of much use if the defendant has no money or assets to pay you or files for bankruptcy. In this event, any judgment you obtain against the defendant will probably be uncollectible.

3. Proving Breach of Contract

Proving a breach of contract case is not complicated—and is particularly easy when you have a written agreement. You must show the following four things:

- *You have a contract:* First, you must show that you have a legally binding contract with the other party. All you have to do is present your written and signed contract to the court.

- *You performed:* You must also prove that you did what was required of you under the terms of the contract. If you did everything you were supposed to do up to the time of the defendant's breach, you'll have no problems with this element.

- *The defendant breached the contract:* You must show that the defendant failed to meet his or her contractual obligations. For example, a client who sues an IC would usually need to prove that the contractor failed to do the agreed-upon work or did work of poor quality. This is usually the heart of the dispute.

HOW IS THE DEFENDANT'S BUSINESS ORGANIZED?

Your ability to collect a court judgment may depend on the way the defendant's business is legally organized.

- If the defendant is a sole proprietorship—that is, individually owns the business—he or she is personally liable for any debts the business owes you. This means that you can go after the proprietor's own personal assets—as well as those of the business—to satisfy the debt. For example, both the proprietor's business and personal bank accounts may be tapped to pay you.

- If the defendant is a partner in a partnership, you can go after the personal assets of all the general partners. If the partnership is a limited partnership, you can't touch the assets of the limited partners.

- If the defendant is a corporation, you could have big problems collecting a judgment. Normally, you can't go after the personal assets of a corporation's owners, such as the personal bank accounts of the shareholders and officers, for breach of contract. Instead, you're limited to collecting from the corpor-

ation's assets. If the corporation is insolvent or goes out of business, there may be no assets to collect. The only likely exception would be if the corporation's owners have personally guaranteed the contract.

- If the defendant is a limited liability company, generally its owners won't be personally liable for any debts the business incurred, just as if it were a corporation.

Often, you can tell how a potential defendant's business is legally organized just by looking at its name on the contract. If it's a corporation, usually its name will be followed by the words Incorporated, Corporation, Company, or Limited—or the abbreviations Inc., Corp., Co., or Ltd. Partnerships often have the words Partnership or Partners in their name, but not always. A limited liability company will usually have the words Limited Liability Company or Limited Company—or the abbreviations L.C., LLC, or Ltd. Co.—in its name. Sole proprietors often use their own names, but they don't have to do so. They may use fictitious business names or dbas ("doing business as" names) that are completely different from their own names.

- *You suffered damages:* Finally, you must show that you suffered an economic loss as a result of the defendant's breach of contract—that is, that you lost money because the defendant didn't perform as promised. (See the discussion of contract damages in Section E4, below.)

Example: Ted, a self-employed graphic designer, contracted with the Acme Sandblasting Company to redesign its logo and newsletter. Ted completed the work, but Acme refused to pay him. After informal negotiations failed, Ted sued Acme in court. To win his case, Ted should produce the written contract with Acme, a decent-looking sample of the redesigned newsletter, and a letter from someone with expertise in the field stating that the work met or exceeded industry standards. Ted would also be wise to try to rebut any points the client is likely to make. For example, if the design work was a few weeks late, Ted would want to present a good excuse, such as the fact that Acme asked for time-consuming changes.

4. What You Can Collect If You Win

If you sue for breach of contract, forget about collecting the huge awards you've heard about people getting when they're injured in accidents and sue for personal injuries.

Damages for breach of contract are limited by law. You can also get the money you lost because the other party didn't fulfill the contract. For example, if you are a contractor who performs all of the work required by the contract, and if the hiring firm refuses to pay you the agreed-upon $1,000 for your work, your damages would be $1,000. In some circumstances, you can also get money to pay for profits you would have made if the other party had fulfilled the contract, or you may be able to get money to repay you for any expenditures you made in anticipation of the other party doing as agreed in the contract. In addition, you can get attorneys fees and expenses if you have a contract clause saying so.

Certain kinds of damages are not available in a contract lawsuit, including:

- *Punitive damages:* Punitive damages are designed to punish wrongdoers for malicious acts.
- *Damages for pain and suffering:* Even though a contract breach may cause you a lot of anxiety and sleepless nights, you cannot ask the other side to pay for your emotional damages.
- *Specific performance:* If an IC fails to do or complete the work called for in the contract, can the client get a court to or-

der the IC to do it? The answer is no. This type of contract-breach remedy is called specific performance, and it's not available for contracts for personal services—that is, a court won't order someone to work. This is considered to be a form of slavery. (In contrast, a court can order a client to pay an IC money—this is not specific performance.)

5. How Long You Have to Sue—Statutes of Limitation

You can't wait forever to file a lawsuit. All states impose specific time periods on how long a plaintiff may wait to sue. The time periods for breach of contract cases are quite long—ranging from three to six years in most states. The limitations period begins to run from the date you knew or should have known of the breach of contract. However, there is no good reason to wait this long to file your suit.

6. Suing in Small Claims Court

Most business contract suits are filed in state court. All states have a special court designed to handle disputes that involve only a small amount of money. These are called small claims courts. The amount for which you can sue in small claims court varies from state to state, but it usually ranges from $2,500 to $7,500.

Small claims court has the same advantages as arbitration: It's usually fast and inexpensive. You don't need a lawyer to go to small claims court. Indeed, some states don't allow lawyers to represent people in small claims court.

You begin a small claims lawsuit by filing a document called a complaint or statement of claim. These forms are available from your local small claims court clerk and are easy to fill out. You may also be asked to attach a copy of your written agreement. You then notify the other side (client or IC), now known as the defendant, of your lawsuit. Depending on your state, the notice can be delivered by certified mail or by a process server. Many defendants settle when they receive a complaint because they don't want to go to court. A hearing date is then set. At the hearing, you present your case to a judge or court commissioner under rules that encourage a minimum of legal and procedural formality. Be sure to bring all your documentation to court, including your invoices, agreement, and correspondence with the defendant.

 For detailed guidance on how to represent yourself in small claims court, see *Everybody's Guide to Small Claims Court* (national and California editions), by Ralph Warner (Nolo).

7. Suing in Other Courts

If your claim exceeds the small claims court limit for your state, you'll need to file your lawsuit in another court. Most business lawsuits are handled in state courts. Every state has its own trial court system with one or more courts that deal with legal disputes between people and businesses. These courts are more formal than small claims courts, and the process usually takes longer. You may be represented by a lawyer, but you don't have to be. Many people have handled their own cases successfully in state trial courts. ■

General Independent Contractor Agreements

This chapter guides you in creating an independent contractor agreement that you can use for almost any type of service. Two separate agreements are provided: The first is designed to be used by hiring firms; the second is for use by the independent contractor.

The agreements presented in the rest of this book are tailored for specific occupations or professions. If you find that none of them meets your needs, use the generic agreements in this chapter.

A. Agreement for Hiring Firm to Use

You should use this agreement if you're hiring an IC to work for you. The agreement is designed to protect your interests as much as possible without being unduly one-sided or unfair to the IC. It also helps establish that the worker is an IC, not your employee.

The entire text of the general independent contractor agreement for use by the hiring firm is on the CD-ROM. You can modify individual clauses to tailor the text to your specific needs. However, be sure to read the following discussion before you make any major changes.

The general independent contractor agreement for use by the hiring firm is also reproduced as a tear-out form in Appendix II. Use this form if you don't have access to a computer. It contains blank spaces for you to fill in by hand or typewriter.

Although we provide a tear-out form for those of you who either don't have computers or who would prefer not to use the CD-ROM, don't just tear out the form and use it as your IC agreement if you can avoid doing so. A fill-in-the-blank IC agreement is the least persuasive type of contract when you are trying to convince a government agency that the worker is an IC and not an employee. The best way to create your agreement is to use the CD-ROM version and tailor it to your needs. If you can't use the CD-ROM version, then retype the version at the back of the book and tailor it to your needs. Use the fill-in-the-blank form only as a last resort.

Call up the agreement on your computer or tear out the form agreement in Appendix II, and read it along with these instructions and explanations.

Title of agreement. You don't have to have a title for an IC agreement, but if you want one you should call it "Independent Contractor Agreement." Never use "Employment Agreement" as a title.

Names of independent contractor and hiring firm. When you first write down the name of the IC, it's best to refer to the IC by the IC's full name. If an IC is incorporated, use the corporate name, not the IC's own name. For example, write "John Smith Incorporated" instead of just "John Smith."

If the IC is unincorporated and is doing business under a fictitious business name, use that name. A fictitious business name or assumed name is a name that sole proprietors or partners use to identify their business, other than their own names. For example, if consultant Al Brodsky calls his one-man marketing research business "ABC Marketing Research," use that name. This shows you're contracting with a business, not a single individual. Never refer to an IC as an employee or to yourself as an employer.

For the sake of brevity, it is usual to identify the IC by shorter names in the remainder of the agreement. You can use an abbreviated version of the IC's full name—for example, "ABC" for "ABC Marketing Research." Or you can refer to the IC simply as "Contractor."

Refer to yourself initially by your full company name and subsequently by a short version of the name, or as "Client" or "Firm."

Include the addresses of the principal places of business of the IC and yourself. If you or the IC have more than one office or workplace, the principal place of business is the main office or workplace.

1. Clause 1. Services to Be Performed

(Choose Alternative A or B.)

☐ ALTERNATIVE A

Contractor agrees to perform the following services:

_____ .

☐ ALTERNATIVE B

Contractor agrees to perform the services described in Exhibit A attached to this Agreement.

The agreement should describe in as much detail as possible what the contractor is expected to do. You must word the description carefully to show only the results you expect the IC to achieve. Don't tell the IC how to achieve those results—this would indicate that you have the right to control how the IC performs the agreed-upon services. Having a right of control is the hallmark of an employment relationship. (See Chapter 2 for a discussion of why it is important that your relationship with the IC not look like an employment relationship.)

While you shouldn't dictate how your IC is to perform the work, it is perfectly fine for you to establish very detailed specifications for the IC's finished work product. The specs should describe only the end results the IC must achieve, not how

DESCRIBING AN INDEPENDENT CONTRACTOR'S JOB

Work descriptions that support IC status	Work descriptions that suggest the worker is an employee
Contractor will deliver the cargo described on the attached manifest to 123 Grub St., Tucson, Arizona, no later than 9:00 a.m. on March 1, 20XX.	Contractor will load cargo as instructed and leave at 9:30 a.m. Use Route 66 until arrival at Poisonby. Take the shortcut described on the attached map to avoid traffic. Spend night in Laughlin at Lazy 8 Motel. Leave Laughlin at 8:00 a.m. Arrive at Tucson by 9:00 a.m.
Contractor will paint all walls and ceilings of the offices located at 222 Main St., Suite 101, using two coats of flat white latex paint. All old paint will be removed, and wood surfaces will be stripped to bare wood before the new paint is applied.	Contractor will paint all walls and ceilings of the offices located at 222 Main St., Suite 101. Contractor will apply paint to the ceilings first and then the floor trim using brushes, not rollers. The walls will be painted last. Contractor will remove all old paint by hand scraping. Contractor will wear gloves and a white paint uniform, and will clean all brushes and rollers at the end of each day.
Contractor agrees to prepare an index of Client's book "History of Sparta."	Contractor will prepare an alphabetical three-level index of Client's book "History of Sparta." Contractor will first prepare 3" x 5" index cards listing every index entry beginning with Chapter One. After each chapter is completed, Contractor will deliver the index cards to Client for Client's approval. When index cards have been created for all 50 chapters, Contractor will create a computer version of the index using Complex Software Version 7.10. Contractor will then print out and edit the index and deliver it to Client for approval.

to obtain those results. See "Describing an Independent Contractor's Job," above, for examples of work descriptions that show IC status or employee status.

You can include the description of the work to be done in the main body of the agreement or, if it's lengthy, on a separate attachment. These options are identified as Alternative A (when you include the explanation of services in the contract) and Alternative B (when the explanation of the services is attached to the contract).

2. Clause 2. Payment

(Choose Alternative A or B and optional clause if desired.)

☐ ALTERNATIVE A

In consideration for the services to be performed by Contractor, Client agrees to pay Contractor $_____ according to the terms of payment set forth below.

☐ ALTERNATIVE B

In consideration for the services to be performed by Contractor, Client agrees to pay Contractor at the rate of $_____ per hour according to the terms of payment set forth below.

(Optional: Check if applicable.)

☐ Unless otherwise agreed upon in writing by Client, Client's maximum liability for all services performed during the term of this Agreement shall not exceed $_____.

There are two common ways an independent contractor may be paid: by a fixed fee or by unit of time.

a. Fixed fee

Paying an IC a fixed amount for the entire job instead of an hourly or daily rate will help you establish the worker's IC status in the eyes of the

government. Remember, a fact of life when people work as ICs is the risk that they might underestimate the job and end up losing money. On the other hand, the IC may earn a substantial profit if the project is completed quickly. The opportunity to earn a profit or suffer a loss is a very strong indicator of IC status and is something employees ordinarily do not have.

This method of payment has an added attraction—it ensures that you know exactly how much the IC's work will cost.

Choose Alternative A and insert the amount if you'll pay the IC a fixed fee.

b. Payment by the hour

Many independent contractors—for example, lawyers, accountants, and plumbers—customarily charge by the hour, day, or other unit of time. Paying an IC by the hour does not support the worker's independent contractor status, but you can get away with it if it's a common practice in the field in which the IC works. Government auditors would likely not challenge the IC status of workers in these occupations as long as there are other strong indicators that they are in business for themselves—for example, they have recurring business expenses such as office rent, they work for more than one client, they have a significant investment in equipment or facilities, and so forth.

Choose Alternative B and insert the amount of the payment if you will pay the IC by the hour or other unit of time.

If you pay by the hour, you may wish to place a cap on the IC's total compensation. This may be a particularly good idea if you're unsure about the IC's reliability or efficiency. Insert the amount of the cap in this optional provision, or delete this language if you don't want to have a cap.

3. Clause 3. Terms of Payment

(Choose Alternative A, B, C, or D.)

☐ ALTERNATIVE A

Upon completion of Contractor's services under this Agreement, Contractor shall submit an invoice. Client shall pay Contractor the compensation described within _____ [15, 30, 45, 60] days after receiving Contractor's invoice.

☐ ALTERNATIVE B

Contractor shall be paid $_____ upon the signing of this Agreement and the remainder of the compensation described above upon completion of Contractor's services and submission of an invoice.

☐ Alternative C

Client shall pay Contractor according to the schedule of payments set forth in Exhibit ___ attached to this Agreement.

☐ Alternative D

Contractor shall invoice Client on a monthly basis for all hours worked pursuant to this Agreement during the preceding month. Invoices shall be submitted on Contractor's letterhead specifying an invoice number, the dates covered in the invoice, the hours expended, and the work performed (in summary) during the invoice period. Client shall pay Contractor's fee within _____ days after receiving Contractor's invoice.

Clause 3 of the agreement specifies how and when you'll pay the IC. Because the IC is running an independent business, the IC should submit an invoice setting forth the amount you have to pay. Keep the IC's invoices in your files—they are evidence, should you ever need it, that you have dealt with the worker as an IC, not an employee.

There are different ways to structure the payments you make to the IC. You can pay the entire amount upon completion, or pay a down payment upfront, with the balance due when the project is finished. Or you can set up a schedule for installment payments. Each method is explained below.

This agreement does not require you to pay a late fee or any other late penalty if your payments to the IC are late. But, of course, paying late is not a good way to keep up good relations with a valuable IC. And, if you don't pay at all, the IC can sue you for the amount due.

a. Full payment upon completion of work

Choose Alternative A if you'll pay the IC the full amount of a fixed fee upon completion of the work. Indicate how soon you'll pay the IC after you receive the invoice. Thirty days is a common payment period, but you can choose a longer or shorter period.

b. Divided payments

Some ICs will insist on a partial down payment of their fixed fee before they begin work. Choose Alternative B if you agree to this arrangement. Insert the amount of the fixed fee you'll pay when the agreement is signed. Many ICs will ask for anywhere from one-fourth to one-half of the amount due in advance. However, this is always a matter for negotiation. Ordinarily, it's best for the hiring firm to pay an IC as little as possible in advance. The less you pay the IC up front, the more incentive the IC will have to complete the job quickly and get paid the balance due. Moreover, if the IC's work turns out to be unsatisfactory, you'll have more leverage to get the IC to redo the job if you haven't given the IC a large advance payment.

c. Installment payments for complex projects

ICs may balk at waiting for a fixed fee for complex or long-term projects due to difficulties in accurately estimating how long the job will take. One way to deal with this problem is to break the job down into phases or milestones and pay the IC a fixed fee upon completion of each phase. If, after one or two phases are completed, it looks like the fixed fee won't be enough to complete the entire project, you can always renegotiate the agreement. Or, if you prefer, you can mutually agree to cancel the agreement and hire someone else to complete the project. This type of arrangement is far more supportive of an IC relationship than hourly payment, since the IC still has some risk of loss.

To set up this arrangement, draw up a schedule of installment payments tying each payment to the IC's completion of specific services, then attach it to the agreement. The main body of the agreement simply refers to the attached payment schedule. Choose Alternative C for this option.

Example: The Splashy Swimming Pool Company hires Dave, a freelance computer programmer, to develop a specialized accounting system. Splashy and Dave agree that the lengthy project will be divided into four phases or modules—accounts receivable, accounts payable, order processing, and invoicing. Splashy will pay Dave upon completion of each module. Splashy and Dave draw up the following schedule and attach it to their agreement:

Schedule of Payments

Client shall pay Consultant according to the following schedule of payments:

Phase 1. $5,000 when an invoice is submitted and the following services are completed: Accounts receivable module is completed and accepted by Client.

Phase 2. $5,000 when an invoice is submitted and the following services are completed: Accounts payable module is completed and accepted by Client.

Phase 3. $5,000 when an invoice is submitted and the following services are completed: Order processing module is completed and accepted by Client.

Phase 4. $5,000 when an invoice is submitted and the following services are completed: Invoicing module is completed and accepted by Client.

d. Hourly payment agreements

Even ICs who are paid by the hour or other unit of time should submit invoices. Never automatically pay an IC weekly or biweekly the way you pay employees. It's best to pay ICs no more than once a month, since this is how businesses are normally paid. Choose Alternative D if you're paying the IC by the hour. Indicate how soon you'll pay the IC after you receive the invoice.

4. Clause 4. Expenses

(Choose Alternative A or B.)
☐ ALTERNATIVE A
Contractor shall be responsible for all expenses incurred while performing services under this Agreement. This includes license fees, memberships, and dues; automobile and other travel expenses; meals and entertainment; insurance premiums; and all salary, expenses, and other compensation paid to employees or contract personnel the Contractor hires to complete the work under this Agreement.
☐ ALTERNATIVE B
Client shall reimburse Contractor for the following expenses that are directly attributable to work performed under this Agreement:

_____.

Contractor shall submit an itemized statement of Contractor's expenses. Client shall pay Contractor within 30 days after receipt of each statement.

Estimating expenses and factoring in these costs when bidding on a job is a strong indicator that a person is in business. For this reason, usually an IC should not be reimbursed for expenses. Make clear to the IC that you will not reimburse expenses and that the IC's set fee or hourly rate is all the money the IC will get. Choose Alternative A if the IC will pay all expenses.

However, for certain types of ICs, it is customary for the client to pay some expenses. For example, attorneys typically charge their clients separately for photocopying charges, deposition fees, and travel. Similarly, a sales or marketing consultant may charge for mileage and phone calls. Where there is an otherwise clear IC relationship and payment of expenses is customary in the IC's trade or business, you can probably get away with doing it. Choose Alternative B and list exactly what expenses you will reimburse if you'll reimburse the IC for expenses.

Be careful to agree to reimburse only expenses that are directly attributable to the IC's work for you. Examples of expenses that you may agree to reimburse include:

- travel expenses other than normal commuting, including airfares, rental vehicles, and highway mileage in company or personal vehicles at a given number of cents per mile
- telephone, fax, online, and telegraph charges
- postage and courier services, and
- printing and reproduction.

Never reimburse an IC for normal fixed overhead costs, such as office rent or commuting to and from the IC's office. These are expenses that should be covered by the business revenues as a whole, not by individual clients.

5. Clause 5. Materials, Equipment, and Office Space

Contractor will furnish all materials, tools, and equipment used to provide the services required by this Agreement.

(Optional: Check if applicable.)

☐ Client agrees to rent to Contractor the following office space and facilities:

_____

The rental will begin on _____ [date] and will end on the earlier of (1) the date Contractor completes the services required by this Agreement; (2) _____ [date]; or (3) the date a party terminates this Agreement. The rental amount is $_____ per ☐ day, ☐ week, ☐ month.

Providing workers with tools and equipment is an indicator of employee status. Therefore, it's not advisable to provide an IC with materials, tools, or equipment. For example, if you're hiring an IC to drive a truck, don't provide the truck. The IC should buy or lease the truck elsewhere. This clause simply states that the IC will provide all tools, equipment, materials, and supplies needed to accomplish the project.

If you must break this rule and provide the IC with tools or equipment, it's best for the agreement to be silent on the matter. Delete this clause entirely and renumber the remaining clauses.

a. Leasing equipment (Optional)

If you must provide your IC with tools or equipment, you may be able to do so in a way that will, in fact, support your client's status as an IC. If the tools, equipment, or materials have substantial value, either sell or lease the items to the IC.

Charge the IC the same amount you'd charge anyone else. If you lease equipment to the IC, enter into a written lease agreement. If you sell the IC equipment, use a written bill of sale. You can find forms for selling or leasing equipment in *Legal Forms for Starting & Running a Small Business*, by Fred Steingold (Nolo).

b. Providing ICs with office space (Optional)

Sometimes it's more convenient for a hiring firm to have an IC work at its offices than at the IC's own office. If you wish to provide the IC with office space on your premises, you should charge rent. Only employees are ordinarily provided with free offices. Charging rent will not only make you some money, it will also help establish that the worker is an IC because the expense will give the worker a risk of loss—if the IC doesn't earn enough to pay for this and other expenses, the IC risks losing money.

If you're providing the IC just a desk to use for a small charge, you can include the simple optional rental clause in your IC agreement. If you include this clause, clearly define exactly what office space and facilities you're providing the

IC—for example, will the IC be allowed to use your copier and fax machine? The rental amount should be a fixed amount, not based on how much money the IC earns.

On the other hand, if you're going to provide the IC with a substantial amount of space, it's advisable to use a more detailed full-fledged commercial lease agreement. You can find a form for such a lease in *Legal Forms for Starting & Running a Small Business*, by Fred Steingold (Nolo).

6. Clause 6. Independent Contractor Status

Contractor is an independent contractor, and neither Contractor nor Contractor's employees or contract personnel are, or shall be deemed, Client's employees. In its capacity as an independent contractor, Contractor agrees and represents, and Client agrees, as follows (check all that apply):

☐ Contractor has the right to perform services for others during the term of this Agreement.

☐ Contractor has the sole right to control and direct the means, manner, and method by which the services required by this Agreement will be performed.

☐ Contractor has the right to perform the services required by this Agreement at any place or location and at such times as Contractor may determine.

☐ Contractor has the right to hire assistants as subcontractors or to use employees to provide the services required by this Agreement.

☐ The services required by this Agreement shall be performed by Contractor, Contractor's employees, or contract personnel, and Client shall not hire, supervise, or pay any assistants to help Contractor.

☐ Neither Contractor nor Contractor's employees or contract personnel shall receive any training from Client in the professional skills necessary to perform the services required by this Agreement.

☐ Neither Contractor nor Contractor's employees or contract personnel shall be required by Client to devote full time to the performance of the services required by this Agreement.

One of the most important functions of an independent contractor agreement is to help establish that the worker is an IC, not your employee. The key to doing this is to make clear that the IC, not the hiring firm, has the right to control how the work will be performed.

The language in this clause addresses most of the factors that the IRS and other agencies consider in determining whether an IC controls how the work is done. (A complete explanation of these factors is in Chapter 2.)

⚠ When you draft your own agreement, include only those provisions in Clause 6 that apply to your particular situation. The more that apply, the more likely the worker will be viewed as an IC.

7. Clause 7. Business Permits, Certificates, and Licenses

Contractor has complied with all federal, state, and local laws requiring business permits, certificates, and licenses required to carry out the services to be performed under this Agreement.

People who are in business for themselves are often required to have various types of business permits, certificates, or licenses. License requirements may be imposed by both your state and local government.

A few states, including Alaska and Washington, require all businesses to obtain state business licenses. Most states don't require general business licenses, but they all require licenses for ICs who work in certain occupations—for example, lawyers, doctors, architects, nurses, and engineers must be licensed in every state. Most states require licenses for other occupations as well—for instance, bill collectors, building contractors, and real estate salespeople. Many cities and counties also require that ICs have business licenses, even if they work at home.

ICs who lack such licenses may be fined by state or local government entities. As a general rule, no action is taken against hiring firms who hire unlicensed ICs. Nevertheless, it is in your interest to make sure the IC has any required licenses. A worker who doesn't have such licenses and permits looks like your employee, not an IC who is running a business.

ICs should obtain necessary licenses and permits on their own. You should neither be involved in the application or certification procedure nor pay for them. This clause simply requires that the IC have all relevant licenses and permits. You need add no information here.

It's a good idea to obtain copies of the IC's licenses and keep them in your file. They will help establish that the worker is an IC if you're audited.

Many states have nonprofit or government-sponsored agencies devoted to small businesses that can provide you with information on whether your state requires ICs to have a license. You can find a state-by-state list of small business resources on the Small Business Administration's website at www.sba.gov. You can also call or write to the national office of the Small Business Administration for a list of state resources at 409 3rd Street, S.W., Washington, DC 20416, 800-827-5722.

Also, many state agencies now have websites that contain information about licensing requirements. A good place to start an Internet search for such information is the State Web Locator at www.infoctr.edu/swl.

8. Clause 8. State and Federal Taxes

> Client will not:
> - withhold FICA (Social Security and Medicare taxes) from Contractor's payments or make FICA payments on Contractor's behalf
> - make state or federal unemployment compensation contributions on Contractor's behalf, or
> - withhold state or federal income tax from Contractor's payments.
>
> Contractor shall pay all taxes incurred while performing services under this Agreement—including all applicable income taxes and, if Contractor is not a corporation, self-employment (Social Security) taxes. Upon demand, Contractor shall provide Client with proof that such payments have been made.

You should never pay or withhold any taxes on an IC's behalf. Doing so is a very strong indicator that the worker is an employee. Indeed, some courts have held that workers were employees based upon this factor alone.

This straightforward provision simply makes clear that the IC must pay all applicable taxes. You need add no information here.

9. Clause 9. Fringe Benefits

> Contractor understands that neither Contractor nor Contractor's employees or contract personnel are eligible to participate in any employee pension, health, vacation pay, sick pay, or other fringe benefit plan of Client.

Do not provide ICs or their employees with fringe benefits that you extend to your own employees, such as health insurance, pension benefits, childcare allowances, or even the right to use employee facilities like an exercise room. This clause makes clear that the IC will receive no such benefits. You need add no information here.

10. Clause 10. Workers' Compensation

> Client shall not obtain workers' compensation insurance on behalf of Contractor or Contractor's employees. If Contractor hires employees to perform any work under this Agreement, Contractor will cover them with workers' compensation insurance to the extent required by law and provide Client with a certificate of workers' compensation insurance before the employees begin the work.
>
> **(Optional: Check if applicable.)**
>
> ☐ Contractor shall obtain workers' compensation insurance coverage for Contractor. Contractor shall provide Client with proof that such coverage has been obtained before starting work.

Each state has its own workers' compensation system that is designed to provide replacement income and medical expenses for employees who suffer work-related injuries or illnesses. To fund

> ### LIABILITY INSURANCE: A MUST FOR HIRING FIRMS
>
> Employees who are covered by workers' compensation cannot sue their employers for damages when they are injured on the job. They are limited to the benefits workers' compensation provides. This is not the case when you hire ICs. That is, ICs can sue you for work-related injuries. For this reason, no matter how small your business, if you hire ICs it is vital that you obtain general liability insurance to protect yourself against personal injury claims by people who are not your employees.
>
> A general liability insurer will defend you in court if an IC, customer, or any other nonemployee claims you caused or contributed to an injury. The insurer will also pay damages or settlements up to the policy limits. Such insurance can be cheaper and easier to obtain than workers' compensation coverage for employees, so you can still save on insurance premiums by hiring ICs rather than employees. If you don't have general liability insurance already, contact an insurance broker or agent to obtain a policy.
>
> Each state has its own workers' compensation law with its own definition of who is an employee and who qualifies as an IC. When you first hire a worker, it's up to you to determine whether the worker is an employee or IC under your state's law. However, your state's workers' compensation agency enforces this law. It may audit you and impose fines and penalties if it determines you've misclassified employees as ICs.
>
> Most states classify workers for workers' compensation purposes using the right-of-control test. This is the same test the IRS and many other government agencies use. (See Chapter 2 for a discussion of classification tests.) However, some states use different tests. You should familiarize yourself with your state's test. For a detailed discussion of these state tests, see *Working With Independent Contractors,* by Stephen Fishman (Nolo). You can also contact your state workers' compensation agency for information.

this system, employers in all but two states—New Jersey and Texas, where workers' compensation is optional—are required to pay for workers' compensation insurance for their employees, either though a state fund or a private insurance company. Employees do not pay for workers' compensation insurance.

You need not provide workers' compensation coverage for a worker who qualifies as an IC under your state workers' compensation law. This can result in substantial savings.

a. Make sure the IC's employees are covered

If the IC has employees, it's very important for the IC to provide them with workers' compensation coverage if required by your state law. If the IC fails to do so, the IC's employees may be considered to be your employees for workers' compensation purposes under your state workers' compensation law, even though they may not be considered your employees for other purposes, such as tax treatment.

If an IC's employees lack workers' compensation coverage and you have employees whom you have insured, your workers' compensation insurer will probably require you to provide coverage under your own policy for these additional workers. You'll have to pay an additional workers' compensation premium.

To avoid having to insure someone else's employees, you should require ICs to provide you with a certificate of insurance establishing that their employees are covered by workers' compensation insurance. A certificate of insurance is issued by the workers' compensation insurer and is written proof that the IC has a workers' compensation policy. Make sure the policy is in effect when the IC's employees will be performing services for you. If it has expired, do not take the IC's word for it that the policy has automatically renewed. Demand a current certificate. To be extra cautious, get the IC's promise in writing that you will be notified immediately if the policy has, for whatever reason, been canceled.

b. Requiring the IC to carry workers' compensation coverage

Usually, ICs who have no employees will not have workers' compensation coverage for themselves. Business owners don't have to provide compensation coverage for themselves, only for employees. Nevertheless, you may want to require ICs to obtain their own workers' compensation coverage.

Your own workers' compensation insurer (or liability insurer) may impose this condition on you. Workers' compensation insurers are worried that uninsured ICs who are injured on the job may claim they are really employees to try to recover benefits from the hiring firm's workers' compensation insurer. And liability insurers don't want to be stuck with the bill if an uninsured IC is injured on your property. (They'll want to know that there is another insurer—the IC's own—to share the bill.) Check your workers' compensation or general liability policy, or ask your agent, to find out whether you are required to obtain certificates of insurance from ICs who work for you, establishing that the ICs themselves are insured.

Choose the optional clause if you want the worker to provide workers' compensation coverage.

11. Clause 11. Unemployment Compensation

> Client shall make no state or federal unemployment compensation payments on behalf of Contractor or Contractor's employees or contract personnel. Contractor will not be entitled to these benefits in connection with work performed under this Agreement.

Federal law requires that all states provide most employees with unemployment compensation, called unemployment insurance in some states. Employers are required to contribute to a state unemployment insurance fund. Employees make no contributions (except in Alaska, where small employee contributions are withheld from em-

ployees' paychecks by their employers). An employee who is laid off or fired for reasons other than serious misconduct is entitled to receive unemployment benefits from the state fund.

Unemployment compensation is only for employees. ICs cannot collect it. Firms that hire ICs don't have to pay unemployment compensation taxes for them. This is one of the significant benefits of classifying workers as ICs, as unemployment compensation taxes typically amount to hundreds of dollars per year for each employee.

Each state has its own unemployment compensation law administered by a state agency, often called the department of labor. Like the determination of who is and is not an employee for purposes of workers' compensation, explained above for Clause 10, each state's law defines who is and who is not an employee for unemployment compensation purposes. When you first hire a worker, it's up to you to decide whether the worker is an employee or IC for unemployment compensation purposes. But again, your state's unemployment compensation agency may review your classification decisions. If it determines you've misclassified an employee as an IC, it will impose fines and penalties.

Unfortunately, you're more likely to be audited by a state unemployment compensation auditor than by any other type of government auditor, including the IRS. This is because ICs often file claims for unemployment compensation when their services for a hiring firm are completed. The ICs claim they should have been classified as employees all along and so are entitled to unemployment compensation payments.

Many states use the right-of-control test discussed in Chapter 2 to determine whether a worker is an IC or employee for unemployment compensation purposes. You should familiarize yourself with your state's test for determining worker status for unemployment compensation purposes. For a detailed discussion of these state tests, refer to *Working With Independent Contractors,* by Stephen Fishman (Nolo). You can also contact your state unemployment compensation agency for information.

If the worker qualifies as an IC under your state's unemployment compensation law, do not pay unemployment compensation taxes for that person. The IC will not be entitled to receive unemployment compensation benefits when the work is finished or the agreement terminated. Clause 11 makes this clear. You need add no information.

12. Clause 12. Insurance

> Client shall not provide any insurance coverage of any kind for Contractor or Contractor's employees or contract personnel. Contractor shall obtain and maintain a broad form Commercial General Liability Insurance policy providing for coverage of at least $_____ for each occurrence. Before commencing any work, Contractor shall provide Client with proof of this insurance and with proof that Client has been made an additional insured under the policy.
>
> Contractor shall indemnify and hold Client harmless from any loss or liability arising from performing services under this Agreement.

An IC should have a liability insurance policy just like any other business; you need not provide it. This type of coverage insures the IC against personal injury or property damage claims by others. For example, if an IC accidentally injures a bystander or one of your employees while performing services for you, the IC's liability policy will pay the costs of defending a lawsuit and money damages up to the limits of the policy coverage.

This clause requires the IC to obtain and maintain a commercial liability insurance policy providing a minimum amount of coverage. You need to state how much coverage is required. $500,000 to $1 million in coverage is usually adequate. But you might want the IC to have even more coverage if the IC is engaged in a hazardous activity. If you're not sure how much coverage to ask for, ask your own insurance broker or agent to advise you.

The insurance clause also requires the IC to add you as an additional insured under the policy. This step is a low-cost endorsement the IC can have added easily to the liability policy. Being added as an additional insured under the IC's policy means that the person who is an additional insured (you) will have all of the benefits of the policy's protections. If you are sued along with the IC in a lawsuit by someone injured by the IC's negligence, the IC's policy will provide defense and pay any claim or verdict up to the limits of the policy. It's to your advantage to involve the IC's insurance company instead of your own if you're named in a lawsuit arising from the IC's negligence.

Do not confuse the term "named insured" with "additional insured." If you are added to a policy as a "named insured," the protections of the policy extend beyond you to your partners, officers, employees, agents, and affiliates. Because the coverage is so broad, it may cost as much as 50% of the original premium to add a named insured to an IC's policy. Understand that being added as an additional insured is cheap (or free) for a good reason—it covers only the individual whose name appears on the certificate.

The clause also requires the IC to indemnify you—that is, to personally repay you—if somebody whom the IC injures decides to sue you. You may be wondering why you need this if you're already an additional insured under the IC's liability policy. Well, there may be circumstances under which you're unable to collect from the IC's insurer—for example, the insurer may deny coverage for some reason. Or the claim may exceed the policy limits of the IC's insurance. In this event, the indemnity clause requires the IC to personally repay you. This gives you additional protection but, of course, will be useful only if the IC has the money to repay you (which is why you want the IC to have insurance in the first place).

13. Clause 13. Term of Agreement

> This agreement will become effective when signed by both parties and will terminate on the earlier of:
> * the date Contractor completes the services required by this Agreement
> * _____ [date], or
> * the date a party terminates the Agreement as provided below.

"Term of the agreement" simply refers to when it begins and ends. Unless the agreement provides a specific starting date, it begins on the date it's signed. This clause establishes that the agreement begins—becomes effective—only after both you and the IC sign. If you and the IC sign on different dates, the agreement begins on the date the last person signed. Until the agreement is signed by both you and the IC, you're not bound by it.

The start date for the agreement marks the moment when you and the IC are legally bound by it—that is, you are supposed to perform as promised. The IC doesn't necessarily have to begin work on the start date. Both you and the IC are legally bound as of the start date even if the IC hasn't actually commenced work.

An independent contractor agreement should have a definite ending date. Ordinarily, this will mark the final deadline for the IC to complete services. However, even if the project is lengthy, the end date should not be too far in the future. (A good outside time limit is six to 12 months.) A longer term makes the agreement look like an employment agreement, not an independent contractor agreement.

If the work is not completed at the end of the term, you can negotiate and sign a new agreement. For now, insert the termination date in this clause. This is the outside termination date. The agreement will end automatically when the contractor completes the services (which could be before the termination date) or if either party terminates the agreement as provided below.

Your obligations don't cease when the work is done. Even after the work is finished, you and the IC are still bound by some of the agreement's terms. You're still required to pay the IC, for example. And the IC will be bound by any warranties made and by any duty of confidentiality imposed by the agreement. (See Section A17 for the confidentiality clause.)

14. Clause 14. Terminating the Agreement

> (Choose Alternative A or B.)
> ☐ ALTERNATIVE A
> With reasonable cause, either Client or Contractor may terminate this Agreement, effective immediately upon giving written notice.
> Reasonable cause includes:
> * a material violation of this Agreement, or
> * any act exposing the other party to liability to others for personal injury or property damage.
> ☐ ALTERNATIVE B
> Either party may terminate this Agreement at any time by giving _____ days' written notice to the other party of the intent to terminate.

If all goes swimmingly, your agreement with the IC will end when the work is completed. But what if you are dissatisfied with the IC's lack of progress, work habits, or interim results—how should you end the relationship? The circumstances under which you may terminate the agreement are important because they can have a big impact on whether a government auditor views the worker as an IC or an employee.

You should not retain the right to fire or terminate an IC for any reason or no reason at all—a right you do have with most employees. This type of unfettered termination right indicates an employment relationship. Instead, you should be able to terminate the agreement only if you have a justifiable business reason (or reasonable cause) to do so, as explained below.

a. Termination for reasonable cause

You should not terminate the IC unless you have reasonable cause for doing so. There are two types of reasonable cause. The first is a serious violation of the agreement by the IC. It would include, for example, the IC's failure to produce results or meet the deadline specified in the agreement.

> **Example:** The Solid Construction Company hires Joe to paint some houses Solid is constructing. Joe agrees to paint five houses per month for the next six months. At the end of two months, Joe has painted only two houses. As a result, Solid is unable to sell the unpainted houses. Joe's substantial failure to deliver the agreed-upon results in the time allotted constitutes reasonable cause for Solid to terminate its agreement with him.

You can also terminate the agreement if the IC does something that exposes you (or threatens to expose you) to liability for personal injury or property damage—for example, if the IC's negligence injures your employees, damages your property, or damages someone else's property. This is a matter of plain fairness—since you stand to lose money if you are sued or have to replace your own damaged property, you shouldn't have to stand idly by and watch the damage mount up. In short, you may not want to continue dealing with an IC who is extremely careless.

Consider terminating a careless contractor. No law requires you to terminate an IC who is careless and causes damage. But keep in mind that if you continue a relationship with an IC whom you know to be trouble, and if someone is injured or property is damaged by the IC's actions, you may find yourself sued along with the IC by the injured party.

Use this power to terminate wisely. If you rashly fire an IC who performs adequately and otherwise satisfies the terms of the agreement,

you'll be liable for breaking the agreement. The IC can sue you in court and obtain a judgment against you for money damages.

Choose Alternative A to preserve your limited termination rights.

b. Termination with notice

Some hiring firms just can't live with the limited termination rights described above. They demand the right to terminate an IC for any reason. (This is called an "at-will" provision.) Such a right is not supportive of the worker's IC status. However, if most of the other aspects of your relationship strongly suggest that the worker is an IC, you may be able to get away with it.

For example, you can get away with unfettered termination rights if you're paying the IC by the project instead of by the hour and if the IC has recurring business expenses, has multiple clients, and employs assistants. All these factors point strongly to an IC relationship and will likely outweigh the fact that the agreement contains an at-will termination provision.

On the other hand, this provision may provide the final nail in the coffin if other factors strongly suggest that the IC is your employee: For example, if the IC is paid by the hour, works on your premises even though the work could be done elsewhere, works only for you, and performs all the services personally, the IC will look more like an employee than an IC. Making the IC subject to at-will firing may simply reinforce this conclusion.

Choose Alternative B to have the right to terminate the agreement for any reason upon written notice and insert how many days' notice you'll give the IC. The notice term can be anywhere from a few days to 30 days or more, depending upon the length of the contract term. The longer the term, the more notice you should give.

Note that this clause is reciprocal—that is, both you and the contractor have the right to terminate the agreement upon notice for any reason. If you wish, you can delete the words "Either party" at the beginning of the clause and replace them with "Client." However, some contractors may view such a one-way termination right as unfair and refuse to agree to it.

15. Clause 15. Exclusive Agreement

> This is the entire Agreement between Contractor and Client.

Normally, business contracts contain a provision stating that the written agreement embodies the complete agreement between those involved. In other words, the provision states that there are no other written agreements or oral understandings floating around that also represent the understandings reached between you and the IC. It means that neither you nor the IC can later bring up side letters, oral statements, or other material not covered by or attached to the contract. This clause prevents either side from claiming that promises not contained in the written contract were made and broken.

Because of this provision, you must make sure that all documents containing any important representations made by the IC are attached to the agreement as exhibits. This may include the IC's proposal or bid, sales literature, side letters, and so forth. If they aren't attached, they won't be considered part of the agreement. And if you and the IC have reached some oral agreements, make sure to write them down and attach them to the contract.

16. Clause 16. Modifying the Agreement (Optional)

> **(Optional: Check box if applicable.)**
>
> ☐ This Agreement may be modified only by a writing signed by both parties.

No contract is engraved in stone. You and the IC can always agree to modify or amend your contract. You can even agree to call the whole thing off and cancel your agreement.

Unless your contract is one that must be in writing to be legally valid (see "Some Agreements Must Be in Writing," in Chapter 1, for a discussion of this issue), usually it can be modified by an oral agreement. In other words, you need not write down the changes. But it's a good idea to put any changes in writing. This avoids disputes arising from misunderstandings or even intentional shiftiness. (See Chapter 1, Section D, for more on modifying contracts.)

Although not absolutely necessary, it's a good idea to require all changes to be in writing and signed by you and the IC before they become binding. You can still negotiate changes by phone or fax, but they won't change the agreement until you and the IC sign a formal modification. This eliminates even the possibility of orally amending the agreement. This approach requires more time and paperwork, but it provides you with the security of knowing that it will be impossible for the IC to argue successfully that the agreement was changed orally.

a. Significant changes require mutual give-and-take

To be legally enforceable, a contract modification that significantly increase's one party's burdens or reduces its benefits must be supported by what lawyers call "consideration"—that is, the party must receive something of value in return for agreeing to the change in the original agreement. For example, if you and the IC agree that the IC will perform additional services not required by your original agreement, the IC's promise to do extra work for you will be enforceable only if you give something of value in return (usually additional money). Similarly, your agreement to modify the original contract so as to give the IC much more time to complete the work would be enforceable only if the IC gave you something of value in return—for example, a reduction in the IC's fee.

SOME ORAL MODIFICATIONS WILL STICK

Even if your agreement requires all modifications to be set forth in writing, as we suggest, some oral modifications may be legally enforceable in some circumstances. This happens when the party seeking to enforce the oral modification has relied on the other party's oral representations and would be harmed unfairly if the oral modification weren't upheld. Basically, courts will enforce the modification if it would be unfair not to do so. For example, if an IC relies on your oral promise to pay the IC more than the original agreement price in exchange for additional work, you will be legally bound by your promise if the IC actually performs the work in reliance on your oral agreement to pay more.

---

b.   Your duty to seriously consider a proposed change

Neither you nor the IC is legally required to accept a proposed change to your contract. But because you are legally obligated to deal with each other fairly and in good faith, you can't simply refuse all reasonable requests for modifications without attempting to reach a resolution. The duty to deal in good faith means that you can't use the writing as a set of blinders or a wholesale excuse to refuse to reconsider.

A form you can use to amend your agreement can be found on the CD-ROM. You'll need to fill in the appropriate changes.

Appendix II contains a blank, tear-out form that you can use to amend your agreement. You'll need to fill in the appropriate information.

17.   Clause 17. Confidentiality (Optional)

(Optional: Check box and complete if applicable.)

☐ Contractor acknowledges that it will be necessary for Client to disclose certain confidential and proprietary information to Contractor in order for Contractor to perform duties under this Agreement. Contractor acknowledges that any disclosure to any third party or any misuse of this proprietary or confidential information would irreparably harm Client. Accordingly, Contractor will not disclose or use, either during or after the term of this Agreement, any proprietary or confidential information of Client without Client's prior written permission except to the extent necessary to perform services on Client's behalf.

Proprietary or confidential information includes:

• the written, printed, graphic, or electronically recorded materials furnished by Client for Contractor to use

• any written or tangible information stamped "confidential," "proprietary," or with a similar legend or any information that Client makes reasonable efforts to maintain the secrecy of

• business or marketing plans or strategies, customer lists, operating procedures, trade secrets, design formulas, know-how and processes, computer programs and inventories, discoveries and improvements of any kind, sales projections, and pricing information

• information belonging to customers and suppliers of Client about whom Contractor gained knowledge as a result of Contractor's services to Client, and

• other: _____.

Contractor shall not be restricted in using any material that is publicly available, already in Contractor's possession, known to Contractor without restriction, or rightfully obtained by Contractor from sources other than Client.

Upon termination of Contractor's services to Client, or at Client's request, Contractor shall deliver to Client all materials in Contractor's possession relating to Client's business.

> Contractor acknowledges that any breach or threatened breach of this clause will result in irreparable harm to Client for which damages would be an inadequate remedy. Therefore, Client shall be entitled to equitable relief, including an injunction, in the event of such breach or threatened breach of this clause). Such equitable relief shall be in addition to Client's rights and remedies otherwise available at law.

ICs often learn about a hiring firm's trade secrets during the course of their work. A trade secret is information or know-how that is not generally known by others and that provides its owner with a competitive advantage in the marketplace. The information can be an idea, written words, a formula, process or procedure, technical design, customer list, marketing plan, or any other secret that gives the owner an economic advantage.

If a trade secret owner takes reasonable steps to keep the confidential information or know-how secret—for example, does not publish it or otherwise make it freely available to the public—the laws of most states will protect the owner from disclosures of the secret by:

- the owner's employees
- people who agree not to disclose it (for example, an IC who signs a contract with a confidentiality clause)
- industrial spies, and
- competitors who wrongfully acquire the information.

This clause identifies confidential information as including:

- the written, printed, graphic, or electronically recorded materials you furnish the IC and any materials you mark "confidential" or with a similar legend
- your business plans, customer lists, operating procedures, trade secrets, design formulas, know-how and processes, com-

puter programs and inventories, discoveries and improvements of any kind, and

- information belonging to your customers and suppliers about whom the IC gained knowledge as a result of the IC's services for you.

Because ICs whom you hire today may end up working for a competitor tomorrow, it's very important that you make clear to ICs that you expect them to protect your trade secrets—that is, not to disclose them to others without your permission.

Be aware, however, that not everything qualifies as a trade secret. For example, the general knowledge, skills, and experience an IC acquires while working for you are not trade secrets and therefore are not covered by this nondisclosure provision. Other things that don't qualify as your trade secrets include:

- material that is publicly available—for example, published in a trade journal
- material already in the IC's possession or known to the IC without restriction before the IC began working for you, and
- material the IC obtains from sources other than you.

As you can see, it can be difficult to tell what is an IC's general knowledge and what is your trade secret. To protect yourself as much as possible, include Clause 17's nondisclosure provision in the agreement. This provision states that the IC may not disclose your confidential information to others without your permission. If the IC violates this clause, you can sue. You may also sue the person or company to whom the IC disclosed your information if you can show that it knew or had reason to know that the IC owed a duty to you to keep the information confidential. You may be able to prevent the competitor from using the information.

If there are any specific items of information you want to make sure the IC knows not to disclose, add them in the "other" section of the clause. The provision also requires the IC to return all your materials to you when the services end.

If the contractor breaches or threatens to breach this confidentiality clause, your main legal remedy will be to sue to prevent the contractor from disclosing or using your confidential information. To do this, you must convince the court that the contractor's disclosure of your confidential information would cause you irreparable harm—harm that money alone might not be able to rectify. The last paragraph in the confidentiality provision is intended to help you prove that such irreparable harm would occur.

18. Clause 18. Resolving Disputes

(Choose Alternative A, B, or C and any desired optional clauses.)

□ ALTERNATIVE A

If a dispute arises under this Agreement, any party may take the matter to court.

(Optional: Check if applicable.)

□ If any court action is necessary to enforce this Agreement, the prevailing party shall be entitled to reasonable attorney fees, costs, and expenses in addition to any other relief to which the party may be entitled.

□ ALTERNATIVE B

If a dispute arises under this Agreement, the parties agree to first try to resolve the dispute with the help of a mutually agreed-upon mediator in

_____

[list city or county where mediation will occur]. Any costs and fees other than attorney fees associated with the mediation shall be shared equally by the parties. If the dispute is not resolved within 30 days after it is referred to the mediator, any party may take the matter to court.

(Optional: Check if applicable.)

□ If any court action is necessary to enforce this Agreement, the prevailing party shall be entitled to reasonable attorney fees, costs, and expenses in addition to any other relief to which the party may be entitled.

□ ALTERNATIVE C

If a dispute arises under this Agreement, the parties agree to first try to resolve the dispute with the help of a mutually agreed-upon mediator in

_____ [list city or county where mediation will occur]. Any costs and fees other than attorney fees associated with the mediation shall be shared equally by the parties. If it proves impossible to arrive at a mutually satisfactory solution through mediation, the parties agree to submit the dispute to a mutually agreed-upon arbitrator in _____

[list city or county where arbitration will occur]. Judgment upon the award rendered by the arbitrator may be entered in any court having jurisdiction to do so. Costs of arbitration, including attorney fees, will be allocated by the arbitrator.

If you and the IC get into a dispute—for example, over the price or quality of the IC's services—you should always attempt to resolve it first through informal negotiations between the two of you. This is by far the easiest and cheapest way to resolve any problem.

Unfortunately, informal negotiations don't always work. If you and the IC reach an impasse, you may need help. This part of the agreement describes how you will handle any disputes you can't resolve on your own. Three alternatives are provided. You need to choose one. (See Chapter 4 for more information on dispute resolution.)

a. Court litigation

Choose this alternative if you want to resolve disputes by going to court. This is the traditional way contract disputes are resolved. It is also usually the most expensive and time-consuming way. Using this clause doesn't prevent you from enjoying the benefits of mediation and arbitration discussed below. If you and the IC agree, you can always submit a dispute to mediation or arbitration instead of going to court. However, you can't force the IC to agree to either of these alternatives.

SMALL CLAIMS COURTS HELP YOU RESOLVE DISPUTES

Litigation in small claims court is less expensive and faster than regular trial court. Small claims courts are designed to help resolve disputes involving a relatively small amount of money. The amount ranges from $2,000 to $7,500, depending on the state in which you live. If your dispute concerns an amount of money over the small claims limit, you can waive the excess and still bring a small claims suit. You don't need a lawyer to sue in small claims court; indeed, lawyers are barred from small claims court in several states.

Alternative A for Clause 18 doesn't require you to file suit in small claims court if your suit is within the jurisdictional amount, but it is always an option—and is usually a good idea.

For a complete discussion of pursuing a case in small claims court, see *Everybody's Guide to Small Claims Court,* by Ralph Warner (Nolo).

The optional language in Alternative A provides that if either person wins a lawsuit based on the agreement—that is, becomes the prevailing party—the loser has to pay the other person's attorney fees and expenses. Without this clause, each side must pay its own expenses. This language can help make filing a lawsuit economically feasible and will give the IC an additional

reason to settle if you have a strong case. And, of course, it will give *you* pause if you are considering filing suit with a weak case.

Note that this "attorney fees and costs" clause covers only those disputes arising under your contract. It will not, for example, cover a lawsuit by the IC against you for personal injuries or sexual harassment. Nor will you be entitled to attorney fees if a contractual dispute with the IC is settled before trial unless you get a concession from the IC that you are the prevailing party. Most suits are settled without trial, and the other side is unlikely to concede that you are entitled to fees and costs. So even if you have an attorney fees clause and end the dispute thinking that you've come out ahead, you may still have to pay your own costs and fees.

Before deciding to include this fees and costs clause, think carefully about the circumstances of this particular IC and this particular job. There may be cases in which you do not want to include an attorney fees provision. An IC who has little or no money won't be able to pay the fees, so the provision will be ineffective. What's worse, an attorney fees provision could help the IC convince a lawyer to file a case against you, since the lawyer will be looking to you for the fee instead of asking the IC to provide a cash retainer up front. If you have substantially more financial resources than the IC or you think you are more likely to break the contract than the IC, an attorney fees provision is not in your interests. If this is your situation, don't include this optional clause in your agreement.

b. Mediation

Choose Alternative B if you want to try mediation before going to court. Mediation, an increasingly popular alternative to full-blown litigation, works like this: You and the IC agree on a neutral third person to try to help you settle your dispute. The mediator tries to help you arrive at a decision but has no power to impose one. In other words, there is no resolution unless both parties agree to it.

Insert the place where the mediation meeting will occur. You'll usually want it in the city or county where your office is located. You don't want to have to travel a long distance to attend a mediation.

If the mediation doesn't help, you still have the option of going to court. If you wish, you may include the optional clause requiring the loser in any court litigation to pay the other party's attorney fees. (See the discussion in Subsection a, above, regarding the practical considerations to consider when contemplating a "fees and costs" clause.)

c.　Arbitration

Choose Alternative C if you want to avoid court altogether. Under this clause, you and the IC agree that you'll first try to resolve your dispute through mediation. If this doesn't work, you must submit the dispute to binding arbitration. Arbitration is like an informal court trial held before an arbitrator rather than a judge or jury.

You and the IC can agree on anyone to serve as the arbitrator. Arbitrators are often retired judges, lawyers, or persons with special expertise in the field involved. Businesses often use private dispute resolution services that maintain a roster of arbitrators. The best known of these is the American Arbitration Association, which has offices in most major cities.

You may be represented by a lawyer in the arbitration, but it's not required. The arbitrator's decision is final and binding—that is, you can't go to court and try the dispute again if you don't like the arbitrator's decision, except in unusual cases where the arbitrator was guilty of fraud, misconduct, or bias.

Be aware that by using this provision, you're giving up your right to go to court. The advantage is that arbitration is usually much cheaper and faster than court litigation. And if your arbitrator is familiar with the subject matter of the dispute,

you may save time and energy by not having to explain lots of background information, as you would if you were speaking to a jury or a judge unfamiliar with your field.

This provision states that the arbitrator's award can be converted into a court judgment. This means that if the losing side doesn't pay the money required by the award, the other party can easily obtain a judgment and enforce it like any other court judgment—for example, by seizing the losing side's bank accounts and property to pay the amount due.

The provision leaves it up to the arbitrator to decide who should pay the costs and fees associated with the arbitration.

You must insert the place where the arbitration hearing will occur. Usually, you'll want it in the city or county where your office is located. You don't want to have to travel a long distance to attend an arbitration.

19.　Clause 19. Applicable Law

> This Agreement will be governed by the laws of the state of _____ .

As you doubtless know, each state has its own system of laws and courts. Most of the time, you and the IC will have offices in the same state, and the work the IC performs will be done in that same state. In this event, it's obvious that your contract should be interpreted according to the law of this state. This law will determine, for example, how long you have to file a lawsuit if the IC breaks the agreement.

However, things aren't so simple if you and the IC are in different states or all or part of the work the IC performs is in another state. For example, a New Jersey business may employ a New York IC to do work on property it owns in Vermont. In this event, if relations go sour, lawyers will engage in a complex legal analysis to determine which state's law should apply to the dispute.

To avoid this lengthy and costly preliminary legal entanglement, it's common for parties to contracts to agree in advance on which state's law should apply in the event of a dispute. In addition to avoiding additional lawyer fees, this allows you (or a savvy IC) to pick the state with laws most favorable to you. Ordinarily, courts enforce such agreements.

In most cases, choosing the law of your home state will be to your advantage, because its law is familiar to both you and your attorney. This clause allows you to do so. Consult an attorney before choosing the law of the IC's home state or the state where the work is to be done.

20. Clause 20. Notices

All notices and other communications in connection with this Agreement shall be in writing and shall be considered given as follows:

- when delivered personally to the recipient's address as stated on this Agreement
- three days after being deposited in the United States mail, with postage prepaid to the recipient's address as stated on this Agreement, or
- when sent by fax or electronic mail, such notice is effective upon receipt provided that a duplicate copy of the notice is promptly given by first class mail, or the recipient delivers a written confirmation of receipt.

When you want to do something important involving the agreement—terminate it, for example—you need to tell the IC about it. This is called giving notice. This provision gives you several options for providing the IC with notice: by personal delivery, by mail, or by fax or electronic mail followed by a confirming letter.

If you give notice by mail, it is not effective until three days after it's deposited in the mail. For example, if you want to end the agreement on 30 days' notice and you mail your notice of termination to the IC, the agreement will not end until 33 days after you mailed the notice.

21. Clause 21. No Partnership

This Agreement does not create a partnership relationship. Contractor does not have authority to enter into contracts on Client's behalf.

Whenever two or more people decide to carry on as co-owners of a business for profit, a partnership is automatically created unless the co-owners form a corporation or a limited liability company. Contrary to what you might think, a partnership may be created without taking any official steps like filing papers with a secretary of state (which must be done to form a corporation or a limited liability company). While you and the IC may want to work closely on your project, you definitely don't want to be the IC's partner. Why? Because partners are liable for each other's debts.

This clause makes it clear to the IC that you are not a partner—and that the IC shouldn't tell people that you are. Make sure the IC doesn't do anything that implies you are a partner, such as putting your name on the IC's business stationery.

This clause will also prove helpful if a creditor of the IC attempts to sue you for the IC's debts by claiming you're a partner. It shows that you never intended to enter into a partnership relationship with the IC and therefore should not be liable for the IC's debts. In other words, the clause will help provide you with a defense against such a creditor's claim.

22. Clause 22. Assignment and Delegation (Optional)

(Choose Alternative A or B, if applicable.)

☐ ALTERNATIVE A

Either Contractor or Client may assign its rights and may delegate its duties under this Agreement.

☐ ALTERNATIVE B

Contractor may not assign or subcontract any rights or delegate any of its duties under this Agreement without Client's prior written approval.

Normally, the person or firm you deal with at the contract-signing stage—the independent contractor—will be the one who does the work, and you'll be the one for whom the work is done. But in an effort to promote flexibility in business relations, the law doesn't automatically insist that the original IC or you stick with the contract throughout its life. Instead, someone else can take over the contract for either of you.

An assignment is the process by which rights or benefits under a contract are transferred to someone else. For example, you might assign the right to receive the benefit of the IC's services to someone else. Such a person is called an assignee. When this occurs, the assignee steps into your shoes. The IC must now work for the assignee, not for you.

> **Example:** Acme, Inc., has signed an IC agreement with Joe, who will perform janitorial services. Acme assigns its rights under the contract to a neighbor in its building called Big Tech, Inc. This means that Joe must perform the work for Big Tech instead of Acme. If Joe fails to do so, Big Tech can sue him for breach of contract.

An IC may also assign the benefits he or she receives under an IC agreement to someone else.

> **Example:** ABC Construction contracts to build a new house for Agnes. To help run its business, ABC borrows money from the Widows & Orphans Bank. As a condition for obtaining a $100,000 loan, the bank requires ABC to assign to it the payments received from Agnes. This means Agnes must pay the bank, not ABC.

Delegation is the flipside of assignment. Instead of transferring benefits under a contract, you transfer the duties. However, the person delegating duties under a contract usually remains responsible if the person to whom the delegation was made fails to perform.

> **Example:** Jimmy, a plumber, has contracted to provide Beta, Inc., with plumbing repair services. Because Jimmy is overloaded with work, he delegates to Mindy his duty to perform plumbing services for Beta. This means that Mindy, not Jimmy, will now do the work. But Jimmy remains liable to Beta if Mindy doesn't perform adequately.

a. Legal restrictions on assignment and delegation

Unless a contract provides otherwise, ordinarily you can freely assign and delegate subject to some important legal limitations. The first limitation is based on simple fairness: You can't assign the benefit of your services to someone else without the IC's consent if it would increase the work the IC must do or otherwise magnify the IC's burdens under the contract. This restriction works both ways—the IC can't delegate duties without your consent if it would decrease the benefits you would receive.

The most significant restrictions are those on delegation of duties. Here, the law protects you from an IC who tries to fob off work onto someone else. An IC may not delegate duties under the contract if it would materially alter the performance you contracted for. Specifically, ICs may not, without your consent, delegate contracts involving:

- personal services including services by IC writers, artists, and musicians, and other services—the performance of which is largely dependent upon an IC's unique abilities, taste, skills, personality, or discretion, and

- contracts for professional services, such as those performed by lawyers, doctors, or architects.

Service contracts with ICs involving more mechanical tasks are ordinarily delegable without the hiring firm's consent. For example, a contract to construct or paint a house would ordinarily be delegable. However, even in these cases, the IC can't delegate duties to a person who lacks the skill or experience to complete the work satisfactorily.

Example: Annie contracts with Frontier Construction to build her a log home. Frontier attempts to delegate its duty to construct the house to Davy, an unlicensed contractor who has never built a log home. Annie is legally entitled to object to this delegation and demand that Frontier do the work itself or delegate the work to a qualified log home builder.

b. How to handle assignment and delegation

You have several options for dealing with assignment and delegation in your agreement.

Option #1: Do nothing. You need not say anything about assignment and delegation in your agreement. Your agreement will be assignable or delegable subject to the restrictions noted above. This is the best option in most cases.

Option #2: Let it happen. If you absolutely do not care who does the work required by the agreement, choose Alternative A to allow an unrestricted right of assignment and delegation. Incidentally, this choice can help you if you are challenged on your treatment of the worker as an IC. It is strong evidence of an IC relationship because it shows you're concerned only with results, not who achieves them. It also demonstrates lack of

control over the IC. Note, however, that if you include this clause in the agreement, the restrictions on assignment and delegation discussed above will not apply—that is, the IC will be able to delegate duties even if they involve personal services.

Option #3: Retain control. There may be some situations in which you really don't want an IC to delegate contractual duties without your consent. If you have hired a particular IC because of his or her special expertise, creativity, reputation for performance, or financial stability, you may not want someone else performing the services, even if the person is technically qualified to do the work. In this event, choose Alternative B to prevent assignment or delegation by the IC under any circumstances without your written approval in advance. Unfortunately, requiring such approval tends to indicate an employment relationship, so don't include it your agreement unless absolutely necessary.

23. Signatures

Client: _____
Name of Client

By: _____
Signature

_____
Typed or Printed Name

Title: _____

Date: _____

Contractor: _____
Name of Contractor

By: _____
Signature

_____
Typed or Printed Name

Title: _____

Taxpayer ID Number: _____

Date: _____

The end of the main body of the agreement should contain spaces for you to sign, write in your title, and enter the date. There should also be space for the IC to sign and provide a taxpayer ID number.

The IRS requires anyone who hires an IC to obtain the person's taxpayer identification number and include it in every contract signed with the IC. This is either the IC's Social Security number or an employer identification number (EIN) obtained from the IRS. If the IC is a corporation, partnership or limited liability company, it will have an EIN, not a Social Security number.

Obtaining and listing the taxpayer ID number is not just a bureaucratic formality. It will become important if you hire an IC who is not a corporation or LLC and you or the IC are audited. If you don't have the number, the IRS will assume that the IC is not reporting or paying taxes on the income you provide. To make up this lost revenue, you'll be required to withhold 31% of all payments over $599 you make to the IC and remit the money to the IRS.

Withholding some of the money earned by the IC and sending it to the IRS (not unlike employee withholding) is called backup withholding. If you fail to backup withhold, the IRS may impose an assessment against you equal to 31% of what you paid the IC. Backup withholding is not required if the IC is incorporated or an LLC. For a detailed discussion of backup withholding, see *Working with Independent Contractors*, by Stephen Fishman (Nolo).

You don't need to worry about backup withholding if you require the IC to list the taxpayer ID number in your IC agreement. Simply keep the agreement on file, and you'll be able to prove to the IRS that you obtained the number. Another approach is to have the IC complete IRS Form W-9, Request for Taxpayer Identification Number, and retain it in your files. You don't need to file it with the IRS. If you're audited, you can show the form to the auditor to prove that you obtained the IC's ID number. Showing the auditor a signed contract containing the number will work just as well.

B. Agreement for Independent Contractor to Use

You should use this agreement if you're working as an IC. It contains a number of provisions favorable to you that a hiring firm wouldn't ordinarily include in an agreement it prepared—for example, requiring the hiring firm to pay a late fee if it doesn't pay you on time.

The entire text of the general independent contractor agreement is on the CD-ROM. You can modify individual clauses to tailor the text to your needs. However, be sure to read the following discussion before you make any major changes.

The agreement is also reproduced as a tear-out form in Appendix III. Use this form if you don't have access to a computer. It contains blank spaces for you to fill in by hand or typewriter.

Although we provide a tear-out form for those of you who either don't have computers or who would prefer not to use the CD-ROM, don't just tear out the form and use it as your IC agreement if you can avoid doing so. A fill-in-the-blank IC agreement is the least persuasive type of contract when you are trying to convince a government agency that you are an IC and not an employee of the hiring firm. The best way to create your agreement is to use the CD-ROM version and tailor it to your needs. If you can't use the CD-ROM version, then retype the version at the back of the book and tailor it to your needs. Use the fill-in-the-blank form only as a last resort.

Call up the agreement on your computer or tear out the form in Appendix III, and read it along with these instructions and explanations.

Title of agreement. You don't need a title for an independent contractor agreement, but if you want one, call it "Independent Contractor Agreement" or "Agreement for Professional Services." Since you are not your client's employee, do not use "Employment Agreement" as a title.

Names of independent contractor and hiring firm. At the beginning of your contract, refer to yourself by your full business name. Later on in the contract, you can use an obvious abbreviation.

If you're a sole proprietor, insert the full name you use for your business. This may be your own name. If you do business under a fictitious business name or an assumed name (sometimes referred to as your "DBA" for "doing business as"), use that name. For example, if Jack Olsen calls his one-person marketing research business "Olsen Marketing Research," he would use that name on the contract. Using a fictitious business name helps show that you're an independent business, not an employee.

If your business is incorporated, use your corporate name, not your own name—for example: "John Smith, Incorporated" instead of "John Smith." If you've formed a limited liability company (LLC) use your full LLC name—for example, "John Smith, LLC." Never refer to yourself as an employee or to the client as an employer.

For the sake of brevity, it is usual to identify yourself and the client by shorter names in the rest of the agreement. You can use an abbreviated version of your full name—for example, "Olsen" for "Olsen Marketing Research." Or you can refer to yourself simply as "Contractor" or "Consultant."

Refer to the client initially by its company name and subsequently by a short version of the name, or as "Client" or "Firm."

Insert the addresses of the principal places of business of both the client and yourself. If you or the client have more than one office or workplace, the principal place of business is the main office.

1. Clause 1. Services to Be Performed

> (Choose Alternative A or B.)
>
> ☐ ALTERNATIVE A
>
> Contractor agrees to perform the following services:
>
> _____
>
> _____
>
> _____
>
> ☐ ALTERNATIVE B
>
> Contractor agrees to perform the services described in Exhibit A, which is attached to and made part of this Agreement.

The first clause in the contract asks you to insert a detailed description of what you are expected to accomplish. Word the description carefully to emphasize the results you're expected to achieve. Don't describe the method by which you will achieve the results. As an independent contractor, it should be up to you to decide how to do the work. The client's control should be limited to accepting or rejecting your final results. The more control the client exercises over how you work, the more you'll look like an employee. (See "Describing an Independent Contractor's Job," in Section A, above, for examples of good and bad work descriptions.)

Use Alternative A if you include the explanation of services in the contract. Choose Alternative B if the explanation of the services is attached to the main contract. Using an attachment page is advisable if the description is extremely lengthy.

2. Clause 2. Payment

(Choose Alternative A or B and optional clause, if desired.)

☐ ALTERNATIVE A

In consideration for the services to be performed by Contractor, Client agrees to pay Contractor $_____.

☐ ALTERNATIVE B

In consideration for the services to be performed by Contractor, Client agrees to pay Contractor at the rate of $_____ per ☐ day, ☐ week, ☐ hour, ☐ month.

(Optional: Check and complete if applicable.)

☐ Contractor's total compensation shall not exceed $_____ without Client's written consent.

Independent contractors can be paid in many different ways. (For a detailed discussion of payment methods, see Nolo's *Working for Yourself: Law & Taxes for Independent Contractors, Freelancers & Consultants,* by Stephen Fishman.) The two most common payment methods are:

- a fixed fee, and
- by unit of time.

a. Fixed fee

In a fixed fee agreement, you charge an agreed amount for the entire project. Clients usually like this arrangement because they know exactly how much they'll have to pay you. A fixed fee can also be very rewarding for you if you work efficiently and if you estimate accurately the time and expense involved in completing the project. How-ever, if you underestimate the time and expense involved, you could end up earning much less than your work is worth. You can charge a fixed fee for any type of service. But before you quote a fee, it's important that the scope of the project be clearly defined.

Choose Alternative A and insert the amount of your fee if you will be paid a fixed fee.

b. Unit of time

Customarily, many independent contractors—for example, lawyers, accountants, and plumbers—charge by the hour, day, or other unit of time. Charging by the hour does not support your independent contractor status, but you can get away with it if it's a common practice in the field in which you work. Choose Alternative B and insert the amount of your payment if you will be paid by the hour or other unit of time.

Be aware that clients often wish to place a cap on the total amount they'll spend on the project when you're paid by the hour. They're afraid you might work slowly in order to earn a larger fee. If the client insists on a cap, make sure it allows you to work enough hours to get the project done. Insert the amount of the cap. If the client does not insist on a cap, delete this language.

For detailed advice on how to determine how much you should charge per hour for your services, see *Working for Yourself: Law & Taxes for Independent Contractors, Freelancers, & Consultants,* by Stephen Fishman (Nolo).

3. Clause 3. Terms of Payment

> (Choose Alternative A, B, C, or D.)
>
> ☐ ALTERNATIVE A
> Upon completing Contractor's services under this Agreement, Contractor shall submit an invoice. Client shall pay Contractor within _____ days from the date of Contractor's invoice.
>
> ☐ ALTERNATIVE B
> Contractor shall be paid $_____ upon signing this Agreement and the remaining amount due when Contractor completes the services and submits an invoice. Client shall pay Contractor within _____ days from the date of Contractor's invoice.
>
> ☐ ALTERNATIVE C
> Contractor shall be paid according to the schedule of payments set forth in Exhibit _____ attached to and made part of this Agreement.
>
> ☐ ALTERNATIVE D
> Contractor shall send Client an invoice monthly. Client shall pay Contractor within _____ days from the date of each invoice.

You need to specify how you will bill the client and be paid. As an IC, you'll have to submit an invoice to the client setting out the amount due. An invoice doesn't have to be fancy or filled with legalese. It should include an invoice number, the dates covered by the invoice, the hours expended if you're being paid by the hour, and a summary of the work performed. There are also several computer accounting programs, such as Quickbooks, that can generate invoices for you.

 An invoice you can use with your clients is on the CD-ROM.

The forms at the back of this book include an invoice. Tear out the original and copy it, saving the original for future use.

a. Payment upon completing work

Choose Alternative A if you will be paid when you complete the work. Insert the time period in which the client is required to pay you after you send the invoice. A period of 30 days is a typical payment term, but it can be shorter or longer if you wish. Note that the time for payment starts to run as soon as you send your invoice, not when the client receives it. This will help you get paid more quickly.

b. Divided payments

You may want something like a down payment for your work—to be paid part of a fixed or hourly fee when the agreement is signed and the remainder when the work is finished. Choose Alternative B if you will be paid in divided payments. Insert the amount of the upfront payment. This amount is, of course, subject to negotiation. Many independent contractors like to receive at least one-third to one-half of a fee before they start work. If the client is new or might have problems paying you, it's wise to get as much money as you can in advance.

c. Fixed-fee installment payments

Choose Alternative C if you will be paid in installments. You may prefer this if the project is long and complex, because you won't have to wait until the project is finished to receive the bulk of your payment. One way to create a payment schedule is to break the job into phases or milestones and receive a fixed fee when each phase is completed. Clients often like this pay-as-you-go arrangement, too.

To do this, draw up a schedule of installment payments tying each payment to your completion of specific services. Usually, it's easier to set forth the schedule in a separate document and attach it to the agreement. The main body of the agreement should simply refer to the attached payment schedule. (See Section A3, above, for an example of such a payment schedule.)

d. Hourly payment for lengthy projects

Choose Alternative D if you're being paid by the hour or other unit of time and the project will last more than one month. Under this provision, you submit an invoice to the client each month setting forth how many hours you've worked. The client is required to pay you within a specific number of days from the date of each invoice. This form of payment is not supportive of your IC status, but you can get away with it if it's a common practice in your field.

4. Clause 4. Late Fees (Optional)

(Check and complete if applicable.)

☐ Late payments by Client shall be subject to late penalty fees of _____ % per month from the due date until the amount is paid.

Many independent contractors charge a late fee if the client doesn't pay within the time specified in the independent contractor agreement or invoice. Charging late fees for overdue payments can get clients to pay on time. If you have a late fee policy, you should state clearly what your late fee is on all your invoices. The late fee is normally expressed as a monthly interest charge.

a. How much can you charge?

Your state might have restrictions on how much you can charge as a late fee. To find out, consult with an attorney.

Alternatively, you can safely charge a late fee that is no higher than what banks charge businesses to borrow money. Find out the current bank interest rate by calling your bank or looking in the business section of your local newspaper.

b. Calculating a late fee

To calculate how much you would actually charge on an overdue account, you'll have to do some math. Your late fee will be based on simple interest—that is, you'll charge the late fee solely on the principal amount you're owed, not on the interest that accrues. (Charging interest on interest, known as compound interest, could be viewed as unreasonable under these circumstances and might not be upheld by a court.)

First, divide the annual interest rate by 12 to determine your monthly interest rate.

> **Example:** Sam decides to start charging clients a late fee for overdue payments. He knows banks are charging 12% interest per year on borrowed money and decides to charge the same. He divides this rate by 12 to determine his monthly interest rate: 1%.

Then, multiply the monthly rate by the amount due to determine the amount of the monthly late fee.

> **Example:** Bogus Corp. is 30 days late paying Sam a $10,000 fee. Sam multiples this amount by his 1% finance charge to determine his late fee, which is $100 (.01 x $10,000). He adds this amount to Bogus' account balance. He adds $100 for every month until Bogus pays its bill.

5. Clause 5. Expenses

(Choose Alternative A or B and optional clause, if desired.)

☐ ALTERNATIVE A

Contractor shall be responsible for all expenses incurred while performing services under this Agreement.

(Optional: Check if applicable.)

☐ However, Client shall reimburse Contractor for all reasonable travel and living expenses necessarily incurred by Contractor while away from Contractor's regular place of business to perform services under this Agreement. Contractor shall submit an itemized statement of such expenses. Client shall pay Contractor within 30 days from the date of each statement.

☐ ALTERNATIVE B

Client shall reimburse Contractor for the following expenses that are directly attributable to work performed under this Agreement:

- travel expenses other than normal commuting, including airfares, rental vehicles, and highway mileage in company or personal vehicles at __ cents per mile

- telephone, fax, online, and telegraph charges

- postage and courier services

- printing and reproduction

- computer services, and

- other expenses resulting from the work performed under this Agreement.

Contractor shall submit an itemized statement of Contractor's expenses. Client shall pay Contractor within 30 days from the date of each statement.

Expenses are the costs you incur that you can attribute directly to your work for a client. They include, for example, the cost of phone calls or traveling done on the client's behalf. Expenses do not include your normal fixed overhead costs such as your office rent, the cost of commuting to and from your office, or the wear and tear on your equipment. They also do not include materials the client provides you to do your work.

a. You pay expenses

Setting your compensation at a level high enough to cover your expenses is simple and safe. On the practical side, it frees you from having to keep records of your expenses. Keeping track of the cost of every phone call or photocopy you make for a client can be a real chore and may be more trouble than it's worth. However, if a project will involve expensive traveling, you may wish to bill the client separately for the cost.

Covering your own expenses will also help bolster your status as an IC. In the recent past, the IRS viewed the payment of a worker's expenses by a client as a sign of employee status. Although the agency now downgrades the importance of this factor, it's not entirely gone. You'll be on more solid ground if you don't ask the client to reimburse you for expenses.

And, even though the IRS has changed its stance, other government agencies may consider payment of a worker's business or traveling expenses to be a strong indication of employment relationship. For this reason, usually it is best that your compensation be high enough to cover your expenses; you should not be reimbursed separately for them.

Choose Alternative A if you are responsible for expenses. Use the Optional paragraph if the client will reimburse you for travel and living expenses while you work on the project.

b. Expenses reimbursed

Some types of independent contractors customarily have expenses paid by their clients. For example, attorneys, accountants, and many independent contractor consultants typically charge their clients separately for photocopying charges, deposition fees, and travel. Where there is an otherwise clear independent contractor relationship and payment of expenses is customary in your trade or business, you can probably get away with doing it.

Choose Alternative B if the client will be responsible for paying your expenses.

6. Clause 6. Materials (Optional)

> **(Check if applicable.)**
>
> ☐ Contractor will furnish all materials and equipment used to provide the services required by this Agreement.

Generally, an independent contractor should provide all the materials and equipment necessary to complete a project. Being provided with tools and equipment is a sign that you're an employee, not an IC. This clause states that you will provide all tools, equipment, materials, and supplies needed to accomplish the project.

If the client is providing you with tools or equipment for free, delete this clause. In this situation, it's best for the agreement not to mention tools and equipment.

If you're buying or leasing tools or equipment from the client, you should enter into a separate purchase or lease agreement. You can find forms for selling or leasing equipment in *Legal Forms for Starting & Running a Small Business*, by Fred Steingold (Nolo).

7. Clause 7. Term of Agreement

> This agreement will become effective when signed by both parties and will terminate on the earlier of:
> - the date Contractor completes the services required by this Agreement
> - _____ [date], or
> - the date a party terminates the Agreement as provided below.

The "term of the agreement" simply refers to when it begins and ends. Unless the agreement provides a specific starting date, it begins on the date it's signed. If you and the client sign on different dates and there is no specified starting date, the agreement begins on the date the last person signed. Normally, you shouldn't begin work until the client signs the agreement, so usually it's best that the agreement not provide a specific start date that might be before the client signs.

The start date for the agreement marks the moment when you and the client are legally bound by it—that is, when you are obligated to perform as promised. You don't necessarily have to begin work on the start date. Indeed, ordinarily you can start work whenever you want; just make sure you meet the contract deadline. Both you and the client are legally bound as of the start date even if you haven't actually commenced work.

An independent contractor agreement should have a definite ending date. Ordinarily, this will mark the final deadline for you to complete your services. However, even if the project is lengthy, the end date should not be too far in the future. A good outside time limit is 12 months. A longer term makes the agreement look like an employment agreement, not an independent contractor agreement. If the work is not completed at the end of 12 months, you can negotiate and sign a new agreement. Insert the termination date in this clause.

8. Clause 8. Terminating the Agreement

> **(Choose Alternative A or B.)**
>
> ☐ ALTERNATIVE A
>
> With reasonable cause, either party may terminate this Agreement effective immediately by giving written notice of termination for cause. Reasonable cause includes:
> - a material violation of this Agreement, or
> - nonpayment of Contractor's compensation after 20 days' written demand for payment.
>
> Contractor shall be entitled to full payment for services performed prior to the effective date of termination.
>
> ☐ ALTERNATIVE B
>
> Either party may terminate this Agreement at any time by giving _____ days' written notice of termination. Contractor shall be entitled to full payment for services performed prior to the date of termination.

Signing a contract doesn't bind you forever, no matter what happens. You and the client can agree to call off your agreement at any time. In addition, contracts typically contain provisions allowing either side to terminate the agreement under certain circumstances without the other side's agreement. This built-in cancellation right is known as "termination."

When a contract is terminated, both you and the client stop performing—that is, you discontinue your work, and the client has no obligation to pay you for any work you may do after the effective date of termination. However, the client is legally obligated to pay you for any work you did prior to the termination date. And you remain liable to the client for any damages it may have suffered due to your failure to perform as agreed before the termination date.

> **Example:** Art, a graphic designer, contracts with Corinth Press to design a book cover. Art agrees to deliver the cover by April 1. Due to the press of other business, Art is unable to complete the assignment. Art and Corinth terminate their agreement, and Corinth hires another designer to prepare the cover. However, because of Art's nonperformance, the book's ship date must be postponed by two weeks. Art is liable to Corinth for its monetary losses, such as the cancellation of advance orders, due to this delay.

It's important to clearly define and limit the circumstances under which you or the client may end the agreement. First, practically speaking, you don't want to live with the uncertainty of knowing that the client can dispense with your services at any time for no good reason.

Second, your status as an IC could be jeopardized if your relationship with the client can be severed at any time. In the past, the IRS viewed a wide-open termination provision giving you, the client, or both of you the right to terminate the agreement at any time as strong evidence of an employment relationship. Although the agency no longer considers this to be such an important factor, it's still wise to place some limits on the client's right to terminate the contract. To be on the safe side, both you and the client should have the power to terminate the agreement only:

- if there is reasonable cause to do so, or
- at most, by giving written notice to the other.

These requirements are explained more fully in Subsections a and b, below.

a. Termination with reasonable cause

Choose Alternative A if you wish to allow either party to end the agreement based on reasonable cause. Termination with reasonable cause means that either you or the client have a good business reason to end the agreement. The standard is an objective one: Would any reasonable person in your situation have a valid business reason to end the agreement? For example, a serious violation of the agreement is reasonable cause to terminate the agreement—but what is considered "serious" depends on the particular facts and circumstances.

Surprisingly, unless your contract provides otherwise, a client's failure to pay you on time may not necessarily constitute reasonable cause for you to terminate the agreement. Our contract provides that late payments are always reasonable cause for terminating the contract. The clause provides that you may terminate the agreement if the client doesn't pay you what you're owed within 20 days after you make a written demand for payment. For example, if you send a client an invoice due within 30 days and the client fails to pay within that time, you may terminate the agreement 20 days after you send the client a written demand to be paid what you're owed. This may give clients another incentive to pay you.

> **Example:** IC Jane Janeway invoices her client on April 1. Her payment is due within 30 days (by April 30). On May 5, Jane still hasn't been paid, so she personally delivers to the client a written notice that she'll cancel the

GENERAL INDEPENDENT CONTRACTOR AGREEMENTS **5/33**

contract if she's not paid within 20 days of the date of the notice (May 25). If the client doesn't pay by the 25th, the agreement is canceled as of that date.

The clause also makes it clear that the client must pay you for the services you performed before the contract was terminated.

b. Termination without cause

Sometimes you or the client just can't live with a limited termination right. Instead, you want to be able to get out of the agreement at any time without incurring liability. For example, if a client's business plans change, it may no longer need your services. Or, you may discover that you have too much work and need to lighten your load.

If you want a broad right to end the work relationship, choose Alternative B. You need to provide at least a few days' notice. Being able to terminate without notice tends to make you *really* look like an employee. Insert how long a notice period you will provide. A period of 30 days is a common notice period, but shorter notice may be appropriate if the project is of short duration.

9. Clause 9. Independent Contractor Status

Contractor is an independent contractor, not Client's employee. Contractor's employees or subcontractors are not Client's employees. Contractor and Client agree to the following rights consistent with an independent contractor relationship (check all that apply).

☐ Contractor has the right to perform services for others during the term of this Agreement.

☐ Contractor has the sole right to control and direct the means, manner, and method by which the services required by this Agreement will be performed.

☐ Contractor has the right to hire assistants as subcontractors, or to use employees to provide the services required by this Agreement.

☐ The Contractor or Contractor's employees or subcontractors shall perform the services required by this Agreement; Client shall not hire, supervise, or pay any assistants to help Contractor.

☐ Neither Contractor nor Contractor's employees or subcontractors shall receive any training from Client in the skills necessary to perform the services required by this Agreement.

☐ Client shall not require Contractor or Contractor's employees or subcontractors to devote full time to performing the services required by this Agreement.

☐ Neither Contractor nor Contractor's employees or subcontractors are eligible to participate in any employee pension, health, vacation pay, sick pay, or other fringe benefit plan of Client.

One of the most important functions of an independent contractor agreement is to help establish that you are an independent contractor, not your client's employee. The key to doing this is to make clear that you, not the client, have the right to control how the work will be performed.

You will need to emphasize the factors the IRS and other agencies consider in determining whether a client controls how the work is done. Of course, if you merely recite what you think the IRS wants to hear but fail to adhere to these understandings in practice, agency auditors won't be fooled. Think of this clause as a reminder to you and your client about how to conduct your business relationship.

You need not insert any information in this provision. Just read each clause. If any clause doesn't apply to your situation, delete it.

10. Clause 10. Local, State, and Federal Taxes

Contractor shall pay all income taxes and FICA (Social Security and Medicare taxes) incurred while performing services under this Agreement. Client will not:

- withhold FICA from Contractor's payments or make FICA payments on Contractor's behalf
- make state or federal unemployment compensation contributions on Contractor's behalf, or
- withhold state or federal income tax from Contractor's payments.

The charges included here do not include taxes. If Contractor is required to pay any federal, state, or local sales, use, property, or value added taxes based on the services provided under this Agreement, the taxes shall be separately billed to Client. Client shall be responsible for any interest or penalties incurred due to late payment or nonpayment of any taxes by Client.

The agreement should address the issue of who pays federal and state income taxes, Social Security taxes, and sales taxes.

a. Income taxes

Your client should not pay or withhold any income or Social Security taxes on your behalf. Doing so is a very strong indicator that you are an employee, not an independent contractor. Indeed, some courts have classified workers as employees based upon this factor alone. Keep in mind that one of the best things about being an independent contractor is that you don't have taxes withheld from your paychecks.

This straightforward provision in your agreement simply makes clear that you'll pay all applicable taxes due on your compensation and, therefore, the client should not withhold taxes from your payments. You need not insert any information here.

b. Sales taxes

A few states require independent contractors to pay sales taxes, even if they provide their clients with services only, not goods. These states include Hawaii, New Mexico, and South Dakota. Many other states require sales taxes to be paid for certain types of services.

Whether or not your state collects sales taxes for your services, include this provision in your agreement, making it clear that the client will have to pay these and similar taxes. States change sales tax laws constantly and more are beginning to look at services as a good source of tax revenue. So this provision could come in handy in the future, even if you don't really need it now.

11. Clause 11. Exclusive Agreement

This is the entire Agreement between Contractor and Client.

When you put your agreement in writing, it will be treated as the last word on the subjects covered if you and the client intend it to be the final and complete expression of your agreement. Clause 11 expresses this intent. The written agreement will take precedence over any written or oral agreements or promises previously made. This means neither you nor the client can bring up earlier side letters, oral statements, or other material not covered by the contract. However, you and the client can still subsequently amend your agreement. (See Clause 12, just below.) Business contracts normally contain a provision like the one in Clause 11.

If you or the client have relied on earlier representations about the job or the nature of the work to be done, make sure that these documents are attached to the agreement as exhibits or that this important information is repeated in the main body of the agreement. For example, a client's request for proposals, your own proposal or bid, or even your sales literature should be attached if these materials explain and round out the agreement. If this material isn't attached, it likely won't be considered part of the agreement.

12. Clause 12. Modifying the Agreement (Optional)

(Check if applicable.)

☐ This Agreement may be modified only by a writing signed by both parties.

It's very common for clients and independent contractors to want to change the terms of an agreement after work has started. For example, you might discover that you underestimated how much time the project will take and need to be paid more to complete it.

Does every change in the contract have to be in writing? Technically, no—just as you can, in theory, make oral contracts for most agreements, you can make oral changes, too. But the practical reasons for writing things down in the first place also apply to changes later on: A written version is harder to forget, misunderstand, or misconstrue. (However, some oral changes will be enforced even if your agreement calls for any changes to be in writing. See "Some Oral Modifications Will Stick," in Section A16, above.)

Although not absolutely necessary, it's a good idea to add this provision to the contract. It requires all amendments be in writing and signed by you and the client before they become binding. You can still negotiate changes by phone or fax, but they won't change the agreement until you and the client sign a formal amendment to the agreement. This approach requires more time and paperwork, but provides you with the security of knowing that it will be almost impossible for the client to legally enforce any claimed oral amendment to the agreement.

 A form you can use to amend your agreement may be found on the CD-ROM. You'll need to fill in the appropriate changes.

 Appendix III contains a blank, tear-out form that you can use to amend your agreement. You'll need to fill in the appropriate information.

a. Significant changes require mutual give-and-take

To be legally enforceable, a contract modification that significantly increases one party's burdens or reduces its benefits must be supported by what lawyers call "consideration"—that is, the party must receive something of value in return for agreeing to the change in the original agreement. For example, if you and the client agree that you will perform additional services not required by your original agreement, your promise will be enforceable only if the client gives you something of value in return—for example, more money. Similarly, an agreement by the client to modify the original contract to give you more time to complete the work will be enforceable only if you give the client something of value in return—for example, a reduction in your fee.

b. Duty to deal in good faith

Neither you nor the client are legally required to accept a proposed change to your contract. But because you are legally obligated to deal with each other fairly and in good faith, you can't simply refuse all reasonable requests for modifications without attempting to reach a resolution. The duty to deal in good faith means that you can't use the writing as a wholesale excuse to refuse to reconsider.

13. Clause 13. Resolving Disputes

(Choose Alternative A, B, or C and any desired optional clauses.)

☐ ALTERNATIVE A

If a dispute arises under this Agreement, any party may take the matter to court.

(Optional: Check if applicable.)

☐ If any court action is necessary to enforce this Agreement, the prevailing party shall be entitled to reasonable attorney fees, costs, and expenses in addition to any other relief to which the party may be entitled.

☐ ALTERNATIVE B

If a dispute arises under this Agreement, the parties agree to first try to resolve the dispute with the help of a mutually agreed-upon mediator in

_____ [list city or county where mediation will occur]. Any costs and fees other than attorney fees associated with the mediation shall be shared equally by the parties. If the dispute is not resolved within 30 days after it is referred to the mediator, any party may take the matter to court.

(Optional: Check if applicable.)

☐ If any court action is necessary to enforce this Agreement, the prevailing party shall be entitled to reasonable attorney fees, costs, and expenses in addition to any other relief to which the party may be entitled.

☐ ALTERNATIVE C

If a dispute arises under this Agreement, the parties agree to first try to resolve the dispute with the help of a mutually agreed-upon mediator in

_____ [list city or county where mediation will occur]. Any costs and fees other than attorney fees associated with the mediation shall be shared equally by the parties. If it proves impossible to arrive at a mutually satisfactory solution through mediation, the parties agree to submit the dispute to a mutually agreed-upon arbitrator in

_____ [list city or county where arbitration will occur]. Judgment upon the award rendered by the arbitrator may be entered in any court having jurisdiction to do so. Costs of arbitration, including attorney fees, will be allocated by the arbitrator.

If you and the client get into a dispute—for example, over the price or quality of your services—you should always attempt to resolve it through informal negotiations between the two of you. This is by far the easiest and cheapest way to resolve any problem.

Unfortunately, informal negotiations don't always work. If you and the client reach an impasse, you may need help. This part of the agreement covers how you and the client will handle any disputes you can't resolve on your own. Three alternatives are provided. You need to choose one. (See Chapter 4 for more information on resolving disputes.)

a. Court litigation

Choose Alternative A if you want to have the option of going directly to court to resolve disputes. This is the traditional way contract disputes are resolved. It is also usually the most expensive and time consuming—unless your dispute can be heard in small claims court as explained in Section A18, above.

If you win a lawsuit based on the agreement, normally you will not be awarded attorney fees unless your agreement requires it. The optional language provides that if either person wins a lawsuit based on the agreement—that is, becomes the prevailing party—the loser is required by this clause to pay the winner's attorney fees and expenses.

If you are short on funds and cannot handle the sometimes considerable up-front charge a lawyer will ask to take your case (called a retainer), this can help make a lawsuit economically feasible. That's because the lawyer may agree to forgo the retainer fee in hopes of winning and getting paid by the other side. There is another advantage: A client who knows that it may have to pay your attorney fees in the event of a dispute has an additional reason to conduct business fairly and to settle if you have a strong case.

However, note that such a clause covers only those disputes arising under your contract. It will not, for example, cover a lawsuit by you against the client for personal injuries or sexual harassment. Nor will you be entitled to attorney fees if a contractual dispute with the client is settled before trial unless you get a concession from the client that you are the prevailing party. Most suits are settled without trial, and a "prevailing party" concession is unlikely. So even if you have an attorney fees clause and feel that you came out on top, you may still have to pay your own fees and costs.

Sometimes an attorney fees provision can work against you. Since it works both ways, it may help your client find an attorney to sue you and make you more anxious to settle. If you think you are more likely to violate the agreement than the client is, an attorney fees provision is probably not a good idea.

b. Mediation

Instead of marching into small claims court or a lawyer's office if you and your client disagree, you might want to try mediation. Mediation, an increasingly popular alternative to full-blown litigation, works like this: You and the client agree on a neutral third person to try to help you settle your dispute. The mediator has no power to impose a decision, only to try to help you arrive at one. In other words, unless both parties agree with the resolution, there is no resolution. If the mediation doesn't help, you still have the option of going to court.

Choose Alternative B if you want to try mediation before going to court. Insert the place where the mediation will occur. You'll usually want it in the city or county where your office is located. You don't want to have to travel a long distance to attend a mediation. If you wish, you may include the optional clause requiring the loser in any court litigation to pay the other party's attorney fees.

c. Arbitration

Choose Alternative C if you want to avoid court altogether. Under this clause, you and the client first try to resolve your dispute through mediation. If this doesn't work, you must submit the dispute to binding arbitration. Arbitration is like an informal court trial without a jury, and the dispute is decided by an arbitrator rather than a judge.

Arbitrators are often retired judges, lawyers, or persons with special expertise in the field involved. Businesses often use private dispute resolution services that maintain a roster of arbitrators. The best known of these is the American Arbitration Association, which has offices in most major cities.

You may be represented by a lawyer in the arbitration, but it's not required. The arbitrator's decision is final and binding—that is, you can't go to court and try the dispute again if you don't like the arbitrator's decision, except in unusual cases where the arbitrator was guilty of fraud, misconduct, or bias.

By using this provision, then, you're giving up your right to go to court. The advantage is that arbitration is usually much cheaper and faster than court litigation. And if your arbitrator is someone familiar with the subject matter of the dispute, you may save time and energy by not having to explain as many background issues as you would if you were speaking to a judge or jury.

This provision also states that the arbitrator's award can be converted into a court judgment. This means that if the losing side doesn't pay the money required by the award, the other party can easily obtain a judgment and enforce it like any other court judgment—for example, by seizing the losing side's bank accounts and property to pay the amount due.

The provision leaves it up to the arbitrator to decide who should pay the costs and fees associated with the arbitration.

You must insert the place where the arbitration hearing will occur. You'll usually want it in the city or county where your office is located. You don't want to have to travel a long distance to attend an arbitration.

14. Clause 14. Limiting Your Liability to the Client (Optional)

(Check if applicable.)

☐ This provision allocates the risks under this Agreement between Contractor and Client. Contractor's pricing reflects the allocation of risk and limitation of liability specified below.

Contractor's total liability to Client under this Agreement for damages, costs, and expenses shall not exceed $_____ or the compensation received by Contractor under this Agreement, whichever is less. However, Contractor shall remain liable for bodily injury or personal property damage resulting from grossly negligent or willful actions of Contractor or Contractor's employees or agents while on Client's premises to the extent such actions or omissions were not caused by Client.

NEITHER PARTY TO THIS AGREEMENT SHALL BE LIABLE FOR THE OTHER'S LOST PROFITS OR SPECIAL, INCIDENTAL, OR CONSEQUENTIAL DAMAGES, WHETHER IN AN ACTION IN CONTRACT OR TORT, EVEN IF THE PARTY HAS BEEN HAS BEEN ADVISED BY THE OTHER PARTY OF THE POSSIBILITY OF SUCH DAMAGES.

Unfortunately, we live in the most litigious society in the world. If something goes wrong with your work, you might end up getting sued by the client and having to pay money damages. If the client suffers an economic loss because of your conduct—including lost profits—you could be required to pay that amount.

Example: Sy Berman is hired to create a custom software system to automate a new biotechnology laboratory. The client tells Sy that un-less the software is in place by April 1, the lab will lose a valuable government contract potentially worth hundreds of thousands of dollars. Sy fails to deliver the software on time, and the lab loses the contract. Sy could be liable not only for the cost of the lab obtaining new software somewhere else, but for the lost profits the lab would have earned had it not lost the government contract.

The money damages a court or arbitrator might require you to pay could far exceed the amount the client actually paid you for your work and could even bankrupt you. Will your insurance policy protect you here? General liability insurance provides coverage for some types of lawsuits—for example, it will protect you if you accidentally damage a client's property. However, such insurance may not protect you if a client sues you because of a problem with your work performance—for example, if you miss a deadline or your work product doesn't satisfy your contract's requirements. Talk to your insurance agent to learn the specifics of your coverage.

Errors and omissions, or E & O, coverage protects you against claims for damages caused by an alleged error or omission in the way you perform your services. (For a detailed discussion of insurance for ICs, see Nolo's *Working for Yourself: Law & Taxes for Independent Contractors, Freelancers & Consultants,* by Stephen Fishman.) However, E & O coverage is expensive, and not all ICs can afford it or obtain it. Even if you have this coverage, your insurance may not cover the entire amount if your client's damages are large. (You're covered only up to the limits of your policy.) And your premiums may go up or your policy may be canceled if your insurer has to pay out substantial damages to an injured client.

For all these reasons, it is advisable for independent contractors to attempt to limit their liability to the client. This is particularly wise if problems with your work or services could cause the client substantial injuries or economic losses.

Understandably, clients are often reluctant to limit your liability to them. Whether a potential client will accept such a clause or you should even ask for one in the first place depends on how badly the client wants to hire you and how much you need the work. If your services are in high demand and you can pick and choose among many eager potential clients, you may be successful in getting them to agree to this provision. ICs in this happy situation often tell a client that they will have to increase the fee for their services if they're required to assume unlimited liability. They may even provide the client with two different fee schedules: one with and one without limited liability. Some ICs increase their fees by a factor of 10, 100, or even 1000 if they are required to assume unlimited liability.

On the other hand, if you're in a very competitive field and are hungry for work, you may not be able to get clients to limit your liability at all. Indeed, it may not even be wise to bring up the issue.

In any event, this optional clause is designed to be as palatable as possible to your clients. This clause limits your total liability for any damages to the client to a set dollar amount or to no more than you were paid, whichever is less. You need to insert the dollar amount. However, you will remain liable for bodily injury or personal property damage to the client resulting from your gross negligence or intentional acts. This is designed to reassure the client. Without such a provision, you could burn down the client's office and have to pay only the amount of your fee as damages.

In addition to a cap on direct damages you have to pay the client, the clause also relieves you entirely of any liability for lost profits or other indirect damages. To help get the client to accept the clause, this is made mutual. That is, you can't get indirect damages from the client either. Such a limitation is very common in business contracts.

15. Clause 15. Notices

All notices and other communications in connection with this Agreement shall be in writing and shall be considered given as follows:

- when delivered personally to the recipient's address as stated on this Agreement
- three days after being deposited in the United States mail, with postage prepaid to the recipient's address as stated on this Agreement, or
- when sent by fax or electronic mail, such notice is effective upon receipt provided that a duplicate copy of the notice is promptly given by first class mail, or the recipient delivers a written confirmation of receipt.

When you want to do something important involving the agreement, you need to tell the client about it. This is called giving notice. For example, you need to give the client notice if you want to modify the agreement or terminate it.

This notice provision gives you several options for providing the client with notice: by personal delivery, by mail, or by fax or electronic mail followed by a confirming letter. Use whatever method is most convenient.

If you give notice by mail, it is not effective until three days after it's deposited in the mail. For example, if you want to end the agreement on 30 days' notice and you mail your notice of termination to the client, the agreement will not end until 33 days after you mailed the notice.

16. Clause 16. No Partnership

This Agreement does not create a partnership relationship. Neither party has authority to enter into contracts on the other's behalf.

Although you are performing services for the client, you are not in business together. You especially don't want to be considered the client's partner for legal purposes. Why? Because partners are liable for each other's debts.

This clause serves two purposes. First, it helps make clear to the client that it is not your partner and should not represent itself as such. Rather, the client is a separate business that has hired you to perform a service. Make sure the client doesn't do anything to make others think it is your partner—for example, including your name on its business stationery.

This clause will also prove helpful if a creditor of the client attempts to sue you for the client's debts by claiming you're the client's partner. It shows that you never intended to enter into a partnership relationship with the client and therefore should not be liable for its debts. In other words, the clause will help provide you with a defense against such a creditor's claim.

17. Clause 17. Applicable Law

> This Agreement will be governed by the laws of the state of _____ .

As you doubtless know, each state has its own system of laws and courts. Whenever a legal dispute arises, it's necessary to decide which state's law should apply. Most of the time, you and the client will have offices in the same state, and the work you perform will be done in that same state. In this event, it's obvious that your contract should be interpreted according to the law of this state. This law will determine, for example, how long you have to file a lawsuit if the client breaks the agreement.

However, things aren't so simple if you and the client are in different states, or if all or part of the work you perform is in another state. For example, a New York hiring firm may contract with a New Jersey IC to do work on a Vermont property. If this business arrangement ends up in court, lawyers will undertake a complex legal analysis to determine which state's law should apply, costing you time and money before the essence of the dispute is even reached.

To avoid this preliminary legal entanglement, it's common for parties to contracts to agree in advance on which state's law should apply in the event of a dispute. In addition to avoiding additional lawyer fees, this allows you (or a savvy client) to pick the state with laws most favorable to you. Ordinarily, courts enforce such agreements.

In most cases, choosing the law of your home state will be to your advantage, since its law is familiar to both you and your attorney. This clause allows you to do so. Consult an attorney before choosing the law of the client's home state or the state where the work is to be done.

18. Clause 18. Assignment and Delegation (Optional)

> **(Check if applicable.)**
>
> ☐ Either Contractor or Client may assign its rights and may delegate its duties under this Agreement.

Normally, the person or firm you deal with at the contract-signing stage will be the one for whom the work is done. But, in an effort to promote flexibility in business relations, the law doesn't automatically insist that you or the original hiring firm stick with each other throughout the life of the contract. Instead, subject to important restrictions, either of you may assign your rights or delegate your duties under the agreement.

An "assignment" is the process by which rights or benefits under a contract are transferred to someone else. For example, a hiring firm might assign the right to receive the benefit of your services to someone else. Such a person is called an assignee. When this occurs, the assignee steps into the hiring firm's shoes. Now you must work for the assignee, not the original hiring firm.

Example: Brutus, Inc., has signed an IC agreement with Joe to perform janitorial services. Brutus assigns its rights under the contract to a neighbor in its office complex called Big Tech, Inc. This means that Joe must perform the work for Big Tech instead of Brutus. If Joe fails to do so, Big Tech can sue him for breach of contract.

You, too, may assign the benefits you receive under an IC agreement to someone else. In other words, even though you continue to do the work, you can designate that someone else receive the compensation.

Example: Sturdee Construction contracts to build a new house for Agnes. To help run its business, Sturdee borrows money from the Widows & Orphans Bank. As a condition for obtaining a $100,000 loan, the bank requires Sturdee to assign to it the payments it receives from Agnes. This means Agnes must pay the bank, not Sturdee.

Delegation is the flipside of assignment. Instead of transferring benefits under a contract, you transfer the duties. However, the person delegating duties under a contract usually remains responsible if the person to whom the delegation was made fails to perform.

Example: Jimmy, a plumber, has contracted to provide Delta, Inc., with plumbing repair services. Because Jimmy is overloaded with work, he delegates to Mindy his duty to perform plumbing services for Delta. This means that Mindy, not Jimmy, will now do the work. But Jimmy remains liable to Delta if Mindy doesn't perform adequately.

a. Legal restrictions on assignment and delegation

Unless a contract provides otherwise, ordinarily you can freely assign and delegate subject to some important legal limitations. The most important restriction is a matter of common sense fairness: The hiring firm can't assign the benefit of your services to someone else without your con-

sent if it would increase the work you must do or otherwise magnify your burdens under the contract. Similarly, you can't delegate your duties without the hiring firm's consent if it would decrease the benefits the firm would receive.

There are other significant restrictions on delegation of your contractual duties. If the work you've agreed to do involves personal services, you cannot delegate it to someone else without the client's consent. Examples of personal services include those by professionals such as lawyers, physicians, or architects. Services by IC writers, artists, and musicians are also considered too personal to be delegated without the hiring firm's consent.

Service contracts involving more mechanical tasks can usually be delegated to someone else without the hiring firm's consent. For example, a contract to construct or paint a house would ordinarily be delegable. However, even in these cases, you can't delegate your duties to a person who lacks the skill or experience to complete the work satisfactorily.

Example: Leslie contracts with Woodsy Construction to build her a log home. Woodsy attempts to delegate its duty to construct the house to Davy, an unlicensed contractor who has never built a log home. Leslie is legally entitled to object to this delegation and demand that Woodsy do the work itself or delegate the work to a qualified log home builder.

b. How to handle assignment and delegation

Generally, you will want to have the option to delegate your duties or assign your rights under the agreement, even if you're performing personal services. This way, for example, you can get someone else to do all or part of the work if you don't have the time.

This assignment clause allows an unrestricted right of assignment and delegation. If you include this clause in the agreement, the restrictions on assignment and delegation discussed above will not apply. If a client balks at including this clause, point out that it can help the client if a government agency challenges your status as an IC. It is strong evidence of an IC relationship because it shows the client is only concerned with results, not who achieves them. It also demonstrates lack of control over you.

If the client still refuses to agree to this, delete the clause and leave the agreement silent on the subject. Your agreement will be assignable or delegable subject to the restrictions noted in Subsection a, above.

You Can Usually Assign Your Right to Payment

While it makes sense to restrict your ability to delegate your obligations under a contract, it makes less sense to restrict your ability to receive the benefits (your fee). After all, what does it matter to the client whether it pays you or someone to whom you've assigned this benefit?

The short answer is, it doesn't. But there is widespread confusion among clients (and their lawyers) about the meaning of a non-assignment clause. Many will read this clause as restricting your ability *both* to delegate your duties and to assign your benefits under the contract. Fortunately, courts generally construe such provisions narrowly to prevent you from delegating your duties, but not from assigning your rights. For example, you could assign your right to payment from the client to a third party.

---

19.   Signatures

Client: _____
              Name of Client

By: _____
              Signature

_____
          Typed or Printed Name

Title: _____

Date: _____

Contractor: _____
              Name of Contractor

By: _____
              Signature

_____
          Typed or Printed Name

Title: _____

Taxpayer ID Number: _____

Date: _____

The end of the main body of the agreement should contain spaces for you to sign, write in your title, and add the date. Make sure the person signing the agreement has the authority to do so. (See Section A23 for information about who has the authority to sign contracts.)

The IRS requires all ICs to provide their taxpayer identification number to their clients, who are supposed to keep it in their files. Your taxpayer ID number is either your Social Security number or an employer identification number (EIN) you obtained from the IRS. (EINs must be used if you have formed a corporation, partnership, or limited liability company.)

Be sure to give the client your taxpayer ID number. Any client who doesn't get your ID number is legally obligated to withhold and remit to the IRS 31% of all payments over $599 made to you. This is called backup withholding. Why is this? Because the IRS assumes that if you don't give a client your ID number, you're not going to pay taxes on the compensation you receive from the client. Backup withholding is not required if you are incorporated.

Many clients are unaware of this law and will fail to backup withhold when required. If the IRS audits such a client and catches the error, it will impose an assessment against it equal to 31% of what the client paid you. Obviously, this will make the client very unhappy.

It's very easy to avoid problems with backup withholding. You can just write the correct number at the end of the agreement where you sign your name. This will serve as proof that you gave the client your ID number. Alternatively, you can fill out IRS Form W-9, Request for Taxpayer Identification Number. The client should retain the form in its files. It need not be filed with the IRS. ■

Consultant Agreements

This chapter contains agreements to be used when you are hiring or working as a consultant. Who, exactly, is a "consultant"? Aside from the old joke ("Being a consultant means never having to say you're unemployed"), anyone can be a consultant. It simply means that the person gives expert or professional advice for a fee. This can include any type of skilled professional who provides advice and assistance to businesses, such as engineers, programmers, marketers, and finance experts.

A consultant agreement for use by the hiring firm is provided in Section A. An agreement for people working as consultants is in Section B.

A. Consultant Agreement for Use by Hiring Firm

You should use this agreement if you're hiring a consultant to work for you. The agreement is designed to protect your interests as much as possible without being unduly one-sided. It will also help establish that the worker is an independent contractor, not your employee.

Many consultants are hired to create or contribute to the creation of intellectual property—for example, computer software, important business documents, marketing plans, inventions, and trademarks. Disputes can easily develop over who owns such material, especially if it's valuable. For example, in one well-known case, a swimming pool installation firm hired a computer consultant to develop an accounting and inventory program. Ownership of the program was never discussed, so when the consultant finished the program, he thought he had the right to resell it to other swimming pool companies. The client disagreed and a court battle ensued. In the end, the court held that the consultant, not the client, owned the program (*Aymes v. Bonelli,* 980 F.2d 857 (2d Cir. 1992)).

This agreement avoids ownership disputes like this by providing that the consultant assigns—transfers—all ownership rights in such material to you.

Another important reason to use this agreement is to protect your trade secrets. Consultants often have access to your most valuable confidential information, such as marketing plans, product information, and financial information. You probably don't want the consultant to use this information or reveal it to your competitors. This agreement requires the consultant to keep your trade secrets confidential.

A consultant's agreement for use by the hiring firm is on the CD-ROM. You can modify individual clauses to tailor the text to your specific needs.

A tear-out form of a consultant's agreement for use by the hiring firm is in Appendix II. Use this form if you don't have access to a computer. It contains blank spaces for you to fill in by hand or typewriter. Copy the original and save it for future uses.

Although we provide a tear-out form for those of you who either don't have computers or who would prefer not to use the CD-ROM, don't just tear out the form and use it as your IC agreement if you can avoid doing so. A fill-in-the-blank IC agreement is the least persuasive type of contract when you are trying to convince a government agency that the worker is an IC and not an employee. The best way to create your agreement is to use the CD-ROM version and tailor it to your needs. If you can't use the CD-ROM version, then retype the version at the back of the book and tailor it to your needs. Use the fill-in-the-blank form only as a last resort.

Call up the agreement on your computer or tear out the form agreement in Appendix II, and read it along with the following instructions and explanations. Many of the provisions in this agreement are identical to those in the general independent contractor agreement in Chapter 5. In these cases, this discussion refers you to the appropriate section of Chapter 5.

Title of agreement. Call the agreement "Consulting Agreement" or "Agreement for Professional Services." Never use "Employment Agreement" as a title.

Names of consultant and hiring firm. Never refer to the consultant as an employee or to yourself as an employer.

Initially, it's best to refer to consultants by their full name. If the consultant is incorporated, use the corporate name, not the consultant's own name. For example, write "John Smith Incorporated" instead of just "John Smith." If the consultant has formed a limited liability company (LLC), use the company name—for example, "John Smith, LLC." If the consultant is a member of a partnership, use the partnership name—for example, "Smith, Weiss, and Fong."

The consultant may instead be doing business under a fictitious business name. A fictitious business name or assumed name (sometimes referred to as a "DBA" for "doing business as") is a name that sole proprietors or partners use to identify their business other than their own names. For example, if consultant Al Brodsky calls his one-man marketing research business "ABC Marketing Research," he should use that name. This shows you're contracting with a business, not a single individual.

For the sake of brevity, it is usual to identify yourself and the consultant by shorter names in the remainder of the agreement. You can use an abbreviated version of the consultant's full name—for example, "ABC" for "ABC Marketing Research." Or you can refer to the consultant simply as "Consultant." Refer to yourself initially by your company name and subsequently by a short version of the name or as "Client" or "Firm."

Put the short names in quotes after the full names. Also include the addresses of the principal places of business of the consultant and yourself. If you or the consultant have more than one office or workplace, the principal place of business is the main office or workplace.

1. Clause 1. Services to Be Performed

(Choose Alternative A or B.)

☐ ALTERNATIVE A

Consultant agrees to perform the following consulting services on Client's behalf:

_____

_____

_____

☐ ALTERNATIVE B

Consultant agrees to perform the consulting services described in Exhibit A attached to this Agreement.

The agreement should describe in as much detail as possible what the consultant is expected to do. You must word the description carefully to show only what results the consultant is expected to achieve. Don't tell the consultant how to achieve those results—this indicates that you have the right to control how the consultant performs the agreed upon services. Such a right of control is the hallmark of an employment relationship. (See Chapter 2 for a discussion of why you do not want the consultant to look like an employee.)

It's perfectly acceptable for you to establish very detailed specifications for the consultant's finished work product. But the specs should describe only the end results the consultant must achieve, not how to obtain those results. For examples of good and bad work descriptions, see Chapter 5, Section A1.

You can include the description in the main body of the agreement or, if it's lengthy, on a separate attachment.

Use Alternative A to explain the services in the contract. Choose Alternative B if you attach the description of the services to the main contract.

2. Clause 2. Payment

There are two common ways an independent contractor may be paid: with a fixed fee and by unit of time. See Chapter 5, Section A2, for a detailed discussion of each method.

3. Clause 3. Terms of Payment

This portion of the agreement states when the consultant will be paid. See Chapter 5, Section A3, for a discussion of this clause.

4. Clause 4. Expenses

This provision concerns whether you will reimburse the consultant for expenses, such as travel costs, photocopying, and telephone expenses. See Chapter 5, Section A4, for a discussion of this clause.

5. Clause 5. Materials

This provision states that you won't provide the consultant with any equipment or materials to do the work. See Chapter 5, Section A5, for a discussion of this clause.

6. Clause 6. Independent Contractor Status

This clause is intended to make clear to a government auditor that the consultant is an IC, not your employee. See Chapter 5, Section A6, for a discussion of this clause.

7. Clause 7. Business Permits, Certificates, and Licenses

The consultant should have all business permits, certificates, and licenses needed to perform the work. Lack of such licenses and permits makes a worker look like an employee, not an IC running an independent business. Consultants should obtain necessary licenses and permits on their own; you should not pay for them. For example, consulting engineers should have their own licenses, if licenses are required by your state's law, and should pay any applicable license fees.

This provision requires that the consultant have all necessary licenses and permits. You need add no information here. See Chapter 5, Section A7, for more information about this provision.

8. Clause 8. State and Federal Taxes

Do not pay or withhold any taxes on a consultant's behalf. Doing so is a very strong indicator that the worker is an employee. Indeed, some courts have held that workers were employees based upon this factor alone.

This straightforward provision simply makes clear that the consultant must pay all applicable taxes. You need add no information here.

9. Clause 9. Fringe Benefits

Do not provide either consultants or their employees with fringe benefits that you provide your own employees, such as health insurance, pension benefits, childcare allowances, or even the right to use employee facilities such as an exercise room. This clause makes clear that the consultant will receive no such benefits. You need add no information here.

10. Clause 10. Workers' Compensation

This clause is about workers' compensation coverage for the consultant and his or her employees. See Chapter 5, Section A10, for a complete discussion of this clause.

11. Clause 11. Unemployment Compensation

This clause concerns unemployment compensation coverage for the consultant. See Chapter 5, Section A11, for a complete discussion of this clause.

12. Clause 12. Insurance

Client shall not provide any insurance coverage of any kind for Consultant or Consultant's employees or contract personnel. Consultant shall obtain and maintain a broad form Commercial General Liability Insurance policy providing for coverage of at least $_____ for each occurrence. Before commencing any work, Consultant shall provide Client with proof of this insurance and with proof that Client has been made an additional insured under the policy. Consultant shall indemnify and hold Client harmless from any loss or liability arising from performing services under this Agreement.

(Optional: Check if applicable.)

☐ Consultant shall obtain professional liability insurance coverage for malpractice or errors or omissions committed by Consultant or Consultant's employees during the term of this Agreement. The policy shall provide for coverage of at least $_____ for each occurrence. Before commencing any work, Consultant shall provide Client with proof of this insurance.

For your own protection and peace of mind, you'll want to insist that the consultant maintain certain types and amounts of liability insurance.

a. General liability insurance

A general liability insurance policy insures the consultant against personal injury or property damage claims by others. For example, if a consultant accidentally injures a bystander or one of your employees while performing services for you, the consultant's liability policy will pay the costs of defending a lawsuit and pay damages up to the limits of the policy coverage.

Clause 12 requires the consultant to obtain and maintain a commercial liability insurance policy providing a minimum amount of coverage. You need to state how much coverage is required. How much coverage should the consultant have?

At least as much as you think it likely that the consultant might be sued for as a result of being negligent while performing services on your behalf; $500,000 to $1 million in coverage is usually adequate. If you're not sure how much coverage to ask for, ask your own insurance broker or agent to advise you.

The insurance clause also requires the IC to add you as an additional insured under the policy. This step is a low-cost endorsement the IC can easily have added to the liability policy by simply asking for it. An additional insured has all of the benefits of the policy's protections. If you and the IC are sued by someone injured by the IC's negligence, the IC's policy will provide defense and pay any claim or verdict up to the limits of the policy.

The consultant should also agree to indemnify—that is, repay—you if somebody the IC injures sues you and collects from you. The indemnity clause gives you the right to collect in turn from the IC.

b. Professional liability insurance

Consultants are often required to obtain another type of insurance called professional liability insurance, also known as errors and omissions, or E & O, coverage. General liability insurance won't cover claims for professional negligence—that is, claims for damages caused because of an error or omission in the way a consultant performs. Such policies specifically exclude coverage for these types of claims. This is where E & O coverage comes in.

It's a good idea for a consultant to have E & O coverage because it will pay for the costs of defending a professional malpractice lawsuit and pay damages up to the limits of the policy coverage.

E & O coverage ensures that there will be money available from an insurer to pay any judgment or settlement you obtain against the consultant if professional negligence causes you damages. A consultant who is uninsured may not have the money or assets to pay off a judgment or settlement. In this event, any judgment you obtain may be worthless.

Unfortunately, E & O coverage is expensive—many consultants don't have it and may balk at obtaining it. If you insist, the consultant may want to charge you more. It's up to you to decide whether it's worth the risk to hire an uninsured consultant. This provision requiring E & O coverage is optional. If you don't want to require it, delete this clause from your agreement.

13. Clause 13. Term of Agreement

The term of the agreement refers to when it begins and ends. This clause provides that the agreement ends when the consultant's services are completed or on a specified date, whichever comes first. See Chapter 5, Section A13, for more information.

14. Clause 14. Terminating the Agreement

This clause states the circumstances in which you or the consultant can terminate the agreement before the term ends. See Chapter 5, Section A14, for more information.

15. Clause 15. Exclusive Agreement

This clause makes clear that this contract represents the entire agreement between you and the consultant. See Chapter 5, Section A15, for more information.

16. Clause 16. Modifying the Agreement (Optional)

This clause provides that the agreement can only be modified by a writing signed by both parties. See Chapter 5, Section A16, for more information.

17. Clause 17. Intellectual Property Ownership (Optional)

(Check if applicable.)

☐ Consultant assigns to Client all patent, copyright, trademark, and trade secret rights in anything created or developed by Consultant for Client under this Agreement. Consultant shall help prepare any papers that Client considers necessary to secure any patents, copyrights, trademarks, or other proprietary rights at no charge to Client. However, Client shall reimburse Consultant for reasonable out-of-pocket expenses incurred.

Consultant must obtain written assurances from Consultant's employees and contract personnel that they agree with this assignment.

Consultant agrees not to use any of the intellectual property mentioned above for the benefit of any other party without Client's prior written permission.

A consultant's work need not be limited to giving advice or solving problems. Consultants are often hired to create intellectual property—for example, writings, music, designs, or inventions. Unless you include a specific provision assigning or transferring the intellectual property rights to the consultant's work to you, you can never be sure that you will own the work you pay the consultant to create. See Chapter 11, Section A6, for a detailed discussion of intellectual property ownership rules.

This clause gives to you ownership of any intellectual property that the consultant creates on your behalf.

18. Clause 18. Confidentiality (Optional)

While working for you, a consultant may have access to your valuable trade secrets—for example, customer lists, business plans, business methods, and techniques not known by your competitors. It is reasonable for you to include a nondisclosure provision in the agreement. See Chapter 5, Section A17, for a detailed discussion.

19. Clause 19. Resolving Disputes

This clause gives you three options for resolving disputes with the consultant: court, mediation followed by arbitration, or binding arbitration. You need to choose one. See Chapter 5, Section A18, for a detailed discussion of these three options for dispute resolution.

20. Clause 20. Applicable Law

See Chapter 5, Section A19, for a discussion of this clause.

21. Clause 21. Notices

This clause sets forth the procedures you must use to give the consultant notice—for example, to terminate the agreement. See Chapter 5, Section A20, for a discussion of this clause.

22. Clause 22. No Partnership

This clause is intended to make clear that you're not the consultant's partner. See Chapter 5, Section A21, for a discussion of this clause.

23. Clause 23. Assignment and Delegation (Optional)

> **(Check if applicable.)**
>
> ☐ Consultant may not assign or subcontract any rights or obligations under this Agreement without Client's prior written approval.

The general rules governing assigning rights and delegating duties under a contract are discussed in Chapter 5, Section A22. You're probably not too concerned about the consultant assigning benefits under the agreement—this would probably just mean that you'll pay someone other than the consultant for the work.

However, delegating the consultant's duties is another story. You're probably hiring the consultant whom you think is the best person for the job. The consultant may have unique talents or experience that others lack. You likely don't want the consultant to fob off the job onto somebody else who may not be as talented.

Fortunately, the law protects you against this. The discussion in Chapter 5, Section A22, points out that under general contract law principles, if the work an IC has agreed to do involves personal services, the IC cannot delegate it to someone else without your consent. In other words, the IC must perform the services personally or have them performed by employees under the IC's supervision.

> **Example:** A wealthy patron of the arts hires Mike Angelo, a renowned artist, to paint murals on the ceilings in his home. Because the job is long and tedious, Mike would like to assign whole rooms to his students, but his patron will have none of it, declaring that he wants Mike Angelo's work and none other.

Given these legal restrictions already in place, you don't really need to say anything restricting delegation by the consultant in your agreement. You can simply delete the optional delegation clause and rely on these legal rules. This is the best option in most cases.

However, there may be situations in which you fear a consultant may not fully understand the restrictions discussed above. To make clear that you do not want anyone else to perform the consultant's duties, you may wish to include the optional clause preventing delegation by the consultant under any circumstances without your written approval in advance.

Unfortunately, a government auditor may view this clause as an indicator of an employment relationship. It highlights the fact that you want the consultant to perform the services personally, which tends to show you want to control who does the work. Typically, such auditors aren't aware of all the niceties of contract law, so they may not understand that you're merely restating what the law already provides. For this reason, include this clause only if you think it's really necessary.

24. Signatures

This is where you and the consultant sign the agreement. You also need to obtain the consultant's taxpayer ID number. See Chapter 5, Section A23, for an explanation of how to sign the agreement.

B. Agreement for Use by Consultant

Use this agreement if you're working as a consultant. It contains a number of provisions favorable to you that a hiring firm wouldn't ordinarily include. For example, this form of the consultant's contract lets you charge a late fee if the hiring firm doesn't pay you on time, and it resolves intellectual property ownership questions to your advantage.

The entire text of the consulting agreement for use by the consultant is on the CD-ROM. You can modify individual clauses to tailor the text to your specific needs. Be sure to read the explanations of the clauses carefully before you change them.

A tear-out form of the consulting agreement for use by the consultant is in Appendix III. Use this form if you don't have access to a computer. It contains blank spaces for you to fill in by hand or typewriter.

Although we provide a tear-out form for those of you who either don't have computers or who would prefer not to use the CD-ROM, don't just tear out the form and use it as your IC agreement if you can avoid doing so. A fill-in-the-blank IC agreement is the least persuasive type of contract when you are trying to convince a government agency that you are an IC and not an employee of the hiring firm. The best way to create your agreement is to use the CD-ROM version and tailor it to your needs. If you can't use the CD-ROM version, then retype the version at the back of the book and tailor it to your needs. Use the fill-in-the-blank form only as a last resort.

Call up the agreement on your computer or tear out the form in Appendix III, and read it along with the following instructions and explanations.

Title of agreement. You don't need a title for a consulting agreement, but if you want one, call it "Consulting Agreement" or "Agreement for Professional Services." Since you are not your client's employee, do not use "Employment Agreement" as a title.

Name of consultant. At the beginning of your contract, refer to yourself by your full business name. Later on in the contract, you can use an obvious abbreviation. Or you can refer to yourself simply as "Contractor" or "Consultant."

If you're a sole proprietor, insert the full name you use for your business. This can be your own name or a fictitious business name or assumed name you use to identify your business. For example, if consultant Al Brodsky calls his one-person marketing research business "ABC Marketing Research," he would use that name on the contract. Using a fictitious business name helps show you're a business, not an employee.

If your business is incorporated, use your corporate name, not your own name—for example: "John Smith, Incorporated" instead of "John Smith."

If you've formed a limited liability company (LLC) use your full LLC name—for example, "John Smith, LLC."

Never refer to yourself as an employee or to the client as an employer.

Name of hiring firm. Refer to the client initially by its company name and subsequently by a short version of the name or "Client" or "Firm."

Addresses. Insert the addresses of the principal places of business of you and the client. If you or the client have more than one office or workplace, the principal place of business is the main office.

1. Clause 1. Services to Be Performed

Insert a detailed description of what you are expected to accomplish. Word the description carefully to emphasize the results you're expected to achieve. Don't describe the method by which you will achieve the results. As an independent contractor, it should be up to you to decide how to do the work. The client's control should be limited to accepting or rejecting your final results. The more control the client exercises over how you work, the more you'll look like an employee. For examples of good and bad work descriptions, see Chapter 5, Section A1.

Use Alternative A if you explain the services in the contract. Choose Alternative B if you attach the description of the services to the main contract.

2. Clause 2. Payment

The two most common payment methods for consultants are payment of a fixed fee by unit of time. These are discussed in detail in Chapter 5, Section B2.

3. Clause 3. Terms of Payment

This portion of the agreement deals with how and when you'll be paid. See Chapter 5, Section B3, for a discussion of this clause.

4. Clause 4. Late Fees (Optional)

See Chapter 5, Section B4, for a discussion of this clause.

5. Clause 5. Expenses

See Chapter 5, Section B5, for a discussion of this clause.

6. Clause 6. Materials

See Chapter 5, Section B6, for a discussion of this clause.

7. Clause 7. Intellectual Property Ownership (Optional)

(Choose Alternative A or B.)

☐ ALTERNATIVE A

Consultant grants to Client a royalty-free nonexclusive license to use anything created or developed by Consultant for Client under this Agreement ("Contract Property"). The license shall have a perpetual term and Client may not transfer it. Consultant shall retain all copyrights, patent rights, and other intellectual property rights to the Contract Property.

☐ ALTERNATIVE B

Consultant assigns to Client all patent, copyright, and trade secret rights in anything created or developed by Consultant for Client under this Agreement. This assignment is conditioned upon full payment of the compensation due Consultant under this Agreement.

Consultant shall help prepare any documents Client considers necessary to secure any copyright, patent, or other intellectual property rights at no charge to Client. However, Client shall reimburse Consultant for reasonable out-of-pocket expenses.

When you're an employee, your employer automatically owns any intellectual property you create as part of your job—for example, important business documents, marketing plans, graphics, designs, music, inventions, or trademarks. This is not the case, however, when you're a consultant. As a consultant, you initially own any intellectual property you create. (See Chapter 11, Section A6, for a detailed discussion of intellectual property ownership rules.)

This clause gives you the opportunity to specify what, if any, rights in your work you will transfer to the client. There are many options. Typically, your client will want to own all the intellectual property rights in your work, but you don't have to agree to this. For example, you could retain sole ownership and grant the client a license to use your work. The only limit on how you deal with ownership of your work is your own imagination and your bargaining power. We discuss two possibilities in this section.

a. You retain ownership

Choose Alternative A if you want to grant the client a nonexclusive license to use the intellectual property you create. Under this arrangement, you keep ownership of your work and merely give the client a nonexclusive license to use it. A nonexclusive license means that you own your work but the client has permission to use it for as long as it wants. The client cannot sell your work to others, but you can.

The license granted in Clause 7 is royalty-free. This means that the sum the client paid you for your services is the only payment you will receive. The client will make no additional payments for the license, no matter how often or how long it uses the material.

> **Example:** The Wily Widget Co. hires efficiency expert and consultant Susan Stanley to advise it on how to speed up the manufacture of widgets. Susan's consulting agreement with Wily provides that it will obtain a nonexclusive license to use Susan's work product.
>
> In the course of her study, Susan devises a new machine tool for widget manufacturing. Wily has only a nonexclusive license to use the tool. Wily may not manufacture it or permit others to do so. Susan may apply for a patent for the tool, license use of the tool to others to use, or manufacture it herself.

b. You transfer ownership to client

Some clients may insist on owning all of the intellectual property rights in your work product. This means that the client, not you, will own your work. Obviously, you should try to charge the client more if you transfer all your rights to it, particularly if you think the rights could be quite valuable. Choose Alternative B if you want to transfer all ownership to the client.

This clause makes the transfer contingent upon your being paid all your compensation from the client—that is, if you're not paid, the transfer does not go into effect.

The clause also obligates you to help prepare any documents the client needs to obtain any copyright, patent, or other intellectual property rights at no charge to the client. This would probably amount to no more than signing a patent or copyright registration application. However, the client is required to reimburse you for your expenses.

8. Clause 8. Consultant's Reusable Materials (Optional)

(Check and complete if applicable.)

☐ Consultant owns or holds a license to use and sublicense various materials in existence before the start date of this Agreement ("Consultant's Materials"). Consultant's Materials include, but are not limited to, those items identified in Exhibit ____, attached to and made part of this Agreement. Consultant may, at its option, include Consultant's Materials in the work performed under this Agreement. Consultant retains all right, title, and interest, including all copyrights, patent rights, and trade secret rights, in Consultant's Materials. Consultant grants Client a royalty-free nonexclusive license to use any of Consultant's Materials incorporated into the work performed by Consultant under this Agreement. The license shall have a perpetual term and may not be transferred by Client.

Many consultants who create intellectual property for clients have certain materials they use over and over again for different clients. For example, a marketing consultant may use certain marketing materials time and again.

Because these recycled materials end up as part of your final work product, you may lose the legal right to reuse them if you transfer all your ownership rights in your work to the client. To avoid this unintended result, include this clause in your agreement. It provides that you retain ownership of these materials and give the client a nonexclusive license to use them only in the context of your work product and no other. In other words, the client could not strip them out and use them itself.

> **Example:** Ace McGraw, a former professional poker player, runs a consulting firm that advises companies that want to make money on the Internet by establishing online gambling websites. Ace writes a report called *McGraw Report On Online Gaming*. Ace provides the report to many different clients. He wants to make sure that clients that retain his consulting services understand that he owns the copyright in the report and will keep ownership. He includes Clause 8, Consultant's Reusable Materials, in his consulting agreement. He lists *McGraw Report On Online Gaming* as part of his reusable materials to which he retains ownership. By including this clause in his agreement, Ace permits clients who hire him to use the report, but he does not give them any ownership rights.

Like the license the client has for your entire work product, the license to use these reusable parts is royalty-free. This means that the sum the client paid you for your services, which includes the reusable parts, is the sole payment you receive for them. The client will make no additional payments for the license. The license also has a perpetual term, meaning that it will last as long as your copyright, patent, or other intellectual property rights do.

Choose this clause if your work will involve the use of reusable materials and you want to preserve your right to reuse them in connection with the intellectual property you created. If you know what materials you'll use in advance, it's a good idea to list them in an exhibit attached to the agreement, but this isn't absolutely necessary.

9. Clause 9. Term of Agreement

See Chapter 5, Section B7, for a discussion of this clause.

10. Clause 10. Terminating the Agreement

See Chapter 5, Section B8, for a discussion of this clause.

11. Clause 11. Independent Contractor Status

See Chapter 5, Section B9, for a discussion of this clause.

12. Clause 12. Local, State, and Federal Taxes

See Chapter 5, Section B10, for a discussion of this clause.

13. Clause 13. Exclusive Agreement

See Chapter 5, Section B11, for a discussion of this clause.

14. Clause 14. Modifying the Agreement (Optional)

See Chapter 5, Section B12, for a discussion of this clause.

15. Clause 15. Resolving Disputes

See Chapter 5, Section B13, for a discussion of this clause.

16. Clause 16. Limiting Your Liability to the Client (Optional)

See Chapter 5, Section B14, for a discussion of this clause.

17. Clause 17. Notices

See Chapter 5, Section B15, for a discussion of this clause.

18. Clause 18. No Partnership

See Chapter 5, Section B16, for a discussion of this clause.

19. Clause 19. Applicable Law

See Chapter 5, Section B17, for a discussion of this clause.

20. Clause 20. Assignment and Delegation (Optional)

(Check if applicable.)

☐ Either Consultant or Client may assign its rights or may delegate its duties under this Agreement.

The general rules governing assigning rights and delegating duties under a contract are discussed in Chapter 5, Section A22. Ordinarily, you are free to assign your benefits under the agreement— usually this means you can assign your right to payment from the client to someone else. This is likely not an issue that the client will care about.

However, delegation of your duties is a different story. The discussion in Chapter 5, Section B18, points out that if a consultant has agreed to perform personal services, the consultant cannot delegate it to someone else without the client's consent. In other words, you must perform the services personally or have them performed by employees under your supervision.

Ordinarily, a consultant's services would be considered personal services. You're being hired for your particular professional advice. This means you ordinarily cannot delegate your duties under the agreement without the client's consent. This is true even if your agreement does not expressly bar delegation.

However, you may want the flexibility to delegate your duties under the agreement. Luckily, your duties can be made delegable if the client agrees. That's what this clause is designed to accomplish. It allows an unrestricted right of delegation (and assignment as well). If you include this clause in the agreement, the normal restrictions on delegation will not apply.

If a client balks at including this clause, point out that it can help the client if a government agency challenges your status as an IC. It is strong evidence of an IC relationship because it shows the client is concerned only with results, not who achieves them. It also demonstrates lack of control over you.

If your client still refuses to agree to the delegation clause, delete the clause entirely. The normal assignment and delegation rules will apply.

21. Signatures

See Chapter 5, Section B19, for a discussion of this clause. ■

Agreements for Household Services

This chapter contains agreements for part-time household workers. These are people who perform services in and around your home such as gardeners, housekeepers, cooks, and chauffeurs. Use the agreement in Section A if you're hiring a household worker. Use the agreement in Section B if you are a household worker.

These agreements are not designed to used for home childcare workers or nannies. Because ordinarily a parent has the right to control how a home childcare worker performs his or her services, such workers are ordinarily considered employees, not independent contractors. For a detailed discussion, see *Working With Independent Contractors,* by Stephen Fishman (Nolo).

A. Agreement for Use by Hiring Party

Often, household workers don't use written client agreements and may be reluctant to sign one. For this reason, this agreement is as simple and short as possible. It contains a series of boxes for you to check off to indicate exactly what services the worker is required to perform.

The text of a household worker agreement for use by the hiring party is on the CD-ROM. You can modify individual clauses to tailor the text to your specific needs.

A tear-out form for a household worker agreement for use by the hiring party is in Appendix II. Use this form if you don't have access to a computer. It contains blank spaces for you to fill in by hand or typewriter.

Although we provide a tear-out form for those of you who either don't have computers or who would prefer not to use the CD-ROM, don't just tear out the form and use it as your IC agreement if you can avoid doing so. A fill-in-the-blank IC agreement is the least persuasive type of contract when you are trying to convince a government agency that the worker is an IC and not an employee. The best way to create your agreement is to use the CD-ROM version and tailor it to your needs. If you can't use the CD-ROM version, then retype the version at the back of the book and tailor it to your needs. Use the fill-in-the-blank form only as a last resort.

Call up the agreement on your computer or tear out the form agreement in Appendix II, and read it along with the following instructions and explanations.

1. Clause 1. Services to Be Performed

(Check and complete applicable provisions.)
Contractor agrees to perform the following services:

a. Cleaning Interior

☐ Contractor will clean the following rooms and areas: _____

b. Cleaning Exterior

Contractor will clean the following:

☐ Front porch or deck: _____
☐ Back porch or deck: _____
☐ Garage: _____
☐ Pool, hot tub, or sauna: _____
☐ Other exterior areas: _____

c. Gardening

☐ Contractor will perform the following gardening services: _____

d. Other Responsibilities

☐ Cooking: _____
☐ Laundry: _____
☐ Ironing: _____
☐ Shopping and errands: _____
☐ Other: _____

The single most important part of the agreement is the description of the services the household worker will perform. This description will serve as the yardstick to measure whether the worker's performance is satisfactory. The agreement contains a number of headings under which you can indicate exactly what services the worker will perform.

You can add to these short descriptions if you wish. However, word the description carefully to emphasize the results the worker is expected to achieve. Don't describe the method by which the worker will achieve the results—for example, the agreement can require the worker to mow your lawn every week but should not specify what type of lawn mower the worker will use. As an IC, the worker should be able to decide how to do the work. Your control should be limited to accepting or rejecting the worker's final results.

2. Clause 2. Payment

> (Choose Alternative A or B.)
> ☐ ALTERNATIVE A
> In consideration for the services to be performed by Contractor, Client agrees to pay Contractor $_____.
> ☐ ALTERNATIVE B
> In consideration for the services to be performed by Contractor, Client agrees to pay Contractor at the rate of $_____ per ☐ hour, ☐ day, ☐ week, ☐ month according to the terms of payment set forth below.

Generally, ICs providing household services are either paid a fixed fee or paid by the hour. Use the alternative for whichever payment method applies.

a. Fixed fees

Choose Alternative A and insert the amount of the fee if you'll pay the worker an agreed-upon fee for the work.

b. Payment by Unit of time

Instead of a fixed fee, many IC household workers charge by the hour, day, or other unit of time. Ordinarily, this method is not supportive of a worker's IC status. But you can get away with it here because it's so common for household workers to be paid this way. Choose Alternative B and insert the amount and term of the payment if you'll pay the worker this way.

3. Clause 3. Terms of Payment

> (Choose Alternative A or B.)
> ☐ ALTERNATIVE A
> Upon completion of Contractor's services under this Agreement, Contractor shall submit an invoice. Client shall pay Contractor the compensation described within _____ [15, 30, 45, 60] days after receiving Contractor's invoice.
> ☐ ALTERNATIVE B
> Contractor shall invoice Client on a monthly basis for all hours worked pursuant to this Agreement during the preceding month. Invoices shall specify an invoice number, the dates covered in the invoice, the hours expended, and the work performed (in summary) during the invoice period. Client shall pay Contractor's fee within _____ [15, 30, 45, 60] days after receiving Contractor's invoice.

"Terms of payment" simply refers to how you will be billed by the worker and when you will pay.

a. Payment upon completing work

Choose Alternative A if you're hiring the worker for one time only—that is, for just one job. This paragraph requires the worker to submit an invoice when the work is completed. You must pay within a set number of days after you receive the invoice. A 30-day payment period is common, but you can insert a different period if you wish.

b.   Payment upon invoice

Choose Alternative B if the worker will be providing you with services on an ongoing basis. It requires the worker to submit an invoice every month. Again, you must pay within a set number of days after you receive the invoice. Insert the applicable time period.

4.   Clause 4. Expenses

(Choose Alternative A or B.)

☐ ALTERNATIVE A

Contractor shall be responsible for all expenses incurred while performing services under this Agreement. This includes license fees, memberships, and dues; automobile and other travel expenses; meals and entertainment; insurance premiums; and all salary, expenses, and other compensation paid to employees or contract personnel the Contractor hires to complete the work under this Agreement.

☐ ALTERNATIVE B

Client shall reimburse Contractor for the following expenses that are attributable directly to work performed under this Agreement: _____

_____ .

Contractor shall submit an itemized statement of Contractor's expenses. Client shall pay Contractor within 30 days after receipt of each statement.

Usually, a household worker whom you treat as an IC should not be reimbursed for expenses, such as the costs of materials and equipment and travel expenses. That's because such reimbursement limits the IC's risk of loss and makes the IC look like an employee. Instead, compensate the worker well enough so that the IC can pay the expenses directly. Choose Alternative A if the worker will pay expenses.

However, choose Alternative B if you decide you must reimburse specified expenses. List exactly what expenses will be reimbursed.

5.   Clause 5. Materials (Optional)

(Check if applicable.)

☐ Contractor will furnish all equipment, tools, materials, and supplies needed to perform the services required by this Agreement.

It's best that the worker provide cleaning and other materials and tools, since this strongly supports IC status. Don't use this optional provision if you'll be providing any materials. In this event, it's best for the agreement not to mention the subject of materials.

6.   Clause 6. Independent Contractor Status

Contractor is an independent contractor, not Client's employee. Contractor's employees or contract personnel are not Client's employees. Contractor and Client agree to the following rights consistent with an independent contractor relationship (check all that apply):

☐ Contractor has the right to perform services for others during the term of this Agreement.

☐ Contractor has the sole right to control and direct the means, manner, and method by which the services required by this Agreement will be performed.

☐ Contractor has the right to perform the services required by this Agreement at any place, location, or time.

☐ Contractor has the right to hire assistants as subcontractors or to use employees to provide the services required by this Agreement.

☐ The Contractor or Contractor's employees or contract personnel shall perform the services required by this Agreement; Client shall not hire, supervise, or pay any assistants to help Contractor.

☐ Neither Contractor nor Contractor's employees or contract personnel shall receive any training from Client in the skills necessary to perform the services required by this Agreement.

☐ Client shall not require Contractor or Contractor's employees or contract personnel to devote full time to performing the services required by this Agreement.

These provisions are designed to show government auditors that the worker is an IC, not your employee. Read them over. If any don't apply to your relationship, delete them.

Of course, if you merely recite these factors but fail to adhere to them, agency auditors won't be fooled. Think of this clause as a reminder about how to conduct your business relationship.

You need not insert any information in this provision. Just read each clause and check the ones that apply.

Note that the last paragraph provides that the worker will not be required to work full time performing the services. A household worker who works for you full time will likely be viewed as your employee, not an IC. (For a detailed discussion, see Nolo's *Working With Independent Contractors,* by Stephen Fishman.)

7. Clause 7. Work Schedule

> Contractor shall perform the services at
> _____
> during reasonable hours on a schedule to be mutually agreed upon by Client and Contractor based upon Client's needs and Contractor's availability to perform such services.

Since the worker will be working in and around your home, you need to set up a work schedule. This clause states that you and the worker will set up a mutually agreeable schedule. You don't need to set forth the schedule in the agreement. Instead, simply state where the work will be performed.

8. Clause 8. Business Permits, Certificates, and Licenses

> Contractor has complied with all federal, state, and local laws requiring business permits, certificates, and licenses required to carry out the services to be performed under this Agreement.

The worker should have all business permits, certificates, and licenses needed to perform the work. For example, your city or county may require a business permit for anyone who conducts a business within city limits, or a chauffeur may need a special driver's license. A worker who lacks such licenses and permits looks like an employee, not an IC running an independent business.

The IC should obtain such licenses and permits; you should not pay for them. You need add no information here.

9. Clause 9. State and Federal Taxes

> Client will not:
> - withhold FICA (Social Security and Medicare taxes) from Contractor's payments or make FICA payments on Contractor's behalf
> - make state or federal unemployment compensation contributions on Contractor's behalf, or
> - withhold state or federal income tax from Contractor's payments.
>
> Contractor shall pay all taxes incurred while performing services under this Agreement— including all applicable income taxes and, if Contractor is not a corporation, self-employment (Social Security) taxes. Upon demand, Contractor shall provide Client with proof that such payments have been made.

Do not pay or withhold any taxes on an IC's behalf. Doing so is a very strong indicator that the worker is an employee. Indeed, some courts have held that workers were employees based upon this factor alone.

10.   Clause 10. Workers' Compensation

Client shall not obtain workers' compensation insurance on behalf of Contractor or Contractor's employees. If Contractor hires employees to perform any work under this Agreement, Contractor will cover them with workers' compensation insurance to the extent required by law and provide Client with a certificate of workers' compensation insurance before the employees begin the work.

If the worker qualifies as an IC under your state's workers' compensation law, do not provide the IC with workers' compensation coverage. If the worker has employees, the worker should provide them with workers' compensation coverage. Be sure that you receive proof that the employees are, indeed, covered. (For a detailed discussion, see Chapter 5, Section A10.)

If you're a homeowner, you undoubtedly have homeowner's insurance (lenders require it). Such insurance will most likely cover you if a household worker is injured while working on your premises—for example, trips and falls due to a faulty staircase—and sues you for damages. Your homeowner's insurance will most likely pay for the costs of your legal defense and any damages or settlement you pay the IC up to the policy limits, typically $100,000 to $300,000. Renters can obtain renter's insurance that provides similar coverage. Talk to your insurance agent to make sure you have the appropriate coverage.

11.   Clause 11. Unemployment Compensation

Client shall make no state or federal unemployment compensation payments on behalf of Contractor or Contractor's employees or contract personnel. Contractor will not be entitled to these benefits in connection with work performed under this Agreement.

Do not pay unemployment compensation taxes for a worker who qualifies as an IC under your state's unemployment compensation law. The worker will not be entitled to receive unemployment compensation benefits when the work is finished or the agreement terminated. (For a detailed discussion, see Chapter 5, Section A11.)

12.   Clause 12. Term of Agreement

This agreement will become effective when signed by both parties and will terminate on the earlier of:
- the date Contractor completes the services required by this Agreement
- _____ [date], or
- the date a party terminates the Agreement as provided below.

The term, or duration, of the agreement should be as short as possible. A good outside time limit is six months to one year. A longer term makes the agreement look like an employment agreement, not an IC agreement. You can negotiate and sign a new agreement when the term expires.

Insert the beginning and ending dates for the agreement. The date the agreement begins can be the date you sign it or another date after you sign.

13.   Clause 13. Terminating the Agreement

(Choose Alternative A or B.)

☐ ALTERNATIVE A

With reasonable cause, either Client or Contractor may terminate this Agreement, effective immediately upon giving written notice.

Reasonable cause includes:
- a material violation of this Agreement, or
- any act exposing the other party to liability to others for personal injury or property damage.

☐ ALTERNATIVE B

Either party may terminate this Agreement any time by giving 30 days' written notice to the other party of the intent to terminate.

The circumstances under which you may terminate the agreement are important. You should not have the right to fire or terminate an IC for any reason or no reason at all. This type of unfettered termination right strongly indicates an employment relationship. Instead, you should be able to terminate the agreement only if you have a reasonable cause to do so—or after giving some notice.

a. Termination for reasonable cause

In most situations, you should have the right to terminate the agreement only if you have reasonable cause to do so. There are two types of reasonable cause. The first is a serious violation of the agreement by the worker. It would include, for example, the IC's failure to produce results specified in the agreement.

You can also terminate the agreement if the worker does something to expose you to liability for personal injury or property damage—for example, if the worker's negligence injures you, damages your property, or damages someone else's property. You aren't obligated to terminate the agreement if this happens, but you have the option of doing so. You may not want to continue dealing with a worker who is extremely careless.

If you fire an IC who performs adequately and otherwise satisfies the terms of the agreement—in other words, if you fire without having reasonable cause—you'll be liable for breaking the agreement. The worker can sue you in court and may be able to obtain a judgment against you for money damages.

Choose Alternative A to have such a limited termination right.

b. Termination with notice

If you can't live with a restricted termination right, your best approach is to add a provision giving you the right to terminate the agreement for any reason upon written notice. Choose Alternative B

to do this. Insert how many days' notice you'll give the worker. The notice term can be anywhere from a few days to 30 days or more, depending upon the length of the contract term. The longer the term, the more notice you should give. This clause is not supportive of the worker's IC status, so use it only if absolutely necessary.

14. Signatures

See Chapter 5, Section A23, for information about signing the agreement.

B. Agreement for Use by Worker

Use this agreement if you're working as a household worker. Probably neither you nor the client want to deal with a lengthy agreement for this type of service, so this agreement is as simple and short as possible.

A household worker agreement for use by the worker is on the CD-ROM. You can modify individual clauses to tailor the text to your specific needs.

A tear-out household worker agreement for use by the worker is in Appendix III. Use this form if you don't have access to a computer. It contains blank spaces for you to fill in by hand or typewriter.

Although we provide a tear-out form for those of you who either don't have computers or who would prefer not to use the CD-ROM, don't just tear out the form and use it as your IC agreement if you can avoid doing so. A fill-in-the-blank IC agreement is the least persuasive type of contract when you are trying to convince a government agency that you are an IC and not an employee of the hiring firm. The best way to create your agreement is to use the CD-ROM version and tailor it to your needs. If you can't use the CD-ROM version, then retype the version at the back of the book and tailor it to your needs. Use the fill-in-the-blank form only as a last resort.

Call up the agreement on your computer or tear out the form agreement in Appendix III. and read it along with the following instructions and explanations.

1. Clause 1. Services to Be Performed

> (Check and complete applicable provisions.)
> Contractor agrees to perform the following services:
>
> ### a. Cleaning Interior
>
> ☐ Contractor will clean the following rooms and areas: _____
>
> ### b. Cleaning Exterior
>
> Contractor will clean the following:
>
> ☐ Front porch or deck: _____
> ☐ Back porch or deck: _____
> ☐ Garage: _____
> ☐ Pool, hot tub, or sauna: _____
> ☐ Other exterior areas: _____
>
> ### c. Gardening
>
> ☐ Contractor will perform the following gardening services: _____
>
> ### d. Other Responsibilities
>
> ☐ Cooking: _____
> ☐ Laundry: _____
> ☐ Ironing: _____
>
> _____
>
> ☐ Shopping and errands: _____
> ☐ Other: _____

The single most important part of the agreement is the description of the services you'll perform for the client. This description will serve as the yardstick to measure whether your performance was satisfactory. The agreement contains a number of headings under which you can indicate exactly what services you will perform.

You can add to these short descriptions if you wish. However, make sure to word the description carefully to emphasize the results you're expected to achieve. Don't describe the method by which you will achieve the results—for example, the agreement can require you to mop the client's tile floors every week but should not specify what type of mop or cleanser you'll use. As an IC, it should be up to you to decide how to do the work. The client's control should be limited to accepting or rejecting your final results.

2. Clause 2. Payment

> (Choose Alternative A or B.)
> ☐ ALTERNATIVE A
> In consideration for the services to be performed by Contractor, Client agrees to pay Contractor $_____.
> ☐ ALTERNATIVE B
> In consideration for the services to be performed by Contractor, Client agrees to pay Contractor at the rate of $_____ per ☐ hour, ☐ day, ☐ week, ☐ month according to the terms of payment set forth below.

Generally, ICs providing household services are paid either a fixed fee or an hourly fee. Use the alternative for whichever payment method applies.

a. Fixed fees

Choose Alternative A to charge a fixed fee. In a fixed fee agreement, you charge an agreed upon amount for the entire household project. Insert the amount of your fee in the blank.

b. Hourly payments

Instead of a fixed fee, many IC household workers charge by the hour, day, or other unit of time. Choose Alternative B and insert your rate of pay to do this.

3. Clause 3. Terms of Payment

> (Choose Alternative A or B.)
> ☐ ALTERNATIVE A
> Upon completion of Contractor's services under this Agreement, Contractor shall submit an invoice. Client shall pay Contractor the compensation described within _____ [15, 30, 45, 60] days after receiving Contractor's invoice.
> ☐ ALTERNATIVE B
> Contractor shall invoice Client on a monthly basis for all hours worked pursuant to this Agreement during the preceding month. Invoices shall specify an invoice number, the dates covered in the invoice, the hours expended, and the work performed (in summary) during the invoice period. Client shall pay Contractor's fee within _____ [15, 30, 45, 60] days after receiving Contractor's invoice.

"Terms of payment" refers to how you will bill the client and be paid. Before you get paid for your work, you should submit an invoice to the client setting out the amount due. An invoice doesn't have to be fancy. It should include an invoice number, the dates covered by the invoice, the hours expended if you're being paid by the hour, and a summary of the work performed.

a. Payment upon completing work

If you are working for the client on only one occasion, use the clause identified as Alternative A. This paragraph requires you to submit an invoice when your work is completed. The client must pay you within a set number of days after you send the invoice. The time for payment begins to run when you send the invoice, not when the client receives it. This will help you get paid more quickly. Insert the payment period. A 30-day payment period is common, but you can require faster payment if you and the client agree.

b. Payment for ongoing services

If you'll provide services on an ongoing basis, use the clause in Alternative B. It requires you to submit an invoice every month. Again, the client must pay you within a set number of days after you send the invoice.

4. Clause 4. Late Fees (Optional)

> (Check and complete if applicable.)
> ☐ Late payments by Client shall be subject to late penalty fees of _____ % per month from the due date until the amount is paid.

Many independent contractors charge a late fee if the client doesn't pay within the time specified in the independent contractor agreement or invoice. See Chapter 5, Section B4, for a detailed discussion of this optional provision.

5. Clause 5. Expenses

> (Choose Alternative A or B.)
> ☐ ALTERNATIVE A
> Contractor shall be responsible for all expenses incurred while performing services under this Agreement. This includes license fees, memberships, and dues; automobile and other travel expenses; meals and entertainment; insurance premiums; and all salary, expenses, and other compensation paid to employees or contract personnel the Contractor hires to complete the work under this Agreement.
> ☐ ALTERNATIVE B
> Client shall reimburse Contractor for the following expenses that are attributable directly to work performed under this Agreement: _____
> Contractor shall submit an itemized statement of Contractor's expenses. Client shall pay Contractor within 30 days after receipt of each statement.

Expenses are the costs you incur that you can attribute to your work for a client. They include, for example, the cost of cleaning or gardening supplies you use to perform your services for the client.

It is usually best that you not be separately reimbursed for expenses. Instead, set your fee high enough to cover your expenses. However, where there is an otherwise clear IC relationship, you can probably do it and still be considered an IC.

a. You pay expenses

Choose Alternative A if you'll pay all your own expenses. This not only helps you look like an IC, but it also frees you from having to keep records of your expenses. Keeping track of everything you buy for a client can be a real chore and may be more trouble than it's worth.

b. Client reimburses expenses

Choose Alternative B if the client will reimburse you for any expenses. Specify what expenses will be reimbursed.

6. Clause 6. Materials (Optional)

> (Check if applicable.)
>
> (Choose Alternative A or B.)
>
> ☐ ALTERNATIVE A
>
> Contractor will furnish all equipment, tools, materials, and supplies needed to perform the services required by this Agreement.
>
> ☐ ALTERNATIVE B
>
> Client shall make available to Contractor, at Client's expense, the following equipment and/or materials:
>
> _____ .
>
> These items will be provided to Contractor by
> _____ [date].

Generally, an IC should provide all the materials and equipment necessary to complete a project. However, this might not always be possible. For example, you may need to use a client's specialized cleaning equipment. If the client is going to supply you with anything, use this optional clause and list any items you'll be given. If you need these items by a specific date, specify the deadline as well.

If you need the client to provide you with extensive equipment for a long period, it's advisable to either purchase or lease it from the client. You should draft and sign a separate purchase or lease agreement. You can find forms for selling or leasing equipment in *Legal Forms for Starting & Running a Small Business,* by Fred Steingold (Nolo).

7. Clause 7. Work Schedule

> Contractor shall perform the services at _____ during reasonable hours on a schedule to be mutually agreed upon by Client and Contractor based upon Client's needs and Contractor's availability to perform such services.

Because you will be working in and around the client's home, you need to set up a work schedule. This agreement states that you and the client will set up a mutually agreeable schedule. You don't need to set forth the schedule in the agreement. Instead, simply state where the work will be performed.

8. Clause 8. Term of Agreement

> This agreement will become effective when signed by both parties and will terminate on the earlier of:
> * the date Contractor completes the services required by this Agreement
> * _____ [date], or
> * the date a party terminates the Agreement as provided below.

The "term of the agreement" refers to when it begins and ends. Unless the agreement provides a specific starting date, it begins on the date it's signed. If you and the client sign on different dates, the agreement begins on the date the last person signed. Normally, you shouldn't begin work until the client signs the agreement, so it's usually best that the agreement not provide a specific start date that might fall before the client signs.

An independent contractor agreement should have a definite ending date. A good outside time limit is 12 months. A longer term makes the agreement look like an employment agreement, not an independent contractor agreement. You can negotiate and sign a new agreement at the end of 12 months. Insert the termination date in this clause.

9. Clause 9. Terminating the Agreement

(Choose Alternative A or B.)

☐ ALTERNATIVE A

With reasonable cause, either Client or Contractor may terminate this Agreement, effective immediately upon giving written notice.

Reasonable cause includes:

- a material violation of this Agreement, or
- nonpayment of contractor's compensation after 20 days' written demand for payment.

Contractor shall be entitled to full payment for services performed prior to the effective date of termination.

☐ ALTERNATIVE B

Either party may terminate this Agreement any time by giving _____ days' written notice to the other party of the intent to terminate.

Contrary to what you might think, signing a contract doesn't mean that you're bound to it forever. You and the client can agree to call off your agreement at any time. In addition, contracts typically contain provisions allowing either side to terminate the agreement under certain circumstances without the other side's agreement.

When a contract is terminated, both you and the client stop performing—that is, you discontinue your work, and the client has no obligation to pay you for any work you may do after the effective date of termination. However, the client is legally obligated to pay you for any work you did prior to the termination date. And even if the contract is terminated, you remain liable to the client for any damages it may have suffered due to your failure to perform as agreed before the termination date.

It's important to clearly define the circumstances under which you or the client may end the agreement. In the past, the IRS viewed a termination provision giving either you, the client, or both of you the right to terminate the agreement at any time to be strong evidence of an employment relationship. However, the agency no longer considers this to be such an important factor.

Even so, it's wise to place some limits on the client's right to terminate the contract. It's usually not in your best interest to give a client the right to terminate you for any reason or no reason at all, since the client may abuse that right. Instead, both you and the client should be able to terminate the agreement without legal repercussions only if there is reasonable cause to do so or upon written notice.

a. Termination with reasonable cause

Choose Alternative A if you wish to allow either yourself or the client to end the agreement with reasonable cause, which is either

- a material violation of the agreement, or
- the client's failure to pay your compensation more than 20 days after you've asked for it.

A material violation means that either you or the client have failed to live up to one or more of the agreement's requirements and that this failure is so serious that it is doubtful that the other side will get what it wanted when it signed the agreement in the first place. The client has bargained for you to perform the household services described in the agreement, during the times called for, in an acceptable manner. You've bargained to be timely paid for performing these services.

Examples of such material violations include:

- your complete failure to do the work called for by the agreement
- serious delay by you in completing the work
- your doing the work so badly that the client has to find someone else to do it
- the client's failure to pay you at all, or
- serious delay in your receiving payment from the client.

Example: Martha, a landscape gardener, is hired by Emily to perform gardening services at her home. Martha performs all the gardening services called for by the agreement, but Emily is habitually late in paying Martha. When her payment is two months overdue, Martha decides she has had enough and tells Emily she is terminating the agreement. Martha is legally entitled to do this because such serious delay in payment is a material violation of the agreement.

Because late payment is one of the most common problems household workers run into, the clause includes a special provision about late payments. It provides that a material violation of the agreement occurs if the client fails to pay you what you are owed within 20 days after you make a written demand for it. You may terminate the agreement in this event.

The clause also makes clear that the client must pay you for the services your performed before the contract was terminated.

b.   Termination without cause

Sometimes you or the client just can't live with a limited termination right. Instead, you want to be able to get out of the agreement at any time without incurring liability. For example, you may have too much work and need to lighten your load.

If you want a broader right to end the work relationship, choose Alternative B. You need to provide at least a few days' notice. Being able to terminate without notice tends to make you look like an employee. Insert the notice period you will provide. A period of 30 days is common, but shorter notice may be appropriate if the project is of short duration.

10.   Clause 10. Independent Contractor Status

Contractor is an independent contractor, not Client's employee. Contractor's employees or contract personnel are not Client's employees. Contractor and Client agree to the following rights consistent with an independent contractor relationship (check all that apply):

☐ Contractor has the right to perform services for others during the term of this Agreement.

☐ Contractor has the sole right to control and direct the means, manner, and method by which the services required by this Agreement will be performed.

☐ Contractor has the right to hire assistants as subcontractors or to use employees to provide the services required by this Agreement.

☐ The Contractor or Contractor's employees or contract personnel shall perform the services required by this Agreement; Client shall not hire, supervise, or pay any assistants to help Contractor.

☐ Neither Contractor nor Contractor's employees or contract personnel shall receive any training from Client in the skills necessary to perform the services required by this Agreement.

☐ Client shall not require Contractor or Contractor's employees or contract personnel to devote full time to performing the services required by this Agreement.

One of the most important functions of the agreement is to help establish that you are an independent contractor, not your client's employee. The key to doing this is to make clear that you, not the client, have the right to control how the work will be performed.

You will need to emphasize the factors that the IRS and other agencies consider in determining whether a client controls how the work is done. Of course, if you merely recite what you think the IRS wants to hear but fail to adhere to these understandings, agency auditors won't be fooled. Think of this clause as a reminder to you and your client about how to conduct your business relationship.

You need not insert any information in this provision. Just read each clause and check the ones that apply.

11. Clause 11. Local, State, and Federal Taxes

Client will not:

- withhold FICA (Social Security and Medicare taxes) from Contractor's payments or make FICA payments on Contractor's behalf
- make state or federal unemployment compensation contributions on Contractor's behalf, or
- withhold state or federal income tax from Contractor's payments.

The charges included here do not include taxes. If Contractor is required to pay any federal, state, or local sales, use, property, or value added taxes based on the services provided under this Agreement, the taxes shall be billed separately to Client. Client shall be responsible for paying any interest or penalties incurred due to late payment or nonpayment of any taxes by Client.

The agreement should address federal and state income taxes, Social Security taxes, and sales taxes.

a. Income taxes

Your client should not pay or withhold any income or Social Security taxes on your behalf. Doing so is a very strong indicator that you are an

employee, not an independent contractor. Indeed, some courts have classified workers as employees based upon this factor alone. Keep in mind that one of the best things about being an independent contractor is that you don't have taxes withheld from your paychecks.

This straightforward provision simply makes clear that you'll pay all applicable taxes due on your compensation and, therefore, the client should not withhold taxes from your payments. You need not insert any information here.

b. Sales taxes

Most states impose sales taxes only on goods, not services. However, Hawaii, New Mexico, and South Dakota require independent contractors to pay sales taxes even if they provide their clients with services only. Many other states require sales taxes to be paid only for certain services.

In some states, sales tax is imposed on those who sell goods or services and then have the option of passing the tax along to their customers or clients. In other states, the tax is imposed directly on the customer or client, and the seller is responsible for collecting the tax and remitting it to the state. In a few states, sellers and customers or clients share sales taxes.

This provision makes clear that the client will have to pay these and similar taxes. This will come in handy if your state seeks to charge you sales taxes for selling household services. In most states, this provision is not really necessary because household services aren't taxed. However, states change sales tax laws constantly, and more are beginning to look at services as a good source of sales tax revenue. So this provision could come in handy in the future even if you don't really need it now.

12. Signatures

See Chapter 5, Section B19, for infomation about signing your agreement. ■

8

Agreements for Salespeople

The agreements in this chapter are for salespeople who are paid by commission. Use the agreement in Section B if you're a firm that is hiring a salesperson. Use the agreement in Section C if you're working as a salesperson.

A. Special IRS Rules for Salespeople

There are special IRS rules governing the employment status of many types of salespeople. In some cases, a salesperson may qualify automatically as an IC for IRS purposes. Other salespeople are automatically considered to be employees. Hiring firms and salespeople need to be familiar with these rules. The legal principles explained below apply to salespeople working in every state.

1. Direct Sellers of Consumer Products

Direct sellers are salespeople who sell consumer products to people in their homes or at a place other than an established retail store—for example, at swap meets. The products they sell include tangible personal property that is used for personal, family, or household purposes—for example, vacuum cleaners, cosmetics, encyclopedias, and gardening equipment. They also include intangible consumer services or products such as home study courses and cable television services.

The IRS automatically classifies a direct seller as an IC if:

- the salesperson's pay is based solely on sales commissions, not on the number of hours worked, and
- there is a written contract with the hiring firm providing that the salesperson will not be treated as an employee for federal tax purposes.

Note carefully, however, that these salespeople will lose this automatic IC status if they perform duties besides selling consumer products on commission.

If these requirements are satisfied, a hiring firm need not pay or withhold Social Security, Medicare, or federal income taxes or pay federal unemployment taxes for a direct seller. This is true regardless of whether the salesperson would qualify as an IC under the normal IRS rules. (See Chapter 2.) In effect, hiring firms that meet these requirements have nothing to fear from the IRS. This benefits direct sellers as well, because they can avoid tax withholding.

> **Example:** Larry sells Bibles door to door for the Holy Bible Co. He is paid a 20% commission on all of his sales. This is his only remuneration from Holy. He has a written contract with Holy that provides that he will not be treated as an employee for federal tax purposes. Holy need not pay Social Security, Medicare, or federal unemployment taxes or withhold federal income taxes for Larry. It's up to Larry to pay his own self-employment taxes.

Unfortunately, only the IRS confers this special status. State agencies—such as your state unemployment tax agency, workers' compensation agency, and state tax department—don't have a similar rule. Direct sellers must qualify as ICs under your state's usual laws for a hiring firm to treat them that way for purposes of workers' compensation and unemployment insurance. If they do qualify, the hiring firm doesn't have to provide them with workers' compensation or unemployment insurance coverage. If they don't qualify, such coverage must be provided; this is so even though they are automatically considered ICs for federal law purposes. Each state has its own rules for determining a worker's status—and different state agencies may apply different tests. (See Chapter 2 for a discussion of these various tests.)

2. Business-to-Business Salespeople

Some types of salespeople are considered automatically to be employees for IRS purposes. This means that they cannot be treated as independent contractors for purposes of Social Security and Medicare taxes and federal unemployment taxes, even if they meet the IRS' usual definition of an IC.

a.   What is a statutory employee?

Congress has determined that certain workers, including some salespeople who otherwise would qualify as ICs, are automatically considered to be employees for purposes of Social Security, Medicare, and federal unemployment tax. This is so even though the workers might qualify as ICs under normal IRS tests for worker status. These workers are called statutory employees. This group of statutory employees includes certain home workers and food, beverage, and laundry distributors.

Statutory employees have a dual status. They are employees for certain purposes and ICs for others. In effect, they wear two hats. If a salesperson qualifies as a statutory employee, the hiring firm must pay half of the person's Social Security and Medicare taxes and withhold the other half from his or her paychecks, and pay federal unemployment taxes. This is the same as for any other employee.

However, statutory employees remain ICs for employment benefit purposes. This means they are not eligible to participate in any benefit plans—such as pensions or insurance coverage—that the hiring firm provides for regular employees.

These statutory employee rules apply only to Social Security, Medicare, and federal unemployment taxes. A statutory employee may still qualify as an IC under state workers' compensation and unemployment compensation rules. (See Chapter 2 for a discussion of these rules.)

Example: Stan is a liquor salesman who sells solely to restaurants and bars. Eighty percent of his sales are made for the Acme Distillery Co. He qualifies as an IC under the normal IRS rules and the rules of his state. However, Stan also qualifies as a statutory employee because all the factors listed in Subsection c, below, are satisfied. This means that Acme must treat Stan as an employee for purposes of Social Security, Medicare, and federal unemployment taxes. But because Stan qualifies as an IC under his state laws, Acme need not provide him with workers' compensation coverage or pay state unemployment taxes for him.

b.   Obligations towards statutory employees

The hiring firm must give every statutory employee an IRS Form W-2, Wage and Tax Statement. The firm should check the Statutory Employee designation in box 15. The W-2 must show the Social Security and Medicare tax withheld, as well as the Social Security and Medicare income. The hiring firm must also file a copy with the Social Security Administration. Form W-2 and instructions for completing it can be obtained by calling the IRS at 800-TAX-FORM (800-829-3676) or by calling your local IRS office. You can also obtain this information from the agency's website at www.irs.gov.

c.   Salespeople who qualify as statutory employees

Salespeople who qualify as statutory employees are ordinarily traveling salespeople paid on a commission basis. The details of their work and the means by which they cover their territories are not typically dictated to them by others. However, they are expected to work their territories with some regularity, take purchase orders, and send them to the hiring firm for delivery to the purchaser.

A salesperson will be deemed a statutory employee by the IRS only if all the following requirements are met:

- the hiring firm and salesperson agree that the work must be performed personally by the salesperson, not by the salesperson's employees
- the salesperson does not have a substantial investment in equipment and facilities
- the salesperson has a continuing relationship with the hiring firm
- the salesperson works at least 80% of the time for one person or company
- the salesperson sells on behalf of, or turns the orders over to, the hiring firm
- the salesperson sells merchandise for resale or supplies for use in the buyer's business operations, as opposed to goods purchased for personal consumption at home, and
- the salesperson sells only to wholesalers, retailers, contractors, or those who operate hotels, restaurants, or similar establishments. This does not include manufacturers, schools, hospitals, churches, municipalities, or state and federal governments.

Example: Linda sells books to retail bookstores for the Scrivener & Sons Publishing Company. Her territory covers the entire Midwest. She works only for Scrivener and is paid a commission based on the amount of each sale. She turns her orders over to Scrivener's, which ships the books to each bookstore customer. Linda is a statutory employee of Scrivener's. The company must pay Social Security and Medicare taxes for her.

d. Avoiding statutory employee status

Obviously, most hiring firms won't want any salesperson it hires to be a statutory employee. It's also not so great for salespeople who want to be ICs because they'll have taxes withheld from their pay. Plus, their compensation will likely be reduced to make up for the fact that the hiring firm must pay half their Social Security and Medicare taxes. Even worse, some hiring firms might refuse to hire salespeople who qualify as statutory employees.

Fortunately, it's easy to avoid this trap and avoid paying Social Security, Medicare, and federal unemployment taxes. All you need do is sign a written agreement stating that the salesperson has the right to subcontract or delegate the work to others. In other words, if the hiring firm states that the worker does not have to perform the services personally, one of the essential conditions mentioned above in Subsection b will not be met, which will preclude statutory employee status. But note that the agreement must reflect reality—that is, the hiring firm must really intend to give the salesperson the right to delegate, not just say so on paper.

Both agreements below include an optional clause giving the salesperson the right to delegate and thereby avoid statutory employee status.

B. Agreement for Use by Hiring Firm

Use this agreement if you're hiring a commission salesperson to work for you. The agreement is designed to protect your interests as much as possible without being unduly one-sided. It also is intended to help establish that the salesperson is an IC, not your employee.

An agreement for a salesperson for use by the hiring firm is on the CD-ROM. You can modify individual clauses to tailor the text to your specific needs.

A tear-out agreement for a salesperson for use by the hiring firm is in Appendix II. Use this form if you don't have access to a computer. It contains blank spaces for you to fill in by hand or typewriter.

Although we provide a tear-out form for those of you who either don't have computers or who would prefer not to use the CD-ROM, don't just tear out the form and use it as your IC agreement if you can avoid doing so. A fill-in-the-blank IC agreement is the least persuasive type of contract when you are trying to convince a government agency that the worker is an IC and not an employee. The best way to create your agreement is to use the CD-ROM version and tailor it to your needs. If you can't use the CD-ROM version, then retype the version at the back of the book and tailor it to your needs. Use the fill-in-the-blank form only as a last resort.

Call up the agreement on your computer or tear out the form agreement in Appendix II, and read it along with the following instructions and explanations.

Title of agreement. You don't have to have a title for an IC agreement, but if you want one you should call it "Independent Contractor Agreement." Never use "Employment Agreement" as a title.

Names of IC and hiring firm. Never refer to an IC as an employee or to yourself as an employer. Don't use employee-like titles for the IC, such as "sales manager" or "sales supervisor." See Chapter 5, Section A, for guidance on which names to use in your agreement.

1. Clause 1. Services to Be Performed

Contractor agrees to sell the following product or merchandise for Client:

_____

(Optional: Check if applicable.)
☐ Contractor shall seek sales of the product in the homes of various individuals.

Describe in the agreement what product, merchandise, or service the salesperson will sell.

If you're hiring the person solely to sell consumer products door to door on a commission-only basis, the person will automatically qualify as an IC under federal tax laws. (See Section A1, above.) In this event, you should include the optional provision stating that the person will seek sales in the homes of various individuals. However, you can't give such a salesperson duties besides selling consumer products on commission, or the worker will lose the automatic IC status.

2. Clause 2. Compensation

> In consideration for the services to be performed by Contractor, Client agrees to pay Contractor a commission on completed sales as follows:
>
> _____
>
> _____
>
> Contractor acknowledges that no other compensation is payable by Client and that all of Contractor's compensation will depend on sales made by Contractor. None of Contractor's compensation shall be based on the number of hours worked by Contractor.

Describe here the commission the salesperson will be paid and when it will be paid.

3. Clause 3. Expenses

See Chapter 5, Section A4, for a discussion of this clause.

4. Clause 4. Materials (Optional)

See Chapter 5, Section A5, for a discussion of this clause.

5. Clause 5. Independent Contractor Status

> Contractor is an independent contractor, not Client's employee. Contractor's employees or contract personnel are not Client's employees. Contractor and Client agree to the following rights consistent with an independent contractor relationship (check all that apply).
>
> ☐ Contractor has the right to perform services for others during the term of this Agreement.
>
> ☐ Contractor shall have no obligation to perform any services other than the sale of the product described here.
>
> ☐ Contractor has the sole right to control and direct the means, manner, and method by which the services required by this Agreement will be performed. Consistent with this freedom from Client's control, Contractor:
>
> > ☐ does not have to pursue or report on leads furnished by Client
> >
> > ☐ is not required to attend sales meetings organized by Client
> >
> > ☐ does not have to obtain Client's pre-approval for orders, and
> >
> > ☐ shall adopt and carry out Contractor's own sales strategy.
>
> ☐ Subject to any restrictions on Contractor's sales territory contained in this Agreement, Contractor has the right to perform the services required by this Agreement at any location or time.
>
> ☐ Contractor has the right to hire assistants as subcontractors or to use employees to provide the services required by this Agreement, except that Client may supply Contractor with sales forms.
>
> ☐ The Contractor or Contractor's employees or contract personnel shall perform the services required by this Agreement; Client shall not hire, supervise, or pay any assistants to help Contractor.
>
> ☐ Neither Contractor nor Contractor's employees or contract personnel shall receive any training from Client in the skills necessary to perform the services required by this Agreement.
>
> ☐ Client shall not require Contractor or Contractor's employees or contract personnel to devote full time to performing the services required by this Agreement.

Even if the salesperson qualifies automatically as an IC for IRS purposes, you will still need to meet the tests of various state agencies—including your state workers' compensation and unemployment compensation agencies. This portion of the agreement is intended to help establish that the salesperson is an IC. Read each of these clauses carefully and check all those that apply.

The key to establishing the salesperson's IC status is to show that you do not have the right to control how the person performs the sales services required by the agreement. (See Chapter 2 for a detailed discussion.) To help make this clear, the agreement provides that the salesperson does not have to pursue or report on any sales leads you provide, is not required to attend sales meetings you organize, need not obtain your pre-approval for any orders, and will adopt and carry out the person's own sales strategy. Make sure that you and all of your employees who deal with the salesperson understand and obey these limitations on your control.

6. Clause 6. Business Permits, Certificates, and Licenses

> Contractor has complied with all federal, state, and local laws requiring business permits, certificates, and licenses required to carry out the services to be performed under this Agreement.

The salesperson should have all business permits, certificates, and licenses needed to perform the work. Most salespeople just need a city or county business license. Some states also require a state business license.

You should not pay for these licenses and permits. It's wise to obtain copies of all the salesperson's licenses, keep them in your file, and check to make sure they're current.

7. Clause 7. State and Federal Taxes

> Client will not:
> - withhold FICA (Social Security and Medicare taxes) from Contractor's payments or make FICA payments on Contractor's behalf
> - make state or federal unemployment compensation contributions on Contractor's behalf, or
> - withhold state or federal income tax from Contractor's payments.
>
> Contractor shall pay all taxes incurred while performing services under this Agreement—including all applicable income taxes and, if Contractor is not a corporation, self-employment (Social Security) taxes. Upon demand, Contractor shall provide Client with proof that such payments have been made.

Do not pay or withhold any taxes on an IC's behalf. Doing so is a very strong indicator that the worker is an employee. Indeed, some courts have held that workers were employees based upon this factor alone.

This straightforward provision simply makes clear that the IC must pay all applicable taxes. You need add no information here.

8. Clause 8. Fringe Benefits

> Contractor understands that neither Contractor nor Contractor's employees or contract personnel are eligible to participate in any employee pension, health, vacation pay, sick pay, or other fringe benefit plan of Client.

Do not provide an IC salesperson with fringe benefits that you provide your own employees, such as health insurance, pension benefits, childcare allowances, or even the right to use employee facilities such as an exercise room. This clause makes clear that the salesperson will receive no such benefits. You need add no information here.

9. Clause 9. Workers' Compensation

See Chapter 5, Section A10, for a discussion of this clause.

10. Clause 10. Unemployment Compensation

See Chapter 5, Section A11, for a discussion of this clause.

11. Clause 11. Insurance

Client shall not provide any insurance coverage of any kind for Contractor or Contractor's employees or contract personnel.

Contractor shall obtain and maintain a broad form Commercial General Liability Insurance policy providing for coverage of at least $_____ for each occurrence and naming Client as an additional insured under the policy.

Contractor shall maintain automobile liability insurance for injuries to person and property, including coverage for all non-owned and rented automotive equipment, providing for coverage of at least $ _____ for each person, $_____ for each accident, and $_____ for property damage.

Before commencing any work under this Agreement, Contractor shall provide Client with proof of this insurance.

Contractor shall indemnify and hold Client harmless from any loss or liability arising from performing services under this Agreement, including any claim for injuries or damages caused by Contractor while traveling in Contractor's automobile and performing services under this Agreement.

An IC salesperson should have a liability insurance policy just like any other business; you need not provide it. This type of coverage insures the IC against personal injury or property damage claims by others. For example, if an IC accidentally injures a bystander while performing services for you, the IC's liability policy will pay the costs of defending a lawsuit and pay damages up to the limits of the policy coverage. However, there are many things such insurance does not cover, such as injuries or damages caused by the IC's intentional acts, fraud, dishonesty, or criminal conduct.

This clause requires the salesperson to obtain and maintain a commercial liability insurance policy providing a minimum amount of coverage. You need to state how much coverage is required. How much coverage should the salesperson have? $500,000 to $1 million in coverage is usually adequate. If you're not sure how much coverage to ask for, ask your own insurance broker or agent to advise you.

The insurance clause also requires the IC to name you as an additional insured under the policy. This is a low-cost endorsement the IC can easily have added to the policy. This will nail down your rights to coverage under the IC's policy if you are sued along with the IC by someone injured by the IC's negligence.

Traveling salespeople usually travel by car. This provision requires the salesperson to have car insurance. The salesperson's personal auto policy probably does not include coverage for regular business use of the car. The salesperson will need to have business use insurance coverage. The cost for this is minimal. If the salesperson has employees, this insurance should cover their use of the car as well.

You need to set forth the amount of coverage the salesperson must have. How much is enough? It's not unreasonable to require $250,000 per person, $500,000 per accident, and $100,000 for property damage.

The clause also requires the salesperson to indemnify—that is, personally repay—you if somebody the IC injures decides to sue you. You may be wondering why you need this if you're already a named insured under the salesperson's liability policy. There may be circumstances under which you're unable to collect from the salesperson's insurer—for example, the insurer may deny coverage for some reason, or the claim may exceed the policy limits of the salesperson's insurance. In this event, the indemnity clause requires the salesperson to repay you personally. This gives you additional protection, but, of course, it will be useful only if the salesperson has the money to repay you (which is why you want the salesperson to have insurance in the first place).

12. Clause 12. Confidentiality

Contractor acknowledges that it will be necessary for Client to disclose certain confidential and proprietary information to Contractor in order for Contractor to perform duties under this Agreement. Contractor acknowledges that any disclosure to any third party or any misuse of this proprietary or confidential information would irreparably harm Client. Accordingly, Contractor will not disclose or use, either during or after the term of this Agreement, any proprietary or confidential information of Client without Client's prior written permission except to the extent necessary to perform services on Client's behalf.

Proprietary or confidential information includes:

- the written, printed, graphic, or electronically recorded materials furnished by Client for Contractor to use

- business or marketing plans or strategies, customer lists, operating procedures, trade secrets, design formulas, know-how and processes, computer programs and inventories, discoveries and improvements of any kind, sales projections, and pricing information

- information belonging to customers and suppliers of Client about whom Contractor gained knowledge as a result of Contractor's services to Client

- any written or tangible information stamped "confidential," "proprietary," or with a similar legend, and

- any information that Client makes reasonable efforts to maintain the secrecy of.

Contractor shall not be restricted in using any material that is publicly available, already in Contractor's possession, known to Contractor without restriction, or rightfully obtained by Contractor from sources other than Client.

Upon termination of Contractor's services to Client, or at Client's request, Contractor shall deliver to Client all materials in Contractor's possession relating to Client's business.

Since salespeople often have to be given access to your valuable customer lists, this provision requires the salesperson to keep such information confidential. It's always wise to include this provision in your agreement, but the extent to which the identities of your customers qualify as a legally protectible trade secret depends on several factors.

A customer list will be a trade secret only if it contains information not generally known in your trade and not already in general use by your competitors. For example, if everyone in your business knows that X Co. is the best customer in the country for your product, X Co.'s identity cannot be a trade secret. But lists of customers whose identities cannot be readily ascertained are protectable—for example, a list of all the people in a conservative bedroom community who regularly purchase pornographic videos by mail would be protectable.

Moreover, you must take reasonable steps to preserve your list's secrecy. One such step is to include this confidentiality provision. Other steps include marking the list "confidential," showing it only to those who really need to see it, and keeping it locked away when not in use. (See Chapter 5, Section A17, for a detailed discussion of trade secrecy.)

13. Clause 13. Term of Agreement

See Chapter 5, Section A13, for a discussion of this clause.

14. Clause 14. Terminating the Agreement

See Chapter 5, Section A14, for a discussion of this clause.

15. Clause 15. Exclusive Agreement

See Chapter 5, Section A15, for a discussion of this clause.

16. Clause 16. Resolving Disputes

See Chapter 5, Section A18, for a discussion of this clause.

17. Clause 17. Applicable Law

See Chapter 5, Section A19, for an explanation of why you should choose the law of a particular state to govern your agreement. Note, however, that this provision applies only to disputes you may have with the salesperson regarding the agreement. It does not restrict other people. For example, if a traveling salesperson gets into a car accident in another state and the victim sues both the salesperson and you, this clause will have no application.

18. Clause 18. Notices

See Chapter 5, Section A20, for a discussion of this clause.

19. Clause 19. No Partnership

See Chapter 5, Section A21, for a discussion of this clause.

20. Clause 20. Assignment and Delegation (Optional)

> **(Check if applicable.)**
>
> ☐ Either Contractor or Client may assign its rights and delegate its duties under this Agreement.

This optional clause permits either you or the salesperson to assign your rights or delegate your duties under the agreement. The general rules about assigning rights and delegating duties under a contract are discussed in Chapter 5, Section A22. Generally, if an IC has agreed to do work that involves personal services, the IC cannot delegate it to someone else without your consent. In other words, the IC must perform the services personally or have them performed by employees under the IC's supervision.

Ordinarily, a judge would consider a salesperson's services to be personal services. You're hiring a salesperson for that individual's particular skill and experience. This means the salesperson ordinarily cannot delegate duties to someone else without your consent. This is true even if your agreement does not expressly bar delegation.

Example: The Techy Computer Co. enters into a sales agreement with Betty, an experienced mainframe computer salesperson, to sell its computers. Betty cannot pass the work to another salesperson without Techy's consent. However, Betty is entitled to have her own employees assist her. The assistance of employees is not considered a delegation of duties.

If you include this clause, you're giving up one of your most important rights under the law of contracts—the right to require the IC to perform the sales services personally. Why would you want to do this? Because by giving up this right and permitting delegation by the IC in your agreement, you eliminate the possibility that the salesperson will be deemed a statutory employee by the IRS. A salesperson who has this delegation right cannot be a statutory employee. (See Section A2, above, for an explanation of why this is important.) But the agreement must reflect reality—that is, you must really intend to give the salesperson the right to delegate, not just say so on paper.

Study Section A2, above, carefully. You may not need to worry about the salesperson being considered a statutory employee by the IRS—for example, because the individual works for you less than 80% of the time. In this event, you need not include the delegation clause. Instead, you can simply let the agreement say nothing on the subject of assignment and delegation, and the normal legal rules will apply.

21. Signatures

See Chapter 5, Section A23, for a discussion of this clause.

C. Agreement for Use by Salesperson

Use this agreement if you're working as a salesperson. Any type of IC salesperson who is paid by commission may use it, except for a real estate salesperson. It contains a number of provisions favorable to you that a hiring firm wouldn't ordinarily put in an agreement it prepared—for example, it requires the hiring firm to pay a late fee if you aren't paid on time and limits your burden to keep the client's trade secrets confidential.

An agreement for use by the salesperson is on the CD-ROM. You can modify individual clauses to tailor the text to your specific needs.

A tear-out agreement for use by the salesperson is in Appendix III. Use this form if you don't have access to a computer. It contains blank spaces for you to fill in by hand or typewriter.

Although we provide a tear-out form for those of you who either don't have computers or who would prefer not to use the CD-ROM, don't just tear out the form and use it as your IC agreement if you can avoid doing so. A fill-in-the-blank IC agreement is the least persuasive type of contract when you are trying to convince a government agency that you are an IC and not an employee of the hiring firm. The best way to create your agreement is to use the CD-ROM version and tailor it to your needs. If you can't use the CD-ROM version, then retype the version at the back of the book and tailor it to your needs. Use the fill-in-the-blank form only as a last resort.

Call up the agreement on your computer or tear out the form in Appendix III, and read it along with the following instructions and explanations.

Title of agreement. You don't need a title for an independent contractor agreement, but if you want one, call it "Independent Contractor Agreement" or "Agreement for Professional Services." Since you are not your client's employee, do not use "Employment Agreement" as a title.

Names of independent contractor and hiring firm. At the beginning of your contract, refer to yourself by your full business name. Later in the contract, you can use an obvious abbreviation.

If you're a sole proprietor, insert the full name you use for your business. This can be your own name or a fictitious business name or assumed name you use to identify your business. For example, if Iona Lott calls her one-person sales business "Lotts a Sales," she would use that name on the contract. Using a fictitious business name helps show that you're a business, not an employee.

If your business is incorporated, use your corporate name, not your own name—for example: "John Smith, Incorporated" instead of "John Smith." If you've formed a limited liability company (LLC) use your full LLC name—for example, "John Smith, LLC."

Do not refer to yourself as an employee or to the client as an employer. For the sake of brevity, it is usual to identify yourself and the client by shorter names in the rest of the agreement. You can use an abbreviated version of your full name—for example, "ABC" for "ABC Sales." Or you can refer to yourself simply as "Contractor" or "Consultant."

Refer to the client initially by its company name and subsequently by a short version of the name or as "Client" or "Firm."

Insert the addresses of the principal places of business of both the client and yourself. If you work out of your home, use that address. If you or the client have more than one office or workplace, the principal place of business is the main office.

1.   Clause 1. Services to Be Performed

Contractor agrees to sell the following product or merchandise for client:

_____

(Optional: Check if applicable.)

☐ Contractor shall seek sales of the product in the homes of various individuals.

Describe what product, service, or merchandise you will sell on the client's behalf.

If you sell consumer products door to door on a commission-only basis, you will automatically qualify as an IC under federal tax laws if you have a signed IC agreement. (See Section A, above, for an explanation of this issue.) In this event, you should include the optional provision stating that you'll seek sales in the homes of various individuals.

2. Clause 2. Compensation

> In consideration for the services to be performed by Contractor, Client agrees to pay Contractor a commission on completed sales as follows:
>
> _____
>
> Contractor acknowledges that no other compensation is payable by Client and that all of Contractor's compensation will depend on sales made by Contractor. None of Contractor's compensation shall be based on the number of hours worked by Contractor.

This agreement is designed to be used by salespeople who are paid by commission only. It's very difficult for any salesperson who is paid a regular salary to qualify as an IC.

Describe the amount and terms of your sales commission in the space provided. The client is required to pay you within the time you specify after you earn your commissions; two weeks is a common time period.

3. Clause 3. Late Fees (Optional)

See Chapter 5, Section B4, for a discussion of this clause.

4. Clause 4. Expenses

> (Choose Alternative A or B.)
> ☐ ALTERNATIVE A
> Contractor shall be responsible for all expenses incurred while performing services under this Agreement. This includes license fees, memberships and dues; automobile and other travel expenses; meals and entertainment; insurance premiums; and all salary, expenses, and other compensation paid to employees or contract personnel the Contractor hires to complete the work under this Agreement.
> ☐ ALTERNATIVE B
> Client shall reimburse Contractor for the following expenses that are directly attributable to work performed under this Agreement: _____.
> Contractor shall submit an itemized statement of Contractor's expenses. Client shall pay Contractor within 30 days after receipt of each statement.

As used here, "expenses" are the costs you incur that you can attribute directly to your work for a client. They include, for example, the cost of phone calls or traveling done on the client's behalf. Expenses do not include your normal fixed overhead costs, such as your office rent or the cost of commuting to and from your office.

Typically, commission salespeople pay their own expenses. It's more supportive of your IC status for you not to be reimbursed for expenses. A better approach is to charge a large enough commission to pay for these items yourself.

Choose Alternative A if you'll pay your own expenses. Choose Alternative B if the client will reimburse you for any expenses and describe what they are.

5. Clause 5. Materials (Optional)

See Chapter 5, Section B6, for a discussion of this clause.

6. Clause 6. Term of Agreement

See Chapter 5, Section B7, for a discussion of this clause.

7. Clause 7. Terminating the Agreement

See Chapter 5, Section B8, for a detailed discussion of the general rules concerning termination. One issue of special importance to salespeople is how they get paid after the contract is terminated. These clauses contain a provision stating that you are entitled to your commissions on all orders received by the company prior to the date of termination.

8. Clause 8. Independent Contractor Status

Contractor is an independent contractor, not Client's employee. Contractor's employees or contract personnel are not Client's employees. Contractor and Client agree to the following rights consistent with an independent contractor relationship (check all that apply):

☐ Contractor has the right to perform services for others during the term of this Agreement.

☐ Contractor shall have no obligation to perform any services other than the sale of the product described here.

☐ Contractor has the sole right to control and direct the means, manner, and method by which the services required by this Agreement will be performed. Consistent with this freedom from Client's control, Contractor:

  ☐ does not have to pursue or report on leads furnished by Client

  ☐ is not required to attend sales meetings organized by Client

  ☐ does not have to obtain Client's pre-approval for orders, and

  ☐ shall adopt and carry out Contractor's own sales strategy.

☐ Subject to any restrictions on Contractor's sales territory contained in this Agreement, Contractor has the right to perform the services required by this Agreement at any location or time.

☐ Contractor has the right to hire assistants as subcontractors or to use employees to provide the services required by this Agreement, except that Client may supply Contractor with sales forms.

☐ The Contractor or Contractor's employees or contract personnel shall perform the services required by this Agreement; Client shall not hire, supervise, or pay any assistants to help Contractor.

☐ Neither Contractor nor Contractor's employees or contract personnel shall receive any training from Client in the skills necessary to perform the services required by this Agreement.

☐ Client shall not require Contractor or Contractor's employees or contract personnel to devote full time to performing the services required by this Agreement.

Even if you qualify automatically as an IC for IRS purposes, you will still need to satisfy various state agencies that you are an IC, not the hiring firm's employee. These agencies include your state workers' compensation and unemployment compensation agencies. This portion of the agreement is intended to help establish that you are an IC. Read each of these clauses carefully and be sure to check only those that apply.

The key to establishing your IC status is to show that the hiring firm does not have the right to control how you perform the sales services required by the agreement. (See Chapter 2 for a detailed discussion.) To help make this clear, the agreement provides, among other things, that you:

- do not have to pursue or report on any sales leads the hiring firm provides
- are not required to attend sales meetings the client organizes
- need not obtain the hiring firm's pre-approval for any orders, and
- will adopt and carry out your own sales strategy.

9. Clause 9. Local, State, and Federal Taxes

See Chapter 5, Section B10, for a discussion of this clause.

10. Clause 10. Exclusive Agreement

See Chapter 5, Section B11, for a discussion of this clause.

11. Clause 11. Confidentiality (Optional)

(Check if applicable.)

☐ During the term of this Agreement and for

_____ ☐ months ☐ years afterward, Contractor will use reasonable care to prevent the unauthorized use or dissemination of Client's confidential information. Reasonable care means at least the same degree of care Contractor uses to protect Contractor's own confidential information from unauthorized disclosure.

Confidential information is limited to information clearly marked as confidential or disclosed orally and summarized and identified as confidential in a writing delivered to Contractor within 15 days of disclosure.

Confidential information does not include information that:

- the Contractor knew before Client disclosed it
- is or becomes public knowledge through no fault of Contractor
- Contractor obtains from sources other than Client who owe no duty of confidentiality to Client, or
- Contractor independently develops.

A trade secret is information or know-how that is not generally known by others and that provides its owner with a competitive advantage in the marketplace. A good example of a trade secret is a company's customer list. A trade secret owner must take reasonable steps to keep such information secret—for example, by not publishing it, restricting the number of employees who have access to it, and requiring outsiders who need to know the information to sign nondisclosure agreements promising not to use or disclose it to others. By taking such steps, a company becomes legally entitled to bring a court action to prevent the confidential information from being improperly used or disclosed.

Because salespeople often have access to valuable confidential information, hiring firms often seek to impose confidentiality restrictions on them. These provisions bar you from disclosing the client's trade secrets—for example, marketing plans, information on products under development, pricing information, manufacturing techniques, or customer lists.

It's not unreasonable for a client to want you to keep its secrets away from the eyes and ears of competitors. Unfortunately, however, many of these provisions are worded so broadly that it is difficult, if not impossible, for you to know how to comply. Vague confidentiality clauses can make it difficult for you to work for other clients without fear of violating your duty of confidentiality.

If, like most ICs, you make your living by performing similar services for many firms in the same industry, you should insist on a confidentiality provision that is reasonable in scope and defines precisely what information you must keep confidential. Such a provision should last for only a limited time—five years at the most.

The optional confidentiality provision in this agreement is designed to be reasonable and understandable. It prevents your unauthorized disclosure of any written material of the client marked "confidential" or information disclosed orally that the client later writes down, marks as "confidential" and delivers to you within 15 days. This enables you to know for sure what material is, and is not, confidential.

12. Clause 12. Resolving Disputes

See Chapter 5, Section B13, for a discussion of this clause.

13. Clause 13. Notices

See Chapter 5, Section B15, for a discussion of this clause.

14. Clause 14. No Partnership

See Chapter 5, Section B16, for a discussion of this clause.

15. Clause 15. Applicable Law

See Chapter 5, Section B17, for a discussion of this clause.

16. Clause 16. Assignment and Delegation (Optional)

> ☐ Either Contractor or Client may assign its rights and delegate its duties under this Agreement.

The rules about assigning rights and delegating duties under a contract are discussed in Chapter 5, Section A22. Generally, if the IC agrees to perform personal services, the IC cannot delegate that work to someone else without the client's consent. In other words, the IC must perform the services personally or have them performed by employees under the IC's supervision.

As a salesperson, your services would likely be considered personal services. You're being hired for your particular skill. This means that you may not be allowed to delegate your duties under the agreement without the client's consent. This is true even if your agreement does not expressly bar delegation.

However, it's often desirable for you to have the flexibility to delegate your duties under the agreement—for example, if you're busy, you can get another salesperson to perform all or part of your sales duties (but you remain liable if your delegate does not perform these duties properly; see Chapter 5, Section A22).

Luckily, your duties can be made delegable if the client agrees. That's what this clause is designed to accomplish. It allows an unrestricted right of delegation (and assignment).

If a client balks at including this clause, point out that it can help the client if a government agency challenges your status as an IC. It is strong evidence of an IC relationship because it shows the client is concerned only with results, not who achieves them. It also demonstrates lack of control over you.

There's one instance where this clause is an absolute necessity: if you could be considered a statutory employee by the IRS. You can't be a statutory employee if your agreement gives you the right to delegate your duties. Including this clause will save both you and the client the money and headaches that go with statutory employee status. (See Section A2, above, for an explanation of statutory employees.)

If the client refuses to give you the right to delegate, delete the clause and let the agreement say nothing on the subject. The normal assignment and delegation rules will apply. This means you'll still be able to assign your benefits under the agreement, but you won't be able to delegate your duties without the client's consent.

17. Signatures

See Chapter 5, Section B19, for information about signing your agreement. ■

Agreements for Accountants and Bookkeepers

This chapter contains IC agreements for accounting and bookkeeping services. Use the agreement in Section A if you're hiring an accountant or bookkeeper. Use the agreement in Section B if you're working as an accountant or bookkeeper.

A. Agreement for Use by Hiring Firm

Use this agreement if you are hiring an accountant or bookkeeper.

An agreement for use by the hiring firm is on the CD-ROM. You can modify individual clauses to tailor the text to your specific needs.

A tear-out agreement for use by the hiring firm is in Appendix II. Use this form if you don't have access to a computer. It contains blank spaces for you to fill in by hand or typewriter.

Although we provide a tear-out form for those of you who either don't have computers or who would prefer not to use the CD-ROM, don't just tear out the form and use it as your IC agreement if you can avoid doing so. A fill-in-the-blank IC agreement is the least persuasive type of contract when you are trying to convince a government agency that the worker is an IC and not an employee. The best way to create your agreement is to use the CD-ROM version and tailor it to your needs. If you can't use the CD-ROM version, then retype the version at the back of the book and tailor it to your needs. Use the fill-in-the-blank form only as a last resort.

Call up the agreement on your computer or tear out the form agreement in Appendix II, and read it along with the following instructions and explanations.

Title of agreement. You don't have to have a title for the agreement, but if you want one you should call it "Independent Contractor Agreement." Do not use "Employment Agreement" as a title.

Names of IC and hiring firm. Never refer to an IC as an employee or to yourself as an employer. Initially, it's best to refer to the accountant or bookkeeper by name. If the person is incorporated, use the corporate name. For example: "John Smith Incorporated" instead of just "John Smith."

If the IC is unincorporated but is doing business under a fictitious business name, use that name. A fictitious business name or assumed name is a name sole proprietors or partners use to identify their business other than their own names. For example, if Alvin calls his one-man bookkeeping service "ABC Bookkeeping Services," you should use that name. This shows you're contracting with a business, not a single individual.

For the sake of brevity, you can identify yourself and the IC by shorter names in the remainder of the agreement. You can use an abbreviated version of the IC's full name—for example, "ABC" for "ABC Bookkeeping Services." Or you can refer to the IC simply as "Contractor," "Accountant," or "Bookkeeper."

Refer to yourself initially by your company name and subsequently by a short version of the name or as "Client" or "Firm."

Put the short names in quotes after the full names. Also include the addresses of the principal places of business of the IC and yourself. If you or the IC have more than one office or workplace, the principal place of business is the main office.

1. Clause 1. Services to Be Performed

See Chapter 5, Section A1, for a discussion of this clause.

2. Clause 2. Payment

> (Choose Alternative A or B and optional clause, if desired.)
>
> ☐ ALTERNATIVE A
>
> In consideration for the services to be performed by Contractor, Client agrees to pay Contractor $_____ according to the terms of payment set forth below.
>
> ☐ ALTERNATIVE B
>
> In consideration for the services to be performed by Contractor, Client agrees to pay Contractor at the rate of $_____ per hour according to the terms of payment set forth below.
>
> **(Optional: Check and complete if applicable.)**
>
> ☐ Unless otherwise agreed upon in writing by Client, Client's maximum liability for all services performed during the term of this Agreement shall not exceed $_____ .

Accountants and bookkeepers either charge by the hour or charge a fixed fee for a specific project—for example, preparing a tax return. Use the clause for whichever payment method applies.

a. Fixed fee

Paying an accountant or bookkeeper a fixed fee for the entire job, rather than an hourly or daily rate, strongly supports a finding of independent contractor status. This way, the person risks losing money if the project takes longer than expected; on the other hand, the person may earn a substantial profit if the project is completed quickly. A worker who has the opportunity to earn a profit or risk a loss is more likely to be considered an IC. This method of payment also ensures that you know exactly how much the IC's work will cost.

Choose Alternative A and insert the amount of the fee if you'll pay a fixed fee.

b. Payment by the hour

Many accountants and bookkeepers charge by the hour, day, or other unit of time. Paying by the hour generally does not support the worker's independent contractor status, but you can get away with it if, as is the case here, it's a common practice in the field in which the IC works. Government auditors would likely not challenge the IC status of an accountant or bookkeeper who charges by the hour, as long as the worker is otherwise running an independent business.

Choose Alternative B and insert the amount of the payment if you will pay the accountant or bookkeeper by the hour or other unit of time.

If you pay by the hour, you may wish to place a cap on the accountant or bookkeeper's total compensation. This may be a particularly good idea if you're unsure how reliable or efficient the person is. Insert the amount of the cap in this provision or delete this language if you don't want to have a cap.

3. Clause 3. Terms of Payment

> (Choose Alternative A, B, or C.)
>
> ☐ ALTERNATIVE A
>
> Upon completion of Contractor's services under this Agreement, Contractor shall submit an invoice. Client shall pay Contractor the compensation described within _____ [15, 30, 45, 60] days after receiving Contractor's invoice.
>
> ☐ ALTERNATIVE B
>
> Contractor shall be paid $_____ upon the signing of this Agreement and the remainder of the compensation described above upon completion of Contractor's services and submission of an invoice.
>
> ☐ ALTERNATIVE C
>
> Contractor shall invoice Client on a monthly basis for all hours worked pursuant to this Agreement during the preceding month. Invoices shall be submitted on Contractor's letterhead specifying an invoice number, the dates covered in the invoice, the hours expended, and the work performed (in summary) during the invoice period. Client shall pay Contractor's fee within _____ days after receiving Contractor's invoice.

This clause explains how and when you'll pay the IC. Because the accountant or bookkeeper is running an independent business, you should receive an invoice setting forth the amount you have to pay.

a. Full payment upon completion of work

Choose Alternative A if you'll pay the full amount of a fixed fee upon completion of the work.

b. Divided payments

Some accountants or bookkeepers will insist on a partial down payment of their fixed fee before they begin work. Choose Alternative B if you agree to this. Insert the amount of the fixed fee you'll pay when the Agreement is signed.

c. Hourly payment agreements

Even accountants or bookkeepers who are paid by the hour or other unit of time should submit invoices. Never automatically pay an accountant or bookkeeper weekly or biweekly the way you pay employees. It's best to pay no more than once a month—this is how businesses are normally paid. Indicate how soon you'll pay the accountant or bookkeeper after you receive the invoice. Thirty days is a common payment period, but you can choose a longer or shorter period.

4. Clause 4. Expenses

> Client shall reimburse Contractor for the following expenses that are attributable directly to work performed under this Agreement:
>
> - travel expenses other than normal commuting, including airfares, rental vehicles, and highway mileage in company or personal vehicles at _____ cents per mile
> - telephone, fax, online, and telegraph charges
> - postage and courier services
> - printing and reproduction
> - computer services, and
> - other expenses resulting from the work performed under this Agreement.
>
> Contractor shall submit an itemized statement of Contractor's expenses. Client shall pay Contractor within 30 days after receipt of each statement.

Usually, paying expenses is an indicator of employee status. In this case, however, it shouldn't be. It is a common practice for accountants and bookkeepers to charge their clients for expenses such as photocopying, long-distance phone calls, and travel expenses. The agreement contains a provision requiring you to reimburse the accountant or bookkeeper for such expenses.

5. Clause 5. Materials

See Chapter 5, Section A5, for a discussion of this clause.

6. Clause 6. Independent Contractor Status

See Chapter 5, Section A6, for a discussion of this clause.

7. Clause 7. Business Permits, Certificates, and Licenses

See Chapter 5, Section A7, for a discussion of this clause.

8. Clause 8. Professional Obligations

> Contractor shall perform all services under this Agreement in accordance with generally accepted accounting practices and principles. This Agreement is subject to the laws, rules, and regulations governing the accounting profession imposed by government authorities or professional associations of which Contractor is a member.

The agreement requires the accountant or bookkeeper to perform the services according to generally accepted accounting practices and principles. Insisting on this level of performance has no affect on the worker's IC status, since you are entitled to require an accountant to follow the rules and regulations of the profession. Including this clause in your agreement puts the IC on notice that you expect the IC to live up to these standards. It also entitles you to sue ICs for breach of contract if they do not live up to them.

9. Clause 9. Insurance

> Client shall not provide any insurance coverage of any kind for Contractor or Contractor's employees or contract personnel. Contractor shall maintain a broad form Commercial General Liability Insurance policy providing for coverage of at least $_____ for each occurrence. Before commencing any work, Contractor shall provide Client with proof of this insurance and proof that Client has been made an additional insured under the policy.
>
> **(Optional: Check and complete if applicable.)**
>
> ☐ Contractor shall maintain an errors and omission insurance policy providing for coverage of at least $_____ for each occurrence. Before commencing any work, Contractor shall provide Client with proof of this insurance.

This portion of the agreement concerns the types and amounts of liability insurance the IC must maintain.

a. General liability insurance

An accountant or bookkeeper should have a general liability insurance policy just like any other business. You need not provide it. This type of coverage insures the consultant against personal injury or property damage claims by others. For example, if the IC accidentally injures a bystander or one of your employees while performing services for you, the IC's liability policy will pay the costs of defending a lawsuit and pay damages up to the limits of the policy coverage.

This clause requires the IC to obtain and maintain a commercial liability insurance policy providing a minimum amount of coverage. You need to state how much coverage is required. $500,000 to $1 million in coverage is usually adequate. If you're not sure how much coverage to ask for, ask your own insurance broker or agent to advise you.

The insurance clause also requires the IC to name you as an additional insured under the policy. This additional insurance is a low-cost endorsement the IC can easily have added to a liability policy. Being added as an insured under the IC's policy will nail down your rights to coverage under the IC's policy if you are named along with the IC in a lawsuit by someone injured by the IC's negligence.

The IC should also agree to indemnify—that is, repay—you if somebody the IC injures decides to sue you.

b. Professional liability insurance

Besides liability coverage, a Certified Public Accountant (CPA) should also have professional liability insurance, also known as errors and omissions, or E & O, coverage. General liability insurance doesn't cover claims for professional negligence—that is, claims for damages caused by an error or omission in the way an accountant performs services. For example, if a CPA's error in preparing your tax return costs you IRS fines and penalties, the CPA's general liability coverage would not help. This is where E & O coverage comes in.

You can always bring a legal action against an IC whose poor performance costs you damages, regardless of whether the IC has an insurance policy. (But if your agreement requires arbitration or mediation, you may not be able to sue in court; see Chapter 5, Section A12.) If your action is successful, a court or arbitrator may order the IC to pay you damages. Unfortunately, such a legal victory may prove worthless if the IC doesn't have the money to pay the damages. This is why you want the IC to have E & O insurance: It will pay any court judgment or settlement you obtain up to the policy limits. This ensures that there will be money available for you to get paid.

Unfortunately, E & O coverage is expensive. As a result, it is not something that bookkeepers, as opposed to CPAs, often have or are willing to obtain.

This provision requiring E & O coverage is optional. If you don't want to require it, delete the clause from your agreement.

10. Clause 10. Term of Agreement

See Chapter 5, Section A13, for a discussion of this clause.

11. Clause 11. Terminating the Agreement

See Chapter 5, Section A14, for a discussion of this clause.

12. Clause 12. Exclusive Agreement

See Chapter 5, Section A15, for a discussion of this clause.

13. Clause 13. Resolving Disputes

See Chapter 5, Section A18, for a discussion of this clause.

14. Clause 14. Applicable Law

See Chapter 5, Section A19, for a discussion of this clause.

15. Clause 15. Notices

See Chapter 5, Section A20, for a discussion of this clause.

16. Clause 16. No Partnership

See Chapter 5, Section A21, for a discussion of this clause.

17. Clause 17. Assignment and Delegation (Optional)

(Check if applicable.)

(Choose Alternative A or B.)

☐ ALTERNATIVE A

Either Contractor or Client may assign rights and may delegate duties under this Agreement.

☐ ALTERNATIVE B

Contractor may not assign or subcontract any rights or delegate any duties under this Agreement without Client's prior written approval.

The general rules governing assigning rights and delegating duties under a contract are discussed in Chapter 5, Section A22. You're probably not too concerned if an accountant or bookkeeper assigns benefits under the agreement—this would probably just mean that you'll pay someone other than the IC for the work.

However, delegating duties is another story. You're probably hiring this person because you think this person is the best person for the job, with unique talents or experience unavailable elsewhere. You don't want the consultant to hand off the job to somebody else who may not be as talented.

Fortunately, the law protects you against this. Under general contract law principles, if an IC has agreed to provide personal services, the IC cannot delegate that work to someone else without your consent. In other words, the IC must perform the services personally or have them performed by employees under the IC's supervision.

A highly skilled CPA's services would ordinarily be considered personal services. This means a CPA ordinarily cannot delegate duties without your consent. This is true even if your agreement does not expressly bar delegation.

However, delegation is probably permissible where you hire a bookkeeper to perform largely mechanical tasks, such as payroll accounting.

You have several options for dealing with assignment and delegation by the IC:

- **Option #1: Do nothing.** You need not say anything about assignment and delegation in your agreement. Your agreement will be assignable or delegable subject to the restrictions noted above. This is the best option in most cases.

- **Option #2: Allow assignment and delegation.** If you absolutely do not care who does the work required by the agreement, choose Alternative A to allow an unrestricted right of assignment and delegation. This clause can help you if you are challenged on your treatment of the worker as an IC. It is strong evidence of an IC relationship because it shows you're concerned only with results, not who achieves them. It also demonstrates lack of control over the IC. However, if you include this clause in the agreement, the restrictions on assignment and delegation discussed above will not apply—that is, the IC will be able to delegate duties even if they involve personal services.

- **Option #3: Expressly forbid delegation by CPAs.** Given the legal restrictions on delegation by CPAs already in place, you don't really need to say anything about restricting delegation by a CPA in your agreement, as explained above in Option #1. You can simply rely on these legal rules. This is the best course in most cases. However, there may be situations in which

you fear a CPA may not fully understand the restrictions discussed above. To make clear to the CPA that delegation of duties is impermissible without your consent, you may wish to choose Alternative B, which prevents delegation by the CPA under any circumstances without your written approval in advance.

Unfortunately, if your agreement is subsequently reviewed by government auditors, they may view this clause as an indicator of an employment relationship. It highlights the fact that you want the CPA to perform the services personally, which tends to show that you want to control who does the work. Typically, such auditors aren't aware of all the niceties of contract law, so they may not understand that you're merely restating what the law already provides. For this reason, you should include this clause only if you really think it's necessary.

- **Option #4: Expressly forbid delegation by bookkeepers.** There may be some situations in which you don't want a bookkeeper to delegate contractual duties without your consent. If you have hired a particular bookkeeper because of special expertise, creativity, reputation for performance, or financial stability, you may not want someone else performing the services, even if the person is technically qualified to do the work. In this event, choose Alternative B to prevent delegation by a bookkeeper under any circumstances without your written approval in advance. Again, include this clause only if absolutely necessary—it tends to indicate that the bookkeeper is your employee, not an IC.

18. Signatures

See Chapter 5, Section A23, for information about signing your agreement.

B. Agreement for Use by IC

Use this agreement if you're an IC performing accounting or bookkeeping services.

The entire text of the agreement for use by an accountant or bookkeeper is on the CD-ROM. You can modify individual clauses to tailor the text to your specific needs.

A tear-out form for use by an accountant or bookkeeper is in Appendix III. Use this form if you don't have access to a computer. It contains blank spaces for you to fill in by hand or typewriter.

Although we provide a tear-out form for those of you who either don't have computers or who would prefer not to use the CD-ROM, don't just tear out the form and use it as your IC agreement if you can avoid doing so. A fill-in-the-blank IC agreement is the least persuasive type of contract when you are trying to convince a government agency that you are an IC and not an employee of the hiring firm. The best way to create your agreement is to use the CD-ROM version and tailor it to your needs. If you can't use the CD-ROM version, then retype the version at the back of the book and tailor it to your needs. Use the fill-in-the-blank form only as a last resort.

Call up the agreement on your computer or tear out the form agreement in Appendix III, and read it along with the following instructions and explanations.

Title of agreement. You don't have to have a title for the agreement, but if you want one you should call it "Independent Contractor Agreement." Do not use "Employment Agreement" as a title.

Names and Addresses. At the beginning of your contract, refer to yourself by your full business name. Later on in the contract, you can use an obvious abbreviation.

If you're a sole proprietor, insert the full name you use for your business. This can be your own name or a fictitious business name or assumed name you use to identify your business. For example, if Sid Sliderule calls his one-person bookkeeping business "Ace Bookkeeping," he would use that name on the contract. Using a fictitious business name helps show that you're a business, not an employee.

If your business is incorporated, use your corporate name, not your own name—for example: "Sid Sliderule, Incorporated" instead of "Sid Sliderule."

If you've formed a limited liability company (LLC), use your full LLC name—for example, "Sid Sliderule, LLC."

Do not refer to yourself as an employee or to the client as an employer. For the sake of brevity, you can identify yourself and the client by shorter names in the rest of the agreement. You can use an abbreviated version of your full name—for example, "Ace" for "Ace Bookkeeping." Or you can refer to yourself simply as "Contractor," "Accountant," or "Bookkeeper."

Refer to the client initially by its company name and subsequently by a short version of the name or as "Client" or "Firm."

Insert the addresses of the principal places of business of both the client and yourself. If you or the client have more than one office or workplace, the principal place of business is the main office.

1. Clause 1. Services to Be Performed

See Chapter 5, Section B1, for a discussion of this clause.

2. Clause 2. Payment

(Choose Alternative A or B and optional provision, if desired.)

☐ ALTERNATIVE A

In consideration for the services to be performed by Contractor, Client agrees to pay Contractor $ _____.

☐ ALTERNATIVE B

In consideration for the services to be performed by Contractor, Client agrees to pay Contractor at the rate of $ _____ per hour.

(Optional: Check and complete if applicable.)

☐ Contractor's total compensation shall not exceed $ _____ without Client's written consent.

Accountants either charge by the hour or charge a fixed fee for a specific project—for example, preparing a tax return. Use the clause for whichever payment method applies.

a. Fixed fees

In a fixed fee agreement, you charge an agreed upon amount for the entire project. Choose Alternative A and insert the amount of the fee to charge a fixed fee.

b. Hourly payments

Instead of a fixed fee, many accountants and bookkeepers charge by the hour, day, or other unit of time. Charging by the hour does not usually help bolster your IC status, but you can probably get away with it since it's a common practice in your field. Choose Alternative B and insert the amount of the charge to charge by the hour.

If you're being paid by the hour or other unit of time, clients will often wish to place a cap on the total amount they'll spend on the project because they're afraid you might work slowly in order to earn a larger fee. An optional paragraph is provided for this. Make sure the cap allows you to spend enough time to satisfactorily complete the project.

3. Clause 3. Terms of Payment

(Choose Alternative A, B, or C.)

☐ ALTERNATIVE A

Upon completing Contractor's services under this Agreement, Contractor shall submit an invoice. Client shall pay Contractor within _____ days from the date of Contractor's invoice.

☐ ALTERNATIVE B

Contractor shall be paid $_____ upon signing this Agreement and the remaining amount due when Contractor completes the services and submits an invoice. Client shall pay Contractor within _____ days from the date of Contractor's invoice.

☐ ALTERNATIVE C

Contractor shall send Client an invoice monthly. Client shall pay Contractor within _____ days from the date of each invoice.

This clause explains how you will bill the client and be paid. To be paid for your work, submit an invoice to the client setting out the amount due. An invoice doesn't have to be fancy or filled with legalese. It should include an invoice number, the dates covered by the invoice, the hours expended (if you're being paid by the hour), and a summary of the work you performed.

a. Payment upon completing work

Choose Alternative A if you will be paid when you complete the work. Insert the time period by which the client is required to pay your fixed or hourly fee after you send the invoice. A period of 30 days is typical, but it can be shorter or longer if you wish. Note that the time for payment starts to run as soon as you send your invoice, not when the client receives it. This will help you get paid more quickly.

b. Divided payments

Choose Alternative B if you will be paid in divided payments—that is, if you opt to be paid part of a fixed or hourly fee when the agreement is signed and the remainder when the work is finished. Insert the amount of the up-front payment. This amount is, of course, subject to negotiation. Many independent contractors like to receive at least one-third to one-half of a fee before they start work. If the client is new or might have problems paying you, it's wise to get as much money in advance as you can.

c. Hourly payment for lengthy projects

Choose Alternative C if you're being paid by the hour or other unit of time and the project will last more than one month. Under this provision, you submit an invoice to the client each month setting forth how many hours you've worked. The client is required to pay you within a specific number of days from the date of each invoice.

4. Clause 4. Late Fees (Optional)

See Chapter 5, Section B4 for a discussion of this clause.

5. Clause 5. Expenses

> Client shall reimburse Contractor for the following expenses that are attributable directly to work performed under this Agreement:
> - travel expenses other than normal commuting, including airfares, rental vehicles, and highway mileage in company or personal vehicles at _____ cents per mile
> - telephone, fax, online, and telegraph charges
> - postage and courier services
> - printing and reproduction
> - computer services, and
> - other expenses resulting from the work performed under this Agreement.
>
> Contractor shall submit an itemized statement of Contractor's expenses. Client shall pay Contractor within 30 days from the date of each statement.

Expenses are the costs you incur that you can attribute directly to your work for a client. They include, for example, the cost of phone calls or traveling done on the client's behalf. Expenses do not include your normal fixed overhead costs such as your office rent or the cost of commuting to and from your office.

Typically, accountants charge their clients for expenses such as photocopying and travel time. The agreement contains a provision requiring the client to reimburse you for such expenses. Paying expenses is usually an indicator of employee status—but in this case it shouldn't be, because it is a common practice in the accounting field.

6. Clause 6. Materials

See Chapter 5, Section B6, for a discussion of this clause.

7. Clause 7. Term of Agreement

See Chapter 5, Section B7, for a discussion of this clause.

8. Clause 8. Terminating the Agreement

See Chapter 5, Section B8, for a discussion of this clause.

9. Clause 9. Independent Contractor Status

See Chapter 5, Section B9, for a discussion of this clause.

10. Clause 10. Professional Obligations

> Contractor shall perform all services under this Agreement in accordance with generally accepted accounting practices and principles. This Agreement is subject to the laws, rules, and regulations governing the accounting profession imposed by government authorities or professional associations of which Contractor is a member.

The agreement provides that you'll perform your services according to generally accepted accounting practices and principles. Obviously, this is what you intend to do in any event, so this clause imposes no additional burdens on you. But it may make the client happy.

11. Clause 11. Local, State, and Federal Taxes

See Chapter 5, Section B10, for a discussion of this clause.

12. Clause 12. Exclusive Agreement

See Chapter 5, Section B11, for a discussion of this clause.

13. Clause 13. Modifying the Agreement (Optional)

See Chapter 5, Section B12, for a discussion of this clause.

14. Clause 14. Resolving Disputes

See Chapter 5, Section B13, for a discussion of this clause.

15. Clause 15. Notices

See Chapter 5, Section B15, for a discussion of this clause.

16. Clause 16. No Partnership

See Chapter 5, Section B16, for a discussion of this clause.

17. Clause 17. Applicable Law

See Chapter 5, Section B17, for a discussion of this clause.

18. Clause 18. Assignment and Delegation

> Either Contractor or Client may assign its rights or delegate any of its duties under this Agreement.

The general rules governing assigning rights and delegating duties under a contract are discussed in Chapter 5, Section B18. That discussion points out that if the work an IC has agreed to do involves personal services, the IC cannot delegate it to someone else without the hiring firm's consent. In other words, the IC must perform the services personally or have them performed by employees under the IC's supervision.

If you're a CPA, your services would ordinarily be considered personal services. You're a highly skilled person who has been hired for your particular professional advice, not that of someone else. This means you ordinarily cannot delegate your duties to the client without the client's consent. This is so even though the agreement does not expressly bar delegation.

However, delegation is probably permissible where you're a bookkeeper hired to perform largely mechanical tasks, such as payroll accounting.

Even if you're a CPA, it's often desirable for you to have the flexibility to delegate your duties under the agreement. Luckily, your duties can be made delegable if the client agrees. That's what this clause is designed to accomplish. It allows an unrestricted right of delegation (and assignment, which is usually not an issue). If you include this clause in the agreement the normal restrictions on assignment and delegation will not apply.

If a client balks at including this clause, point out that it can help the client if a government agency challenges your status as an IC. It is strong evidence of an IC relationship because it shows the client is concerned only with results, not who achieves them. It also demonstrates lack of control over you.

If the client still refuses to agree to this, delete the clause and leave the agreement silent on the subject. The normal assignment and delegation rules will apply. This means you'll still be able to assign your benefits under the agreement, but will be unable to delegate your duties without the client's consent.

19. Signatures

See Chapter 5, Section B19, for information about signing your contract. ∎

Agreements for Software Consultants

The agreements in this chapter are for software consulting and programming services. Use the agreement in Section A if you're hiring a software consultant. Use the agreement in Section B if you're working as a software consultant.

A. Agreement for Use by Hiring Firm

Use this agreement if you're hiring an independent contractor to perform software consulting or programming services. The agreement is designed to protect your interests as much as possible and help establish that the worker is an IC, not your employee.

A software consultant agreement for use by the hiring firm is on the CD-ROM. You can modify individual clauses to tailor the text to your specific needs.

A tear-out form for hiring a software consultant for use by the hiring firm is in Appendix II. Use this form if you don't have access to a computer. It contains blank spaces for you to fill in by hand or typewriter.

Although we provide a tear-out form for those of you who either don't have computers or who would prefer not to use the CD-ROM, don't just tear out the form and use it as your IC agreement if you can avoid doing so. A fill-in-the-blank IC agreement is the least persuasive type of contract when you are trying to convince a government agency that the worker is an IC and not an employee. The best way to create your agreement is to use the CD-ROM version and tailor it to your needs. If you can't use the CD-ROM version, then retype the version at the back of the book and tailor it to your needs. Use the fill-in-the-blank form only as a last resort.

Call up the agreement on your computer or tear out the form agreement in Appendix II, and read it along with the following instructions and explanations.

Title of agreement. You don't have to have a title for an IC agreement, but if you want one you should call it "Independent Contractor Agreement" or "Consulting Agreement." Do not use "Employment Agreement" as a title.

Names of IC and hiring firm. Never refer to an IC as an employee or to yourself as an employer.

Initially, it's best to refer to the consultant by name. If an IC is incorporated, use the corporate name, not the consultant's own name. For example: "K.L.R. Appps, Incorporated" instead of just "K.L.R. Appps." If the consultant is unincorporated but does business under a fictitious business name, use that name. A fictitious business name or assumed name is a name sole proprietors or partners use to identify their business other than their own names. For example, if consultant Pete Programmer calls his software consulting business "Hardy Software," use that name. This shows you're contracting with a business, not a single individual.

For the sake of brevity, you can identify yourself and the IC by shorter names in the remainder of the agreement. You can use an abbreviated version of the consultant's full name—for example, "Hardy" for "Hardy Software." Or you can refer to the IC simply as "Contractor" or "Consultant."

Refer to yourself initially by your company name and subsequently by a short version of the name or as "Client" or "Firm."

Put the short names in quotes after the full names. Also include the addresses of the principal places of business of the IC and yourself. If you or the consultant have more than one office or workplace, the principal place of business is the main office.

1. Clause 1. Services to Be Performed

(Choose Alternative A or B.)

☐ ALTERNATIVE A

Consultant agrees to perform the following services for Client:

_____

_____

_____

☐ ALTERNATIVE B

Consultant agrees to perform the services described in Exhibit A, which is attached to and made part of this Agreement.

The agreement should describe, in as much detail as possible, what the consultant is expected to do. However, the description should tell the consultant only what results to achieve, not how to achieve them. As an IC, it's up to the consultant to decide how to perform the services. (See Chapter 5, Section A1, for examples of good and bad service descriptions.)

It is often helpful to break down the project into discrete parts or stages—often called phases or milestones. This makes it easier for the hiring firm to monitor the consultant's progress and may aid the consultant in budgeting time. For example, the consultant's pay could be contingent upon the completion of each milestone. (Payment schedules are covered below, under "Payment.")

You can include the description in the main body of the agreement. If it's a lengthy explanation, put it on a separate document labeled Exhibit A and attach it to the agreement.

2. Clause 2. Payment

(Choose Alternative A, B, or C and optional clause, if desired.)

☐ ALTERNATIVE A

Consultant shall be paid $_____ upon completion of the work as detailed in Clause 1.

☐ ALTERNATIVE B

Client shall pay Consultant a fixed fee of $_____ , in _____ installments as follows:

(a) $_____ upon completion of the following services: _____

_____

(b) $_____ upon completion of the following services: _____

_____

(c) $_____ upon completion of all remaining work to be performed and the services to be rendered in accordance with the schedule set forth in Clause 1, above, and written acceptance by Client.

☐ ALTERNATIVE C

Consultant shall be compensated at the rate of $_____ per _____ ☐ hour; ☐ day; ☐ week; ☐ month.

(Optional: Check and complete if applicable.)

☐ Unless otherwise agreed upon in writing by Client, Client's maximum liability for all services performed during the term of this Agreement shall not exceed $_____ .

There are a number of ways a consultant may be paid. Choose the alternative that suits your needs.

a. Fixed fee

The simplest payment option is to pay a fixed fee for the entire project (Alternative A). This way, you know exactly how much the work will cost you, and the consultant won't have an incentive to pad the bill by working more hours. Before quoting you a fixed fee, the consultant will need to have a concrete idea of what the project entails.

b.   Installment payments

Paying a consultant a fixed fee for the entire job, rather than an hourly or daily rate, supports a finding of independent contractor status. However, this can pose problems for the consultant due to difficulties in accurately estimating how long the job will take. One way to deal with this problem is to break down the job into phases and pay the consultant a fixed fee upon completion of each phase. If you use this approach, you must describe in the agreement what work the consultant must complete to receive each payment.

Example: The Splashy Swimming Pool Company hires Dave, a software consultant, to develop a computerized accounting system. Splashy and Dave agree that the lengthy project will be divided into four phases or modules—accounts receivable, accounts payable, order processing, and invoicing. Dave will be paid a specified amount upon completion of each module. They draw up the following schedule and attach it to their agreement:

SCHEDULE OF PAYMENTS

Client shall pay Consultant according to the following schedule of payments:

Phase 1. $5,000 when an invoice is submitted and the following services are completed: Accounts receivable module is completed and accepted by Client.

Phase 2. $5,000 when an invoice is submitted and the following services are completed: Accounts payable module is completed and accepted by Client.

Phase 3. $5,000 when an invoice is submitted and the following services are completed: Order processing module is completed and accepted by Client.

Phase 4. $5,000 when an invoice is submitted and the following services are completed: Invoicing module is completed and accepted by Client.

---

c.   Payment by the hour/day/week/month

If a fixed fee for the job is impractical, it doesn't make much difference for IRS purposes whether the consultant is paid by the hour, day, week, or month. It's generally a good idea to place a cap on the consultant's total compensation. This may be a particularly good idea if you're unsure how reliable or efficient the consultant is.

3.   Clause 3. Expenses

(Choose Alternative A or B.)

□ ALTERNATIVE A

Consultant shall be responsible for all expenses incurred while performing services under this Agreement.

□ ALTERNATIVE B

Consultant will not be reimbursed for any expenses incurred in connection with the performance of services under this Agreement, unless those expenses are approved in advance in writing by Client.

The IRS considers the payment of a worker's business or traveling expenses to be a mild indicator of an employment relationship. It's best not to reimburse an IC for expenses. Instead, pay the IC a fee that's high enough to cover the IC's expenses.

Select one of the two alternatives. Alternative A provides you will not pay any of the consultant's expenses. Alternative B provides that you will pay only those expenses you agree in writing to reimburse in advance. Some consultants may insist on such reimbursement, so you may not be able to avoid it.

4.   Clause 4. Invoices

> Consultant shall submit invoices for all services rendered. Client shall pay Consultant within _____ days after receipt of each invoice.

An independent contractor should never be paid weekly, bi weekly, or monthly the way an employee is. Instead, the IC should submit invoices, which you should pay at the same time and in the same manner as you pay other vendors. You need to decide how long you will have to pay the consultant after you receive the invoice. Thirty days is a common period.

5.   Clause 5. Independent Contractor Status

See Chapter 5, Section A6, for a discussion of this clause.

6.   Clause 6. Intellectual Property Ownership

> (Choose Alternative A or B.)
>
> ☐ ALTERNATIVE A
>
> Work Product includes, but is not limited to, the programs and documentation, including all ideas, routines, object and source codes, specifications, flow charts, and other materials, in whatever form, developed solely for Client under this Agreement.
>
> Consultant hereby assigns to Client its entire right, title, and interest, including all patent, copyright, trade secret, trademark, and other proprietary rights, in the Work Product.
>
> Consultant shall, at no charge to Client, execute and aid in the preparation of any papers that Client may consider necessary or helpful to obtain or maintain—at Client's expense—any patents, copyrights, trademarks, or other proprietary rights. Client shall reimburse Consultant for reasonable out-of-pocket expenses incurred under this provision.
>
> ☐ ALTERNATIVE B
>
> Work Product includes, but is not limited to, the programs and documentation, including all ideas, routines, object and source codes, specifications, flow charts and other materials, in whatever form, developed solely for Client under this Agreement.
>
> Client agrees that Consultant shall retain any and all rights Consultant may have in the Work Product.
>
> Consultant hereby grants Client an unrestricted, nonexclusive, perpetual, fully paid-up, worldwide license to use and sublicense the use of the Work Product for the purpose of developing and marketing its products, but not for the purpose of marketing Work Product separate from its products.

Usually, you will want to obtain sole ownership of what the consultant creates, whether it be a computer program, program documentation (instructions on how to use and maintain the program), or other material. However, you can work out other arrangements, if you'd like. Two different ownership alternatives are presented in the contract. Choose only one alternative and delete the other.

a.   Client owns consultant's work product

As discussed in detail in Chapter 2, a hiring firm can never be sure that it will own what it pays an independent contractor to create (the consultant's "work product") unless it obtains an assignment of the consultant's rights. Obtaining ownership of the consultant's work product presents many advantages: You can do anything you want with the material, and the consultant is prevented from giving or selling the same material to your competitors.

If you and the consultant agree that you will own the consultant's work product, you should obtain a written assignment of the consultant's intellectual property rights in such work (Alternative A). Among the most important reasons to obtain an assignment from the IC is that it gives the hiring firm the right to create derivative works based on the original work product—for example, new versions of a program originally created by the consultant.

b.   Consultant owns work product

You and the consultant may agree that the consultant will retain ownership rights in the work product (perhaps in return for receiving reduced monetary compensation). In this event, instead of assigning rights to the client, the consultant will grant you a nonexclusive license to use the work product.

Alternative B grants you a nonexclusive license to use the consultant's work product in any of your products, but prevents you from selling the consultant's work to others. This is a generous

license grant. This means you can use the work product however you want for your own purposes; you just can't sell or license it to others. Meanwhile, the consultant can sell or license the same work product to others. This can result in substantial additional income for the consultant. As a result, it's not unreasonable for you to pay the consultant substantially less than you would have for a full assignment of rights.

7.   Clause 7. Consultant's Materials

> Consultant's Materials means all programs and documentation, including routines, object and source codes, tools, utilities, and other copyrightable materials, that:
>
> - do not constitute Work Product
> - are incorporated into the Work Product, and
> - are owned solely by Consultant or licensed to Consultant with a right to sublicense.
>
> Consultant's Materials include, but are not limited to, the following: _____
>
> _____
>
> _____
>
> _____
>
> Consultant shall retain any and all rights Consultant may have in Consultant's Materials. Consultant hereby grants Client an unrestricted, nonexclusive, perpetual, fully paid-up, worldwide license to use and sublicense the use of Consultant's Materials for the purpose of developing and marketing its products.

A software consultant will often have various development tools, routines, subroutines, and other programs, data, and materials that the consultant brings to the job and that might end up in the final product. The agreement uses the term "consultant's materials" to refer to these items.

You must make sure the consultant grants you a license to use these materials. This clause grants you a nonexclusive license to use them in any of your products, including products other than those the consultant works on. To avoid confusion later on, ask the consultant to list these materials in the agreement.

8. Clause 8. Confidential Information (Optional)

(Check if applicable.)

☐ Consultant agrees that the Work Product is Client's sole and exclusive property. Consultant shall treat the Work Product on a confidential basis and not disclose it to any third party without Client's written consent, except when reasonably necessary to perform the services under this Agreement.

Consultant acknowledges that it will be necessary for Client to disclose certain confidential and proprietary information to Consultant in order for Consultant to perform duties under this Agreement. Consultant acknowledges that any disclosure to any third party or any misuse of this proprietary or confidential information would irreparably harm Client. Accordingly, Consultant will not use or disclose to others without Client's written consent Client's confidential information, except when reasonably necessary to perform the services under this Agreement. Confidential Information includes, but is not limited to:

- the written, printed, graphic, or electronically recorded materials furnished by Client for use by Contractor
- Client's business or marketing plans or strategies, customer lists, operating procedures, trade secrets, design formulas, know-how and processes, computer programs and inventories, discoveries, and improvements of any kind
- any written or tangible information stamped "confidential," "proprietary," or with a similar legend, and

- any written or tangible information not marked with a confidentiality legend, or information disclosed orally to Consultant, that is treated as confidential when disclosed and later summarized sufficiently for identification purposes in a written memorandum marked "confidential" and delivered to Consultant within 30 days after the disclosure.

Consultant shall not be restricted in the use of any material that is publicly available, already in Consultant's possession, known to Consultant without restriction, or rightfully obtained by Consultant from sources other than Client.

Consultant's obligations regarding proprietary or confidential information extend to information belonging to customers and suppliers of Client about whom Consultant may have gained knowledge as a result of Consultant's services to Client.

Consultant will not disclose to Client information or material that is a trade secret of any third party.

The provisions of this clause shall survive any termination of this Agreement.

In the course of doing work for you, the consultant may be exposed to your most valuable trade secrets—for example, your business plans, customer lists, operating procedures, computer source codes, and design formulas. It is reasonable, therefore, for you to include a nondisclosure provision in the agreement. Such a provision states that the consultant may not disclose your trade secrets to others without your permission.

There are two important parts of the confidentiality provision in the agreement. The first paragraph provides that the consultant has a duty not to disclose to others the work the consultant creates for the hiring firm. The only exception is if the disclosure is necessary in order for the consultant to perform the services required by the agreement—for example, the consultant may have to disclose your confidential information to assistants.

The first paragraph should be deleted from the agreement if the consultant retains ownership of the work product. (See Clause 6 for information about work product ownership.)

The second paragraph prevents the unauthorized disclosure by the consultant of any of your confidential information. Confidential information includes any written material that you mark confidential or information disclosed orally that the you later write down, mark as confidential, and deliver to the consultant. An exception is made where the disclosure is necessary to perform the services.

The third paragraph makes it clear that not everything qualifies as a trade secret. For example, the general knowledge, skills, and experience a consultant acquires while working for you are not trade secrets and, therefore, are not covered by this nondisclosure provision. "General knowledge" and "skills" are knowledge and skills that any average software consultant would acquire while performing the services. "Trade secrets" are specialized knowledge that is not generally known and that a software consultant would not know unless you disclosed it.

Other material that won't qualify as your trade secrets includes information that

- is publicly available—for example, published in a trade journal
- is already in the consultant's possession or known to the consultant without restriction before the consultant began working for you, and
- is obtained by the consultant from sources other than you.

9. Clause 9. Term of Agreement

See Chapter 5, Section A13, for a discussion of this clause.

10. Clause 10. Termination of Agreement

- Each party has the right to terminate this Agreement if the other party has materially breached any obligation herein and such breach remains uncured for a period of 30 days after notice thereof is sent to the other party.
- If at any time after commencement of the services required by this Agreement, Client shall, in its sole reasonable judgment, determine that such services are inadequate, unsatisfactory, no longer needed, or substantially not conforming to the descriptions, warranties, or representations contained in this Agreement, Client may terminate this Agreement upon _____ days' written notice to Consultant.

Many consultants and hiring firms want to have the right to terminate their agreements for any reason on short notice. Unfortunately, the IRS considers such an unfettered termination right to be an indicator of an employment relationship. This clause attempts to reach a compromise between the parties' desire to be able to get out of the agreement and the IRS' rules and guidelines for classifying independent contractors.

The clause has two important parts. The first paragraph permits either party to terminate the agreement if the other has breached it and has failed to remedy the breach within 30 days.

The second paragraph permits termination if you are seriously dissatisfied with the consultant's work. (Unlike the situation for which the first paragraph is designed, there needn't be a total breach.)

The second paragraph does not permit the client to terminate the agreement for irrational or whimsical reasons, no matter how sincerely felt. Instead, the client's reaction to the quality of the work is measured against an objective standard of reasonableness. The client may terminate the agreement only if, in the client's reasonable judgment, the consultant's performance is inadequate or unnecessary.

11. Clause 11. Return of Materials

> Upon termination of this Agreement, each party shall promptly return to the other all data, materials, and other property of the other held by it.

This clause requires both you and the consultant to return each other's materials—for example, copies of your source code the consultant may have at the end of the agreement.

12. Clause 12. Warranties and Representations

> Consultant warrants and represents that:
> - Consultant has the authority to enter into this Agreement and to perform all obligations hereunder.
> - The Work Product and Consultant's Materials are and shall be free and clear of all encumbrances including security interests, licenses, liens, or other restrictions except as follows: _____
> _____
> - The use, reproduction, distribution, or modification of the Work Product and Consultant's Materials does not and will not violate the copyright, patent, trade secret, or other property right of any former client, employer, or third party.
> - For a period of _____ days following acceptance of the Work Product, the Work Product will be:
>   - free from reproducible programming errors and defects in workmanship and materials under normal use, and
>   - substantially in conformance with the product specifications.
> - The Work Product shall be created solely by Consultant, Consultant's employees during the course of their employment, or independent contractors who assigned all right, title, and interest in the work to Consultant.

A warranty is a promise or statement regarding the quality, quantity, performance, or legal title of a product or service. Software independent contractor agreements typically contain several warranty provisions.

The first bullet provides that the consultant has the authority to perform as promised. This means, for example, that the consultant is not prevented from working for you because the consultant has previously signed a noncompetition agreement for a previous client. If the consultant becomes unable to perform the services because a prior client complains that it would violate such an agreement, you can sue the consultant for breach of warranty.

In the second bullet, the consultant promises that nothing will prevent the client from obtaining free and clear ownership or license rights to the consultant's work product and material. If the consultant has licensed any materials from third parties that are to be included in the work product, the licenses should be listed here. Note that the consultant has already granted you a license to use any such materials in Clause 7, above, dealing with consultant's materials.

In the third bullet, the consultant promises that the work product and materials will not infringe on others' copyrights, patents, trade secrets, or other intellectual property rights.

In the fourth bullet, the consultant promises that the work product will perform in substantial conformance with the specifications and will be free of reproducible programming errors. You need to state how long this warranty will last—it can be anywhere from 90 days to one year or longer.

Finally, in the fifth bullet, the consultant promises that any workers whom the consultant hires will assign their rights to the consultant so the consultant may in turn assign or license those rights to you.

13. Clause 13. Indemnification

> Consultant agrees to indemnify and hold harmless Client against any claims, actions, or demands, including without limitation reasonable attorney and accounting fees, alleging or resulting from the breach of the warranties contained in this Agreement. Client shall provide notice to Consultant promptly of any such claim, suit, or proceeding and shall assist Consultant, at Consultant's expense, in defending any such claim, suit, or proceeding.

Indemnification is a fancy legal word that means a promise to repay someone for their losses or damages if a specified event occurs. This clause requires the consultant to indemnify you or your business if someone sues or threatens to sue you because the consultant breached any of the warranties made in the agreement. For example, if the consultant copied code from a third party and sold it to you, then that third party sued you for infringement, the consultant would be required to pay for your legal defense and any damages awarded by a court.

14. Clause 14. Assignment and Delegation (Optional)

> **(Check if applicable.)**
>
> ☐ Consultant may not assign or subcontract any rights or obligations under this Agreement without Client's prior written approval.

The general rules governing assigning rights and delegating duties under a contract are discussed in Chapter 5, Section A22. If an IC has agreed to perform personal services, the IC cannot delegate that work to someone else without your consent. In other words, the IC must perform the services personally or have them performed by employees under the IC's supervision.

Ordinarily, a software consultant's services would be considered personal services. This means that the consultant ordinarily cannot delegate duties without your consent. This is true even if your agreement does not expressly bar delegation.

Example: The Big Bang Demolition Co. enters into a consulting agreement with Al, a software consultant, to perform a study on how using computers can make demolition work safer. Al cannot pass the work to another software consultant without Big Bang's consent. However, Al is entitled to have his own employees assist him.

Given these legal restrictions already in place, you need not say anything in your agreement about restricting assignment and delegation by the consultant. You can simply rely on these legal rules. This is the best way to proceed in most cases. If you decide to go this route, delete the optional assignment and delegation clause.

However, there may be situations in which you fear a consultant may not fully understand the restrictions discussed above. To make clear to the consultant that delegation of duties is impermissible without your consent, you may wish to include this optional clause preventing assignment or delegation by the consultant under any circumstances without your written approval in advance.

Unfortunately, if your agreement is subsequently reviewed by government auditors, they may view this clause as an indicator of an employment relationship. It highlights the fact that you want the consultant to perform the services personally, which tends to show that you want to control who does the work. Remember, when government auditors review your agreement, they are looking for indications that the consultant is an employee instead of an IC. Therefore, you should include this clause only if you really think it's necessary.

15.   Clause 15. Insurance

Client shall not provide any insurance coverage of any kind for Consultant or Consultant's employees or contract personnel. Consultant shall obtain and maintain a broad form Commercial General Liability Insurance policy providing for coverage of at least $_____ for each occurrence. Before commencing any work, Consultant shall provide Client with proof of this insurance and with proof that Client has been made an additional insured under the policy.

(Optional: Check and complete if applicable.)

☐ Consultant shall obtain professional liability insurance coverage for malpractice or errors or omissions committed by Consultant or Consultant's employees during the term of this Agreement. The policy shall provide for coverage of at least $_____ for each occurrence. Before commencing any work, Consultant shall provide Client with proof of this insurance.

This portion of the agreement covers the types and amounts of liability insurance the consultant is required to maintain.

a.   General liability insurance

This type of coverage insures the consultant against personal injury or property damage claims by others. (It does not provide coverage for the consultant's own property or injuries to the consultant.) For example, if a consultant accidentally injures a bystander or one of your employees while performing services for you, the consultant's liability policy will pay the costs of defending a lawsuit and pay damages up to the limits of the policy coverage. A consultant should have a general liability insurance policy just like any other business; you need not provide it.

This clause requires the consultant to obtain and maintain a commercial liability insurance policy providing a minimum amount of coverage. You need to state how much coverage is required. How much coverage should the consultant have? $500,000 to $1 million in coverage is usually adequate. If you're not sure how much coverage to require, ask your own insurance broker or agent to advise you.

The insurance clause also requires the consultant to name you as an additional insured under the policy. This additional insurance can easily be added to the consultant's liability policy. Being added as an insured under the consultant's policy will nail down your rights to coverage under the consultant's policy if you are sued along with the consultant by someone injured by the consultant's negligence.

b.   Professional liability insurance

Consultants are also often required to obtain professional liability insurance, also known as errors and omissions, or E & O, coverage. General liability insurance doesn't cover claims for professional negligence—that is, claims for damages caused by an error or omission in the way a consultant performs services. This is where E & O coverage comes in.

From the consultant's point of view, it's a good idea to have E & O coverage in case the consultant is sued—it will pay for the costs of defending a professional malpractice lawsuit and pay damages up to the limits of the policy coverage. However, it's just as important for you that such coverage be in place in case you need to sue the consultant for professional negligence. It ensures that there will be money available from an insurer to pay any judgment or settlement you obtain against the consultant. A consultant who is uninsured may not have the money or assets to pay off a judgment or settlement. In this event, any judgment you obtain may be worthless.

Unfortunately, E & O coverage is expensive. Some software consultants don't have it and may balk at obtaining it. If you insist, the consultant may want to charge you more. It's up to you to decide whether it's worth the risk to hire an uninsured consultant. This provision requiring E & O coverage is optional. If you don't want to require it, delete the clause from your agreement.

16. Clause 16. Resolving Disputes

See Chapter 5, Section A18, for a discussion of this clause.

17. Clause 17. Exclusive Agreement

See Chapter 5, Section A15, for a discussion of this clause.

18. Clause 18. Applicable Law

See Chapter 5, Section A19, for a discussion of this clause.

19. Clause 19. Notices

See Chapter 5, Section A20, for a discussion of this clause.

20. Clause 20. No Partnership

See Chapter 5, Section A21, for a discussion of this clause.

21. Signatures

See Chapter 5, Section A23, for a discussion of this clause.

B. Agreement for Use by Consultant

Use this agreement if you're working as an independent contractor performing software consulting services. The agreement is designed to protect your interests as much as possible. It contains a number of provisions favorable to you that a hiring firm wouldn't ordinarily put in an agreement it prepared—for example, requiring the hiring firm to pay a late fee if it doesn't pay you on time and limiting your liability to the client if something goes wrong.

 The entire text of the software consultant agreement for use by the IC is on the CD-ROM. You can modify individual clauses to tailor the text to your specific needs.

Call up the agreement on your computer or tear out the form agreement in Appendix III, and read it along with the following instructions and explanations.

Title of agreement. You don't need a title for an independent contractor agreement, but if you want one, call it "Independent Contractor Agreement" or "Consulting Agreement." Since you are not your client's employee, do not use "Employment Agreement" as a title.

Names and Addresses. At the beginning of your contract, refer to yourself by your full business name. Later in the contract, you can use an obvious abbreviation. Do not refer to yourself as an employee or to the client as an employer.

If you're a sole proprietor, use the full name you use for your business. This can be your own name or a fictitious business name or assumed name you use to identify your business. For example, if consultant Chip Micro calls his software consulting business "Chip Consulting," he would use that name on the contract. Using a fictitious business name helps show you're a business, not an employee.

If your business is incorporated, use your corporate name, not your own name—for example: "John Smith, Incorporated" instead of "John Smith."

For the sake of brevity, you can identify yourself and the client by shorter names in the rest of the agreement. You can use an abbreviated version of your full name—for example, "ABC" for "ABC Consulting." Or you can refer to yourself simply as "Contractor" or "Consultant." Refer to the client initially by its company name and subsequently by a short version of the name or as "Client" or "Firm."

Include the addresses of the principal places of business of the client and yourself. If you or the client have more than one office or workplace, the principal place of business is the main office.

1. Clause 1. Services to Be Performed

See Chapter 5, Section B1, for a discussion of this clause.

2. Clause 2. Payment

(Choose Alternative A, B, or C and optional provision, if desired.)

☐ ALTERNATIVE A

Consultant shall be paid $_____ upon execution of this agreement and $_____ upon completion of the work as detailed in Clause 1.

☐ ALTERNATIVE B

Client shall pay Consultant a fixed fee of $_____, in _____ installments according to the payment schedule described in Exhibit _____ which is attached to and made part of this Agreement.

☐ ALTERNATIVE C

Consultant shall be compensated at the rate of $_____ per _____ ☐ hour; ☐ day; ☐ week; ☐ month.

(Optional: Check if applicable.)

☐ Unless otherwise agreed upon in writing by Client, Client's maximum liability for all services performed during the term of this Agreement shall not exceed $_____ .

Software consultants can be paid in a variety of ways. Choose the alternative that suits your needs.

a. Fixed fee

In a fixed fee agreement, you charge an agreed upon amount for the entire project. Alternative A requires the client to pay you an initial sum when the work is commenced and the remainder when it's finished.

b. Installment payments

If the project is long and complex, you may prefer to be paid in installments rather than waiting until the project is finished to receive the bulk of your payment. One way to do this is to break the job into phases, or milestones, and be paid a fixed fee when each phase is completed. Clients often like this pay-as-you-go arrangement, too, because each payment you receive is tied to completion of a specific portion of the project. If you don't complete the phase or perform unsatisfactorily, the client doesn't have to pay you.

To do this, draw up a schedule of installment payments, tying each payment to your completion of specific services. It's usually easier to set forth the schedule in a separate document and attach it to the agreement as an exhibit. The main body of the agreement should simply refer to the attached payment schedule.

Following is an example of a schedule of payments. This schedule requires four payments: a down payment when the contract is signed and three installment payments tied to completion of specific portions of the software project (termed "modules"). However, you and the client can have as many payments as you want.

SCHEDULE OF PAYMENTS

Client Makesalot, Inc., shall pay Contractor Chip Micro according to the following schedule of payments:

1. $2,000 when this Agreement is signed.

2. $5,000 when an invoice is submitted and the following services are completed: Accounts receivable module is completed and accepted by Client.

3. $5,000 when an invoice is submitted and the following services are completed: Accounts payable module is completed and accepted by Client.

4. $5,000 when an invoice is submitted and the following services are completed: Order processing module is completed and accepted by Client.

---

c. Payment by unit of time

The majority of software consultants charge by the hour or other unit of time. This is by far the safest way to be paid if it's difficult to estimate exactly how long a project will take. Choose Alternative C if you wish to be paid this way. Being paid this way does not help establish that you're an IC, but it's so common in the software field that it shouldn't pose a problem.

Many clients will wish to place a cap on your total compensation because they're afraid you'll work slowly and pad your bill. If this is the case, include the optional sentence providing for a cap, but make sure it's large enough to allow you to complete the job.

3. Clause 3. Invoices

Consultant shall submit invoices for all services rendered. Client shall pay the amounts due within _____ days of the date of each invoice.

You should always submit an invoice to be paid. Fill in how much time the client will have to pay you after you send your invoice. Thirty days is common, but you can shorten this period if you wish.

4. Clause 4. Late Fees (Optional)

See Chapter 5, Section B4, for a discussion of this clause.

5. Clause 5. Expenses

(Choose Alternative A or B and optional provision, if desired.)

☐ ALTERNATIVE A

Consultant shall be responsible for all expenses incurred while performing services under this Agreement.

(Optional: Check if applicable.)

☐ However, Client shall reimburse Consultant for all reasonable travel and living expenses necessarily incurred by Consultant while away from Consultant's regular place of business to perform services under this Agreement. Consultant shall submit an itemized statement of such expenses. Client shall pay Consultant within 30 days from the date of each statement.

☐ ALTERNATIVE B

Client shall reimburse Consultant for the following expenses that are attributable directly to work performed under this Agreement:

- travel expenses other than normal commuting, including airfares, rental vehicles, and highway mileage in company or personal vehicles at _____ cents per mile

- telephone, fax, online, and telegraph charges

- postage and courier services

- printing and reproduction

- computer services, and

- other expenses resulting from the work performed under this Agreement.

Consultant shall submit an itemized statement of Consultant's expenses. Client shall pay Consultant within 30 days from the date of each statement.

Expenses are the costs you incur that are directly attributable to your work for a client. It would include, for example, the cost of phone calls or traveling done on the client's behalf. Expenses do not include your normal fixed overhead costs such as your office rent or the cost of commuting to and from your office. They also do not include materials the client provides you to do your work. Two options are presented regarding expenses: you can pay them all yourself or have the client reimburse you for them.

a. You pay expenses

Government agencies may consider payment of a worker's business or traveling expenses to be an indicator of an employment relationship. For this reason, usually it is best that you not be separately reimbursed for expenses. Instead, set your fees high enough to cover your expenses.

Setting your compensation at a level that covers your expenses has another advantage as well: It frees you from having to keep records of your expenses. Keeping track of every phone call or photocopy you make for a client can be a real chore and may be more trouble than it's worth. Choose Alternative A to pay your own expenses.

b. Expenses reimbursed

Some projects may involve extensive traveling, long-distance phone calls, and other expenses. You may wish to bill the client separately for these costs. In this event, use Alternative B, which requires the client to reimburse you for these expenses. You need to provide how many cents per mile you'll charge for travel time.

6. Clause 6. Materials

See Chapter 5, Section B6, for a discussion of this clause.

7. Clause 7. Term of Agreement

See Chapter 5, Section B7, for a discussion of this clause.

8. Clause 8. Terminating the Agreement

See Chapter 5, Section B8, for a discussion of this clause.

9. Clause 9. Independent Contractor Status

See Chapter 5, Section B9, for a discussion of this clause.

10. Clause 10. Intellectual Property Ownership

(Choose Alternative A or B and optional provision, if desired.)

□ ALTERNATIVE A

Consultant assigns to Client its entire right, title, and interest in anything created or developed by Consultant for Client under this Agreement ("Work Product") including all patents, copyrights, trade secrets, and other proprietary rights. This assignment is conditioned upon full payment of the compensation due Consultant under this Agreement.

Consultant shall, at no charge to Client, execute and aid in the preparation of any papers that Client may consider necessary or helpful to obtain or maintain—at Client's expense—any patents, copyrights, trademarks, or other proprietary rights. Client shall reimburse Consultant for reasonable out-of-pocket expenses incurred under this provision.

(Optional: Check if applicable, then check applicable provision.)

☐ Client grants to Consultant a nonexclusive

    ☐ irrevocable license to use the Work Product
    OR:

    ☐ license for the term of _____ years to use the
    Work Product.

☐ ALTERNATIVE B

Consultant shall retain all copyright, patent, trade secret, and other intellectual property rights Consultant may have in anything created or developed by Consultant for Client under this Agreement ("Work Product"). Consultant grants Client a nonexclusive worldwide license to use and sublicense the use of the Work Product for the purpose of developing and marketing its products, but not for the purpose of marketing Work Product separate from its products. The license shall have a perpetual term and may not be transferred by Client. This license is conditioned upon full payment of the compensation due Consultant under this Agreement.

Usually, the client will want to obtain sole ownership of what you create, whether it be a computer program, documentation, or other material. However, you can make other arrangements, if you'd like. Two different ownership alternatives are presented in the contract. Choose only one alternative and delete the other.

a. Client owns consultant's work product

Choose Alternative A if you agree that the client will own your work product. This assigns your rights to the client. But the assignment is expressly conditioned upon receipt of all compensation from the client. In other words, no money, no assignment.

Optional sentence. An assignment of all rights means that you may not use the work you performed for the client without the client's permission—for example, you may not include your own work in a program that you later write for someone else unless you get permission from the original client. If the client agrees, this permission can be given, in advance, in the independent contractor agreement. The optional sentence in this clause grants you a nonexclusive license to use the work product you've created for the client.

The license can be limited in any way. For example, the client could prevent you from using your work product when you perform services for the client's competitors.

There's one more wrinkle to the nonexclusive license option. You have to decide whether it's going to be permanent. In legal jargon, you'll have to decide whether it will be irrevocable—that is, last forever—or be limited to a specified time period—for example, one or two years. Choose the appropriate language in the optional clause.

b. Consultant owns work product

You don't have to sign over the rights to the work product you create for the client. You and the client may agree that you will retain your ownership rights in the work product (perhaps in return for receiving less money for the job). In this event, instead of assigning your rights to the client, you grant the client a nonexclusive license to use the work product.

Alternative B grants the client a nonexclusive license to use your work product in any of its products. Importantly, it prevents the client from selling your work to others. You can make this license more restrictive—for example, you could limit the client's use to a particular product. If you want to do this, modify the clause to indicate the restrictions.

11.   Clause 11. Consultant's Reusable Materials

Consultant owns or holds a license to use and sublicense various materials in existence before the start date of this Agreement ("Consultant's Materials"). Consultant may, at its option, include Consultant's Materials in the work performed under this Agreement. (Choose Alternative A or B.)

☐ ALTERNATIVE A

Consultant retains all right, title, and interest, including all copyright, patent rights, and trade secret rights, in Consultant's Materials. Subject to full payment of the consulting fees due under this Agreement, Consultant grants Client a nonexclusive worldwide license to use and sublicense the use of Consultant's Materials for the purpose of developing and marketing its products, but not for the purpose of marketing Consultant's Materials separate from its products. The license shall have a perpetual term and may not be transferred by Client. Client shall make no other commercial use of Consultant's Materials without Consultant's written consent.

(Optional: Check and complete if applicable.)

☐ This license is granted subject to the following terms: _____

☐ Consultant's Materials include, but are not limited to, those items identified in Exhibit ____ , attached to and made part of this Agreement.

☐ ALTERNATIVE B

Consultant retains all right, title, and interest, including all copyright, patent rights, and trade secret rights, in Consultant's Materials. Subject to full payment of the consulting fees due under this Agreement, Consultant grants Client a nonexclusive worldwide license to use Consultant's Materials in the following product(s): _____ .

The license shall have a perpetual term and may not be transferred by Client. Client shall make no other commercial use of Consultant's Materials without Consultant's written consent.

(Optional: Check and complete if applicable.)

☐ Consultant's Materials include, but are not limited to, those items identified in Exhibit ____ , attached to and made part of this Agreement.

A software consultant often has various development tools, routines, subroutines, and other programs, data, and materials that the consultant brings to the job and that might end up in the final product. These items are often known as "consultant's materials."

Unless you want to transfer ownership of these materials to the hiring firm, you should make sure that the independent contractor agreement provides that you retain all your ownership rights in this material. However, the agreement must also give the hiring firm a nonexclusive license to use the background technology that you include in your work product.

There are two clauses to choose from. The first clause permits the client to use your materials in any of its products. The second clause limits the client's use of your materials to the product or products specified in the clause.

a.   Client's license extends to all products

The client would probably prefer to have the right to use your technology in any of its products. Alternative A accomplishes this. In this event, you may want to insist on some sort of payment or royalty provision. If so, include the optional sentence and describe the compensation.

b.   Client's license limited to specific products

Alternative B permits the hiring firm to use the background technology only in a particular product or products. Make sure you describe any such products in detail.

c.   Optional clause: Identifying materials

If you know what your consultant materials will be in advance, it's a good idea to list them in an exhibit attached to the agreement. Include this optional clause if you do this.

If possible, identify your background technology in the source code copies of the programs you deliver to the client and in any printouts of code delivered to the client. You might include a notice like the following where such material appears: "[Your Company Name] CONFIDENTIAL AND PROPRIETARY."

12. Clause 12. Confidentiality

During the term of this Agreement and for _____ ☐ months ☐ years afterward, Consultant will use reasonable care to prevent the unauthorized use or dissemination of Client's confidential information. Reasonable care means at least the same degree of care Consultant uses to protect its own confidential information from unauthorized disclosure.

Confidential information is limited to information clearly marked as confidential, or disclosed orally and summarized and identified as confidential in a writing delivered to Consultant within 15 days of disclosure.

Confidential information does not include information that:

- the Consultant knew before Client disclosed it
- is or becomes public knowledge through no fault of Consultant
- Consultant obtains from sources other than Client who owe no duty of confidentiality to Client, or
- Consultant independently develops.

Since software consultants often have access to their clients' valuable confidential information, hiring firms often seek to impose confidentiality restrictions on them. It's not unreasonable for a client to want you to keep its secrets away from the eyes and ears of competitors. Unfortunately, however, many of these provisions are worded so broadly that they can make it difficult for you to work for other clients without fear of violating your duty of confidentiality.

If, like most software consultants, you make your living by performing similar services for many firms within the computer industry, insist on a confidentiality provision that is reasonable in scope and defines precisely what information you must keep confidential. Such a provision should last for only a limited time—five years at the most.

The optional confidentiality provision in this agreement prevents you from disclosing:

- any written material of the client marked "confidential," or
- information disclosed orally that the client later writes down, marks as "confidential," and delivers to you within 15 days.

These protocols enable you to know for sure what material is, and is not, confidential.

13. Clause 13. Warranties

(Choose Alternative A or B.)

☐ ALTERNATIVE A

THE GOODS OR SERVICES FURNISHED UNDER THIS AGREEMENT ARE PROVIDED AS IS WITHOUT ANY EXPRESS OR IMPLIED WARRANTIES OR REPRESENTATIONS; INCLUDING, WITHOUT LIMITATION, ANY IMPLIED WARRANTIES OF MERCHANTABILITY OR FITNESS FOR A PARTICULAR PURPOSE.

☐ ALTERNATIVE B

Consultant warrants that all services performed under this Agreement shall be performed consistent with generally prevailing professional or industry standards. Client must report any deficiencies in Consultant's services to Consultant in writing within _____ days of performance to receive warranty remedies.

Client's exclusive remedy for any breach of the above warranty shall be the reperformance of Consultant's services. If Consultant is unable to reperform the services, Client shall be entitled to recover the fees paid to Consultant for the deficient services.

THIS WARRANTY IS EXCLUSIVE AND IN LIEU OF ALL OTHER WARRANTIES, WHETHER EXPRESS OR IMPLIED, INCLUDING ANY IMPLIED WARRANTIES OF MERCHANTABILITY OR FITNESS FOR A PARTICULAR PURPOSE AND ANY ORAL OR WRITTEN REPRESENTATIONS, PROPOSALS, OR STATEMENTS MADE PRIOR TO THIS AGREEMENT.

A warranty is a promise or statement regarding the quality, quantity, performance, or legal title of goods or services you provide. There are two main types of warranties: express and implied. An express warranty comes into existence when a software consultant makes a specific promise or representation about a software product—for example, "the software will perform 10 trillion calculations per minute." Express warranties can be made orally or in writing.

Certain warranties are also implied—that is, they automatically exist by operation of law even if no explicit promise was made by the seller or person providing services. These warranties exist even if your written agreement doesn't mention them. Two main implied warranties are involved in software transactions:

- **Implied warranty of merchantability.** This warranty basically means that the consultant promises that the software is fit for its commonly intended use—in other words, it is of at least average quality. This implied warranty applies only to sales of new software or other goods.
- **Implied warranty of fitness for a specific purpose.** If a client is relying on the consultant's expertise to select suitable software, and the consultant is aware that the customer intends to use the software for a particular purpose, the software becomes impliedly guaranteed for that purpose. For instance, suppose a company hires a computer consultant to choose and install a new computer system. The client tells the consultant that it needs the system to process new accounts within 12 hours. If the system fails to perform as required by the client, the client can sue the consultant for breaching the implied warranty of fitness for a specific purpose.

If the software product or service you provide fails to live up to an express or implied warranty, the client can sue you in court for breach of warranty and obtain damages. The client doesn't have to prove that you were negligent—that is, failed to do your work properly. All it has to show is that your goods or services didn't perform the way you said they would either expressly or by implication. This makes it much easier for the client to obtain damages.

Clients often expect software consultants to make some type of express warranty regarding their services. This is an area where your interests and those of the client are diametrically opposed. Naturally, you would prefer to make no warranties at all. But savvy clients may refuse to hire you unless you'll vouch for your work.

Two options are provided below. Subsection a provides that you'll make no warranties at all. If the client won't go for this, you can use the limited warranty in Subsection b.

a. Providing software as is

Many software consultants don't want to provide any warranties at all. To do this, choose Alternative A stating that your software goods or services are provided as is. This means that if your goods or services are unsatisfactory, the client has no basis for a complaint.

The clause also disclaims—or disavows—the implied warranties discussed above.

However, this clause won't excuse you from failure to live up to the project specifications included in your written agreement. Nor will the clause protect you from charges of outright fraud if you lied to the client, nor will it shield you from liability if your software is so defective it injures someone. To be effective, the as-is clause should be printed in capitals so that the client won't overlook it.

b. Limited warranty

Some clients may balk at having no warranty protection at all. In this event, you can use Alternative B to give the client a limited warranty. It requires the client to report in writing any deficiencies in your work within a specified time period (30 to 90 days) You are then required to redo your services or, if you can't do this, pay back the client what it paid you for your faulty work. The concluding paragraph states that this clause is the only warranty you're providing the client and disclaims the implied warranties discussed above. It's a good idea to print such a statement in capitals so that the client won't overlook it.

14.   Clause 14. Limited Liability

- In no event shall Consultant be liable to Client for lost profits of Client or special, incidental, or consequential damages (even if Consultant has been advised of the possibility of such damages).

- Consultant's total liability under this Agreement for damages, costs, and expenses, regardless of cause, shall not exceed the total amount of fees paid to Consultant by Client under this Agreement.

- Client shall indemnify Consultant against all claims, liabilities, and costs, including reasonable attorney fees, of defending any third-party claim or suit, other than for infringement of intellectual property rights, arising out of or in connection with Client's performance under this Agreement. Consultant shall promptly notify Client in writing of such claim or suit, and Client shall have the right to fully control the defense and any settlement of the claim or suit.

Many consultants seek to include a provision in their agreements limiting their total liability to the client. Understandably, clients are often reluctant to limit your liability to them. The best way to get a client to accept it is to make it clear that you will have to increase the fee for your services if you're required to assume unlimited liability. You may even provide the client with two different fee schedules: one with and one without limited liability. Some software consultants increase their fees by a factor of 10, 100, or even 1000 if they are required to assume unlimited liability.

This clause contains three separate liability limiting provisions, all or any of which may be in-

cluded or excluded from the agreement; this is a matter for negotiation between you and the client.

- Under the first paragraph, you are relieved from liability for the client's lost profits or special, incidental, or consequential damages caused by defects in your work. Such damages might include damages for interruption of the client's business or loss of production due to malfunctioning software. These types of damages—lost profits in particular—can far exceed your total compensation and could even send you into bankruptcy.

- The second paragraph, which is perhaps the most important, limits your total liability to the client to the amount of money actually received from the client. Instead, you can set a specific dollar amount.

- The third paragraph requires the client to indemnify you against third-party claims. These are claims for money brought against you by people or entities other than the client. Indemnification is a fancy legal word that means that the client must pay attorney fees and other costs of defending you against such claims. If the defense is unsuccessful or there's a settlement, the clause means the client will pay any monetary damages, too.

Example: Freelance programmer Joe Jones writes a custom software program designed to operate a chemical factory. The software crashes and so does the factory, resulting in a chemical spill costing hundreds of thousands of dollars to clean up. Numerous lawsuits are brought against both the chemical factory and Jones by property owners affected by the spill. Jones has to hire a lawyer to defend him, and the claims are ultimately settled for a six-figure amount. Because Jones's contract with the chemical factory had an indemnfication clause, the factory is required to reimburse Jones for his legal expenses and the costs of the settlement.

15. Clause 15. Taxes

> The charges included here do not include taxes. If Consultant is required to pay any federal, state, or local sales, use, property, or value added taxes based on the services provided under this Agreement, the taxes shall be billed separately to Client. Client shall be responsible for paying any interest or penalties incurred due to late payment or nonpayment of such taxes by Client.

Hawaii, New Mexico, and South Dakota require all independent contractors to pay sales taxes, even if they provide their clients with services only. Many other states require sales taxes to be collected and paid for certain specified services. Contact your state sales tax department for information on your state's requirements.

Even if your state doesn't presently require you to collect sales tax from the client, include this clause in your agreement, which makes it clear that the client will have to pay these and similar taxes. States change their sales tax laws constantly, and more and more are looking at services as a good source of sales tax revenue. This provision could come in handy in the future, even if you don't really need it now.

16. Clause 16. Contract Changes

> Client and Consultant recognize that:
> - Consultant's original cost and time estimates may be too low due to unforeseen events, or to factors unknown to Consultant when this Agreement was made
> - Client may desire a mid-project change in Consultant's services that would add time and cost to the project and possibly inconvenience Consultant, or
> - other provisions of this Agreement may be difficult to carry out due to unforeseen circumstances.
>
> If any intended changes or any other events beyond the parties' control require adjustments to this Agreement, the parties shall make a good faith effort to agree on all necessary particulars. Such agreements shall be put in writing, signed by the parties, and added to this Agreement.

It's very common for clients and software consultants to want to change the terms of an agreement after work has begun. For example, the client might want to make a change in the contract specifications that could require you to do more work for which you should be compensated. Or you might discover that you underestimated how much time the project will take and need to be paid more to complete it and avoid losing money.

This provision recognizes that the agreement may have to be changed. It states that you and the client must write down and sign your changes. Such a contract provision requiring modifications to be in writing is probably not legally necessary—that is, both you and the client can still make changes without writing them down, and a court could rule that the oral changes must be honored. At the very least, it reminds both of you that it's always a good idea to document changes in writing.

Keep in mind that neither you nor the client is ever required to accept a proposed contract change. But because you are obligated to deal with each other fairly and in good faith, you can't simply refuse without attempting to reach a resolution. If you and the client can't agree on the changes, you're required to submit your dispute to mediation. If that doesn't work, you agree to use binding arbitration. This avoids expensive court litigation.

17. Clause 17. Resolving Disputes

See Chapter 5, Section B13, for a discussion of this clause.

18. Clause 18. Exclusive Agreement

See Chapter 5, Section B11, for a discussion of this clause.

19. Clause 19. Applicable Law

See Chapter 5, Section B17, for a discussion of this clause.

20. Clause 20. Notices

See Chapter 5, Section B15, for a discussion of this clause.

21. Clause 21. No Partnership

See Chapter 5, Section B16, for a discussion of this clause.

22. Signatures

See Chapter 5, Section B19, for a discussion of this clause. ■

11

Agreements for Creative Contractors

This chapter contains independent contractor agreements for ICs who create or contribute to the creation of copyrightable works of authorship. This includes:

- written works of all types
- photography
- graphics, illustrations, and artwork, and
- music.

A. Agreement for Use by Hiring Firm

Use this agreement if you're hiring an independent contractor to work for you. The agreement is designed to protect your interests as much as possible without being unduly one-sided. It will help establish that the worker is an IC, not your employee.

An independent contractor agreement for creative ICs for use by the hiring firm is on the CD-ROM. You can modify individual clauses to tailor the text to your specific needs.

A tear-out form for an agreement for creative ICs for use by the hiring firm is in Appendix II. Use this form if you don't have access to a computer. It contains blank spaces for you to fill in by hand or typewriter.

Although we provide a tear-out form for those of you who either don't have computers or who would prefer not to use the CD-ROM, don't just tear out the form and use it as your IC agreement if you can avoid doing so. A fill-in-the-blank IC agreement is the least persuasive type of contract when you are trying to convince a government agency that the worker is an IC and not an employee. The best way to create your agreement is to use the CD-ROM version and tailor it to your needs. If you can't use the CD-ROM version, then retype the version at the back of the book and tailor it to your needs. Use the fill-in-the-blank form only as a last resort.

Call up the agreement on your computer or tear out the form agreement in Appendix II, and read it along with the following instructions and explanations.

Title of agreement. You don't have to have a title for an IC agreement, but if you want one you should call it "Independent Contractor Agreement." Do not use "Employment Agreement" as a title.

Names of IC and hiring firm. Never refer to an IC as an employee or to yourself as an employer. See Chapter 5, Section A, for more about which names to use.

1. Clause 1. Services to Be Performed

(Choose Alternative A or B.)

☐ ALTERNATIVE A

Contractor agrees to perform the following services:

_____

_____

_____

☐ ALTERNATIVE B

Contractor agrees to perform the services described in Exhibit A attached to and made part of this Agreement.

The agreement should describe in as much detail as possible what you expect the contractor to do. You must limit the description to the results the IC is expected to achieve. Don't tell the IC how to achieve those results—this indicates that you have the right to control how the IC performs the agreed-upon services. Such a right of control is the hallmark of an employment relationship. (See Chapter 2 for more about this issue.)

It's perfectly all right for you to establish detailed specifications for the IC's finished work product. And if there is a choice as to how the final product may be presented to you—on hard copy, computer disk, film, and so on—be sure to specify which form will be acceptable. But the specs and end-product instructions should describe only the results the IC must achieve, not how the IC has to do the work.

For example, if you're hiring an IC to write an index, you should describe what the finished index should look like, but you shouldn't tell the IC how to arrive at the end product. A good work description should read like the one shown below. Note how the client has specified the form of the result (a computer disk written in WordPerfect and at least 100 pages long) but not the process to be used in achieving it.

Contractor agrees to prepare an index of Client's book "History of Sparta" of at least 100 single-spaced 8.5" x 11" pages. Contractor will provide Client with a printout of the finished index and a 3.5-inch computer disk version in WordPerfect for Windows format.

Here's an example of a work description that is inappropriate for an IC—it tells the IC how to perform the service, which is a red flag for government auditors.

Contractor will prepare an alphabetical three-level index of Client's book "History of Sparta." Contractor will first prepare 3" x 5" index cards listing every index entry beginning with Chapter One. After each chapter is completed, Contractor will deliver the index cards to Client for Client's approval. When index cards have been created for all 50 chapters, Contractor will create a computer version of the index using Complex Software Version 7.6. Contractor will then print out and edit the index and deliver it to Client for approval.

Describe the form of the end product in terms of your IC's medium. For example, if you're hiring a photographer, you need to tell the IC what type of image to deliver. Will the photographs will be in color or black and white? How many photos will be provided? Will the photographer deliver negatives, prints, or transparencies, or will the photographs be delivered in digital form?

If you're hiring an artist or sculptor, define clearly how large the work will be, what materials will be used, and where the finished product must be delivered.

You can include the description in the main body of the agreement or, if it's lengthy, on a separate attachment. Use Alternative A if you explain the services in the contract. Choose Alternative B if the description is attached to the main contract.

2. Clause 2. Payment

(Choose Alternative A or B and optional clause, if applicable.)

☐ ALTERNATIVE A

In consideration for the services to be performed by Contractor, Client agrees to pay Contractor $_____ according to the terms of payment set forth below.

☐ ALTERNATIVE B

In consideration for the services to be performed by Contractor, Client agrees to pay Contractor at the rate of $_____ per ☐ hour; ☐ day; ☐ week; ☐ month according to the terms of payment set forth below.

(Optional: Check and complete if applicable.)

☐ Unless otherwise agreed upon in writing by Client, Client's maximum liability for all services performed during the term of this Agreement shall not exceed $_____.

This clause contemplates that the IC will be paid either a fixed fee for the work or paid by the hour or other unit of time. See Chapter 5, Section A2, for a detailed discussion of payment methods.

However, there are other ways to pay for creative work. For example, you could pay a creative IC a royalty—that is, a percentage of what you earn by selling copies of the IC's work. For more information on royalty agreements, see *Licensing Art & Design: A Professional's Guide to Licensing and Royalty Agreements,* by Caryn R. Leland (Allworth Press).

3. Clause 3. Terms of Payment

This clause refers to how and when you'll pay the IC. See Chapter 5, Section A3, for a detailed discussion.

4. Clause 4. Expenses

(Choose Alternative A or B.)

☐ ALTERNATIVE A

Contractor shall be responsible for all expenses incurred while performing services under this Agreement.

☐ ALTERNATIVE B

Client shall reimburse Contractor for the following expenses that are directly attributable to work performed under this Agreement: _____

_____

Contractor shall submit an itemized statement of Contractor's expenses. Client shall pay Contractor within 30 days after receipt of each statement.

As a general rule, it's best not to reimburse an IC for expenses. Instead, compensate the IC well enough so that the expenses come out of the IC's own pocket. Choose Alternative A if the IC will pay all expenses.

However, it is customary for clients to reimburse many types of creative ICs for expenses—for example, travel costs, costs to purchase or rent equipment, materials and supplies, charges for models and props, printing and reproduction, and postage or shipping costs. If there is an otherwise clear IC relationship and reimbursement of expenses is common in the IC's field, you can get away with such reimbursement without jeopardizing the worker's status as an IC. Choose Alternative B if you'll reimburse the IC for expenses, then list the expenses you will reimburse.

5. Clause 5. Materials

Contractor will furnish all materials, tools, and equipment used to provide the services required by this Agreement.

You shouldn't provide an IC with materials, tools, or equipment to perform the services. For example, if you're hiring a photographer, don't provide expensive camera equipment. The IC should buy or lease the equipment elsewhere. Providing workers with tools and equipment is a strong indicator of employee status. This clause states that the IC will provide all tools, equipment, materials, and supplies needed to accomplish the project.

If you must break this rule and provide the IC with equipment, materials, or supplies, delete this clause. In this situation, it's best for the agreement not to even mention tools and equipment. You aren't required to highlight the fact that you are giving the worker materials in your agreement. Doing so will just make your life more difficult if you're audited by the IRS or other government agency. Simply delete the materials and equipment clause from the agreement. (Of course, if you're ever questioned about it, you'll have to tell the truth.)

However, there is a way to furnish items to the IC that will counterbalance the suggestion that the IC is really your employee. If the equipment, materials, or supplies that the IC needs have substantial value, either sell or lease the items to the IC. Since employers normally don't sell or lease equipment to their employees, this will support the worker's IC status.

Charge the IC the same amount you'd charge anyone else. If you lease equipment to the IC, enter into a written lease agreement. If you sell the IC equipment, use a written bill of sale. You can find forms for selling or leasing equipment in *Legal Forms for Starting & Running a Small Business,* by Fred Steingold (Nolo).

6. Clause 6. Intellectual Property Ownership

(Choose Alternative A, B, or C.)

☐ ALTERNATIVE A

Contractor assigns to Client Contractor's entire right, title, and interest in anything created or developed by Contractor for Client under this Agreement. Contractor shall help prepare any papers that Client considers necessary to secure any copyrights, trademarks, or other proprietary rights at no charge to Client. However, Client shall reimburse Contractor for reasonable out-of-pocket expenses incurred.

Contractor must obtain written assurances from Contractor's employees and contract personnel that they agree with this assignment.

Contractor agrees not to use any of the intellectual property mentioned above for the benefit of any other party without Client's prior written permission.

☐ ALTERNATIVE B

To the extent that the work performed by Contractor under this Agreement ("Contractor's Work") includes any work of Authorship entitled to protection under the copyright laws, the parties agree to the following provisions.

- Contractor's Work has been specially ordered and commissioned by Client as a contribution to a collective work, a supplementary work or other category of work eligible to be treated as a work made for hire under the United States Copyright Act.
- Contractor's Work shall be deemed a commissioned work and a work made for hire to the greatest extent permitted by law.
- Client shall be the sole author of Contractor's Work and any work embodying the Contractor's Work according to the United States Copyright Act.
- To the extent that Contractor's Work is not properly characterized as a work made for hire, Contractor grants to Client all right, title, and interest in Contractor's Work, including all copyright rights, in perpetuity and throughout the world.

- Contractor shall help prepare any papers Client considers necessary to secure any copyrights, patents, trademarks, or intellectual property rights at no charge to Client. However, Client shall reimburse Contractor for reasonable out-of-pocket expenses incurred.

Contractor agrees not to use any of the intellectual property mentioned above for the benefit of any other party without Client's prior written permission.

☐ ALTERNATIVE C

Contractor assigns to Client the following intellectual property rights in the work created or developed by Contractor under this Agreement: _____

_____

(Check applicable provision.)

☐ The rights granted above are exclusive to Client.
OR:

☐ The rights granted above are nonexclusive.

This clause determines who will own the work the IC creates. Contrary to what you might think, you don't automatically own a work of authorship an IC creates on your behalf, even though you pay the IC. This is because of the U.S. copyright law, which is the federal law governing ownership of works of authorship.

Because this rule of law is a bit surprising, a little copyright law groundwork might be in order. A work of authorship—such as any type of written work, artwork, graphics, photography, sculpture, video, music, and architectural blueprints and designs—is protected automatically by copyright the moment it's created. Unless the work qualifies as a work made for hire (see Section A6, below), the IC will initially own the copyright—you will have no ownership rights at all unless you obtain them from the IC.

Why is it important who owns the copyright? Because only the copyright owner has the exclusive right to copy, distribute, create new works from, perform, and display the work. Anyone who exercises any of these rights without the copyright owner's permission is guilty of copyright infringement and may be sued by the owner.

Unless you obtain all or some of the IC's copyright rights as provided below, you won't own the work the IC creates, even though you paid for it. You'll likely just have a nonexclusive right to use it. But the IC could use it too—for example, by reselling it to your competitors—without your permission.

> **Example:** Mark pays Sally, a freelance photographer, to take some photographs of toxic waste dumps to supplement his treatise on toxic waste management. Sally is an IC. Mark fails to obtain ownership of Sally's copyright rights in the photos. As a result, Sally owns the copyright in the photos. Since Mark paid Sally to take the photos, he has a nonexclusive right to use them in his treatise. However, he can't resell them to others or otherwise exercise any of Sally's copyright rights. Sally may re-sell the photos, reproduce them in a book she's publishing, or exercise any of her other rights without Mark's permission.

Now that you understand the importance of this issue, you'll be able to choose among three options for dealing with intellectual property ownership when you hire an IC. You can:

- obtain an assignment of all the IC's rights (see Subsection a, below)
- characterize the IC's work as a work made for hire, (see Subsection b, below) or
- obtain an assignment of some of the IC's rights (see Subsection c, below).

a. Obtaining an assignment of all the IC's rights

Ordinarily, you will want to obtain ownership of all the intellectual property rights in any work of authorship an IC creates on your behalf. This is very easy to do. You just have to include a clause in your agreement providing that the IC assigns (transfers) to you the IC's "entire right, title, and interest" in the work. By using this clause, you will acquire all the IC's rights in the work. You will have the exclusive right to copy, distribute, per-

form, display, and create new works from the IC's work. The IC will have no rights at all.

However, an assignment cannot last forever. Copyright law provides that a creator of a work of authorship, including an IC, can revoke an assignment 35 to 40 years after it's made and get all of the rights back.

> **Example:** Bill hires Penny, a freelance illustrator, to create illustrations for a children's book. His IC agreement provides that Penny assigns all her rights in the illustrations to Bill. Thirty-five years after the book is published, Penny or her heirs can get her rights back from Bill without paying him anything. Bill and his publisher can continue to use the illustrations in the book, but Penny will have the right to resell or otherwise use them.

In most cases, however, this termination right is superfluous. Few works of authorship have a useful economic life of more than 35 years. Unless Penny, in the example above, has the success of a Maurice Sendak, after 35 years it's likely no one will care if the termination right is exercised.

b. Characterizing the work as a work made for hire

There is another way to ensure that you will own the intellectual property rights in the product made by your IC. You can do this by characterizing the work as a work made for hire.

This awkward term has a rather simple meaning. A work made for hire never belongs to the IC, not even for a moment. Instead, as long as it is clear that you have paid the IC to produce this work for hire, you automatically own all the copyright rights. Indeed, you are considered to be the work's author for copyright purposes, even though you didn't create it. All the creator receives is compensation from you.

If you intend for the IC to create a work made for hire, it is absolutely critical that your agreement state this understanding. Without it, the product will belong to the IC, and at best you'll have only a nonexclusive right to use it.

The advantages of owning the work from the outset are many. As the author, you're entitled to register the work with the U.S. Copyright Office. You own all the exclusive rights that make up a copyright, such as the right to copy and distribute the work. You can exercise these rights yourself, sell or license them to others, or do whatever else you want with them. The person or people you paid to create the work have no say over what you do with the work.

How is this different from obtaining an assignment of all the IC's rights? As a practical matter, it isn't much different. Either way, you own all the IC's rights. The main advantage to using the work-made-for-hire approach is that, unlike the assignment route, it's permanent. The IC will not have the right to terminate the transfer after 35 years.

Before you happily choose the work-made-for-hire option, we must caution you that there are restrictions. Not all works of authorship created by IC may be works made for hire. A work created by an IC can be a work made for hire only if it falls within one of the following nine categories:

- a contribution to a collective work—for example, a work created by more than one author, such as a newspaper, magazine, anthology, or encyclopedia

- a part of an audiovisual work—for example, a motion picture screenplay

- a translation

- supplementary works—for example, forewords, afterwords, supplemental pictorial illustrations, maps, charts, editorial notes, bibliographies, appendixes, and indexes

- a compilation—for example, an electronic database

- an instructional text

- a test

- answer material for a test, and

- an atlas.

You and the IC must also sign an agreement stating that the work is made for hire. The written agreement is critical. Without it you can't have a work made for hire. The agreement should always be signed before the IC starts work. Alternative B fulfills this requirement. It also provides that if, for some reason, the work fails to qualify as a work made for hire, the IC assigns to you the copyright rights in the work. This means that you will still own all the copyright rights in the work, subject to the IC's termination rights down the road.

DON'T USE WORK FOR HIRE IN CALIFORNIA

California law provides that a person who commissions a work made for hire is considered to be the employer of the creator of the work for purposes of the workers' compensation, unemployment insurance, and unemployment disability insurance laws. (Cal. Labor Code § 3351.5(c); Cal. Unemployment Insurance Code § 621 and § 686.) California is the only state with such laws.

These laws are a good reason not to enter into work-made-for-hire agreements with creative ICs in California. Instead, have the IC assign the desired copyright rights to you in advance by using Alternative A.

c. Obtaining an assignment of some of the IC's rights

You don't have to obtain an assignment of all the IC's rights or characterize the work as a work made for hire. Instead, you can obtain only some of the rights and let the IC keep the rest.

To understand how this piecemeal assignment works, understand that a copyright is made up of a bundle of rights, including:

- reproduction rights—the right to make copies of a protected work

- distribution rights—the right to sell or otherwise distribute copies to the public

- rights to create adaptations (or derivative works)—the right to prepare new works based on the protected work, and

- performance and display rights—the right to perform or display in public a protected literary, musical, dramatic, choreographic, pantomime, motion picture, or audiovisual work, whether in person or by means of a device like a television.

Each of these rights can be subdivided and sold separately or together with others. A copyright is infinitely divisible—for example, any of an IC's rights can be divided and sold by:

- time period—a few weeks, months, years, or forever
- geographic area—for example, North America or worldwide
- category of use—for example, advertising or educational uses
- medium of use—for example, magazines, books, television, movies, public performances, and
- language—for example, English or other languages.

Finally, any of these rights can be transferred on an exclusive or nonexclusive basis—that is, you could be the only person entitled to exercise them (exclusive), or the IC may let others do so (nonexclusive).

Why would you want to take less than all the IC's rights? Because you may not need all of them, and the IC might charge you less if you don't take them all.

Use the clause in Alternative C to obtain only some of the IC's rights. You'll need to define clearly what rights you're obtaining.

Example 1: Schooldaze Publishing needs 25 spot illustrations for use in a high school textbook. It contracts with Sally, an IC illustrator, to create them. Schooldaze wants just the illustrations for its textbook; it doesn't care if Sally makes other uses of them. So Schooldaze uses Alternative C in Clause 6. It completes the clause so as to obtain the nonexclusive right to use Sally's illustrations in its textbook. The completed clause looks like this:

Contractor assigns to Client the following intellectual property rights in the work created or developed by Contractor under this Agreement: The right to use the 25 spot illustrations created by Contractor in the textbook entitled High School Civics; *however, it may be exploited in any language or medium now known or later invented, including, but not limited to, print, microfilm, and electronic media, and in translations in all languages.* The right is nonexclusive.

Example 2: Bill, a freelance writer, contracts with *Untrue Adventures Magazine* to write an article about the famed creature Bigfoot. The magazine doesn't need all of Bill's copyright rights in his story; it wants just the exclusive right to publish it for the first time in North America. In magazine industry parlance, such rights are called "first North American serial rights." Once the article has been initially published in *Untrue,* Bill can resell it or otherwise uses it. *Untrue*'s editor uses Alternative C in Clause 6. The completed clause looks like this: *Contractor assigns to Client the following intellectual property rights in the work created or developed by Contractor under this Agreement: One-time North American publication rights in the English language in Client's magazine called* Untrue Adventures Magazine.

Example 3: Phil, a wealthy philanthropist, hires Art, a famous artist, to paint a portrait of his beautiful mansion. Art wants to retain copyright ownership of his painting so he can reproduce it—for example, on postcards or calendars. Phil doesn't care if he owns all the copyright rights in Art's work, but he wants to be the only person who has the legal right to display the original painting in public—for example, in museums or art galleries. Art agrees, but he wants to limit Phil's exclusive right to display his painting to a term of 20 years. Phil uses Alternative C in Clause 6. He completes the clause so as to obtain the exclusive right to display the painting in public. The completed clause looks like this: *Contractor assigns to Client the following intellectual property rights in the work created or developed by Contractor under this Agreement: The right to display the original copy of Contractor's painting in public. This right shall last for 20 years from the date of this Agreement and shall extend worldwide.* The rights granted in this clause are exclusive.

7. Clause 7. Releases (Optional)

(Check if applicable.)

☐ Contractor shall obtain all necessary copyright permissions and privacy releases for materials included in the Images at Client's request. Contractor shall indemnify Client against all claims and expenses, including reasonable attorney fees, due to Contractor's failure to obtain such permissions or releases.

In some cases, it may be necessary for either the IC or you to obtain a publicity/privacy release from a third party—typically, the subject of the IC's creative work. A release is simply an agreement in which a person promises not to sue the person obtaining the release if the person performs an act specified in the release.

A publicity/privacy release is ordinarily necessary when you use a person's name, likeness, or voice for commercial purposes—for example, in a TV commercial or magazine ad. By signing the release, people agree to let you use these elements of their persona. If you don't obtain a release, people might be able to sue you for violating their right to privacy or publicity. For example, if you hire a photographer to create a glossy ad for your apartment complex, you'll want the photographer to obtain releases from the happy tenants who are featured in her pictures.

Usually, the IC will be in the best position to obtain such releases. Indeed, obtaining publicity/privacy releases is standard practice for many creative ICs—for example, commercial photographers customarily obtain releases from their models. Generally, obtaining a release is not difficult—the subject just has to sign a simple form.

 The CD-ROM contains a release.

 You'll find a tear-out form for a release in Appendix II.

This clause requires the IC to obtain any needed releases and to indemnify—repay—you if you're sued because the IC failed to obtain a release.

8. Clause 8. Moral Rights Waiver for Works of Fine Art (Optional)

(Check if applicable.)

☐ Contractor waives any and all moral rights or any similar rights in the work created or developed by Contractor under this Agreement ("Work Product") and agrees not to institute, support, maintain, or permit any action or lawsuit on the grounds that Client's use of the Work Product:

• constitutes an infringement of any moral right or any similar right

• is in any way a defamation or mutilation of the Work Product

• damages Contractor's reputation, or

• contains unauthorized variations, alterations, changes, or translations of the Work Product.

In 1991, Congress enacted the Visual Artists Rights Act, granting artists certain types of moral rights ("VARA"; 17 U.S.C. 106A). Moral rights consist of:

• the right of attribution (the right to be named as author of the work), and

• the right of integrity (the right to prevent the intentional destruction or modification of the work in a way that harms the artist's reputation).

These rights give IC artists the ability to object if they believe someone has distorted or mutilated their work.

The VARA covers only works of fine art: paintings, drawings, sculptures, and photographs produced for exhibition in a limited signed edition of no more than 200 copies. The VARA does not apply to motion pictures or other audiovisual works, electronic publications, or applied artworks such as advertisements or technical drawings. Nor does the VARA apply to works made for hire. So you don't need to include this clause if you've characterized the IC's work as a work made for hire.

An artist's rights under the VARA can be waived by contract. If the VARA might apply to a work you've hired an IC to create, it is a good idea to obtain such a waiver. This way, the contractor will not have the legal right to object under the VARA if you later decide that you need to modify or even destroy the contractor's work. In the absence of a waiver, a contractor-artist could sue you if you altered, mutilated, or destroyed the work without permission—or even if you simply displayed it without proper attribution.

9. Clause 9. Independent Contractor Status

Contractor is an independent contractor, and neither Contractor nor Contractor's employees or contract personnel are, or shall be deemed, Client's employees. In its capacity as an independent contractor, Contractor agrees and represents, and Client agrees, as follows. (check all that apply):

☐ Contractor has the right to perform services for others during the term of this Agreement.

☐ Contractor has the sole right to control and direct the means, manner, and method by which the services required by this Agreement will be performed.

☐ Contractor has the right to perform the services required by this Agreement at any place or location and at such times as Contractor may determine.

☐ Contractor has the right to hire assistants as subcontractors, or to use employees to provide the services required by this Agreement.

☐ The services required by this Agreement shall be performed by Contractor, or Contractor's employees or contract personnel, and Client shall not hire, supervise, or pay any assistants to help Contractor.

☐ Neither Contractor nor Contractor's employees or contract personnel shall receive any training from Client in the professional skills necessary to perform the services required by this Agreement.

☐ Neither Contractor nor Contractor's employees or contract personnel shall be required by Client to devote full time to the performance of the services required by this Agreement.

One of the most important functions of an independent contractor agreement is to help establish that the worker is an IC, not your employee. The key to doing this is to make clear that the IC, not the hiring firm, has the right to control how the work will be performed.

The language in this clause addresses most of the factors the IRS and other agencies consider in determining whether an IC controls how the work is done. (See Chapter 2 for more about this issue.)

When you draft your own agreement, include only those provisions that apply to your situation. The more that apply, the more likely the worker will be viewed as an IC.

10. Clause 10. Business Permits, Certificates, and Licenses

Contractor has complied with all federal, state, and local laws requiring business permits, certificates, and licenses required to carry out the services to be performed under this Agreement.

The IC should have all necessary business permits, certificates, and licenses. A creative IC would likely just need a local business license. A worker who lacks such a license and permits looks like an employee, not an IC running an independent business. This provision requires the IC to have all necessary licenses. You need add no information here.

11. Clause 11. State and Federal Taxes

See Chapter 5, Section A8, for a discussion of this clause.

12. Clause 12. Fringe Benefits

See Chapter 5, Section A9, for a discussion of this clause.

13. Clause 13. Workers' Compensation

See Chapter 5, Section A10, for a discussion of this clause.

14. Clause 14. Unemployment Compensation

See Chapter 5, Section A11, for a discussion of this clause.

15. Clause 15. Contractor's Insurance

See Chapter 5, Section A12, for a discussion of this clause.

16. Clause 16. Client's Liability

> Client shall exercise the same care with regard to any materials or work product belonging to Contractor that are in Client's possession as it would for its own property. However, if any property received from Contractor is lost or damaged, Client's liability shall be limited to the coverage available under its business liability policy in force at the time this Agreement is signed.

If the IC performs all or part of the services on your premises, expensive equipment the IC owns may end up in your possession for a period of time—for example, camera or video equipment, props, sets, or costumes. This clause is designed to limit your liability if such property is damaged while in your possession. Your liability is limited

to the amount of your insurance coverage. This way, the only out-of-pocket cost you'll have is the amount of your deductible.

17. Clause 17. Warranties and Representations

> Contractor warrants and represents that:
> - Contractor is free to enter into this Agreement
> - the work created or developed by Contractor under this Agreement ("Work Product") shall be original or all necessary permissions and releases obtained and paid for, and
> - Contractor's Work Product shall not infringe upon any copyright or other proprietary right of any other person or entity.
> Contractor agrees to indemnify Client for loss, liability, or expense resulting from actual breach of these Warranties.

You want to know that the IC's work will not infringe on anybody else's copyright or other intellectual property rights. In legal jargon, you'll want a "representation and warranty"—which is simply a legally binding promise that the IC is not stepping on anybody's copyright toes. If the promise turns out not to be true, you can sue the IC for breach of contract.

Such a provision makes sense because the IC—the one creating the work—is in the best position to avoid committing copyright infringement. Remember that because the person is an IC, you have no control over how the work is done—for example, you can't prevent the IC from copying from someone else.

If the IC's work does infringe, it's highly likely that you will be sued for infringement along with the IC. This clause is intended to limit your financial exposure to such lawsuits by requiring the IC to indemnify—reimburse—you for the costs involved in defending them.

18.   Clause 18. Term of Agreement

See Chapter 5, Section A13, for a discussion of this clause.

19.   Clause 19. Terminating the Agreement

See Chapter 5, Section A14, for a discussion of this clause.

20.   Clause 20. Exclusive Agreement

See Chapter 5, Section A15, for a discussion of this clause.

21.   Clause 21. Modifying the Agreement (Optional)

See Chapter 5, Section A16, for a discussion of this clause.

22.   Clause 22. Confidentiality (Optional)

See Chapter 5, Section A17, for a discussion of this clause.

23.   Clause 23. Resolving Disputes

See Chapter 5, Section A18, for a discussion of this clause.

24.   Clause 24. Applicable Law

See Chapter 5, Section A19, for a discussion of this clause.

25.   Clause 25. Notices

See Chapter 5, Section A20, for a discussion of this clause.

26.   Clause 26. No Partnership

See Chapter 5, Section A21, for a discussion of this clause.

27.   Clause 27. Assignment and Delegation (Optional)

> **(Check if applicable.)**
>
> ☐ Contractor may not assign or subcontract any rights or obligations under this Agreement without Client's prior written approval.

The general rules about assigning rights and delegating duties under a contract are discussed in Chapter 5, Section A22. Generally, if an IC agrees to perform personal services, the IC cannot delegate that work to someone else without your consent. In other words, the IC must perform the services personally or have them performed by employees under the IC's supervision.

A creative IC's services would definitely be considered personal services. You're hiring the IC for a particular creative ability. This means a creative IC ordinarily cannot delegate duties without your consent. This is true even if your agreement does not expressly bar delegation.

> **Example:** The Ediface Construction Co. enters into an IC agreement with Mario, an IC artist, to create a sculpture for a new skyscraper under development. Mario cannot pass the work to another artist without Ediface's consent. However, Mario is entitled to have his own employees assist him.

Given these legal restrictions already in place, you need not say anything about restricting assignment and delegation by the IC in your agreement. You could simply delete the optional assignment clause and rely on these legal rules. This is the best option in most cases.

However, there may be situations in which you fear the IC may not fully understand the restrictions discussed above. To make it clear to the IC that delegation of duties is impermissible without your consent, you may wish to include this optional clause. Unfortunately, if your agreement is subsequently reviewed by a government auditor, he or she may view this clause as an indicator of an employment relationship. It highlights the fact that you want the consultant to perform the services personally, which tends to show you want to control who does the work. Remember that when government auditors review your agreement, they are looking for any excuse they can find to claim that the consultant is an employee instead of an IC.

For this reason, you should include this clause only if you really think it's necessary.

28. Signatures

See Chapter 5, Section A23, for an explanation of how to sign your agreement.

B. Agreement for Use by Contractor

Use this agreement if you're working as an IC. It contains a number of provisions favorable to you that a hiring firm wouldn't ordinarily put in an agreement it prepared—for example, requiring the hiring firm to pay a late fee if it doesn't pay you on time.

 An agreement for use by creative ICs is on the CD-ROM. You can modify individual clauses to tailor the text to your specific needs.

A tear-out agreement for use by creative ICs is in Appendix III. Use this form if you don't have access to a computer. It contains blank spaces for you to fill in by hand or typewriter.

Although we provide a tear-out form for those of you who either don't have computers or who would prefer not to use the CD-ROM, don't just tear out the form and use it as your IC agreement if you can avoid doing so. A fill-in-the-blank IC agreement is the least persuasive type of contract when you are trying to convince a government agency that you are an IC and not an employee of the hiring firm. The best way to create your agreement is to use the CD-ROM version and tailor it to your needs. If you can't use the CD-ROM version, then retype the version at the back of the book and tailor it to your needs. Use the fill-in-the-blank form only as a last resort.

Call up the agreement on your computer or tear out the form in Appendix III, and read it along with the following instructions and explanations.

Title of agreement. You don't need a title for an independent contractor agreement, but if you want one, call it "Independent Contractor Agreement" or "Agreement for Professional Services." Since you are not your client's employee, do not use "Employment Agreement" as a title.

Names of independent contractor and hiring firm. See Chapter 5, Section B, for information about which names to use in your agreement.

1. Clause 1. Services to Be Performed

(Choose Alternative A or B.)
☐ ALTERNATIVE A
Contractor agrees to perform the following services on Client's behalf:

_____

_____

☐ ALTERNATIVE B
Contractor agrees to perform the services described in Exhibit A, which is attached to this Agreement.

Insert a detailed description of what you are expected to accomplish. Word the description carefully to emphasize the results you're expected to achieve. Don't describe the method by which you will achieve the results. As an independent contractor, it should be up to you to decide how to do the work. The client's control should be limited to accepting or rejecting your final results. The more control the client exercises over how you work, the more you'll look like an employee.

Be sure to define clearly exactly what form the finished product will take. (See Section A1, above, for examples.)

Use Alternative A if you describe the services in the contract. Choose Alternative B if you attach the description to the main contract.

2. Clause 2. Payment

(Choose Alternative A or B and optional provision, if desired.)

☐ ALTERNATIVE A

In consideration for the services to be performed by Contractor, Client agrees to pay Contractor $_____.

☐ ALTERNATIVE B

In consideration for the services to be performed by Contractor, Client agrees to pay Contractor at the rate of $_____ per _____ [hour, day, week, or other unit of time].

(Optional: Check and complete if applicable.)

☐ Contractor's total compensation shall not exceed $_____ without Client's written consent.

This clause contemplates that you will receive a fixed fee for the work or be paid by the hour or other unit of time. (See Chapter 5, Section B2, for a detailed discussion.)

However, there are other payment methods. For example, you could be paid a royalty—that is, a percentage of what the client earns by selling copies of your work. For more information on royalty agreements, see *Licensing Art & Design: A Professional's Guide to Licensing and Royalty Agreements,* by Caryn R. Leland (Allworth Press).

3. Clause 3. Terms of Payment

See Chapter 5, Section B3, for a discussion of this clause.

4. Clause 4. Late Fees (Optional)

See Chapter 5, Section B4, for a discussion of this clause.

5. Clause 5. Expenses

(Choose Alternative A or B and optional provision, if desired.)

☐ ALTERNATIVE A

Contractor shall be responsible for all expenses incurred while performing services under this Agreement.

(Optional: Check if applicable.)

☐ However, Client shall reimburse Contractor for all reasonable travel and living expenses necessarily incurred by Contractor while away from Contractor's regular place of business to perform services under this Agreement. Contractor shall submit an itemized statement of such expenses. Client shall pay Contractor within 30 days from the date of each statement.

☐ ALTERNATIVE B

Client shall reimburse Contractor for the following expenses that are attributable directly to work performed under this Agreement:

• travel expenses other than normal commuting, including airfares, rental vehicles, and highway mileage in company or personal vehicles at _____ cents per mile

• telephone, fax, online, and telegraph charges

• postage and courier services

• printing and reproduction

• computer services, and

• other expenses resulting from the work performed under this Agreement.

Contractor shall submit an itemized statement of Contractor's expenses. Client shall pay Contractor within 30 days from the date of each statement.

Expenses are the costs you incur that you can attribute directly to your work for a client. They include, for example, travel costs, costs to purchase or rent equipment, materials and supplies, charges for models and props, printing and reproduction, and postage or shipping costs. Expenses do not include your normal fixed overhead costs such as your office rent, the cost of commuting to and from your office, or the wear and tear on your equipment. There are two ways to handle expenses.

a. You pay expenses

Setting your compensation at a level high enough to cover your expenses is simple and safe. On the practical side, it frees you from having to keep records of your expenses. Covering your own expenses will help bolster your status as an IC, as well. In the recent past, the IRS viewed a client's payment of a worker's expenses as a sign of employee status. Although the agency now downgrades the importance of this factor, it's not entirely gone. And even though the IRS has changed its stance, other government agencies may consider payment of your expenses to be a strong indication of an employment relationship. You'll be on more solid ground if you don't ask the client to reimburse you for expenses.

Choose Alternative A if you are responsible for expenses. Use the optional paragraph if the client will reimburse you for travel and living expenses while you work on the project.

b. Client reimburses expenses

However, it is customary for clients to reimburse many types of creative ICs for expenses. If there is an otherwise clear IC relationship and reimbursement of expenses is common in your field, you can get away with such reimbursement without jeopardizing your status as an IC. Choose Alternative B if you'll be reimbursed for expenses.

6. Clause 6. Materials

> Contractor will furnish all materials and equipment used to provide the services required by this Agreement.

Generally, an independent contractor should provide all the materials and equipment necessary to complete a project. Being provided with tools and equipment is a sign that you're an employee, not

an IC. This clause states that you will provide all tools, equipment, materials, and supplies needed to accomplish the project.

If the client is providing you with tools or equipment, delete this clause. In this situation, it's best for the agreement not to mention tools and equipment.

7. Clause 7. Intellectual Property Ownership

> (Choose Alternative A or B.)
>
> ☐ ALTERNATIVE A
>
> Contractor hereby licenses to Client the following intellectual property rights in the work created or developed by Contractor under this Agreement:
>
> _____ .
>
> This license is conditioned upon full payment of the compensation due Contractor under this Agreement. Contractor reserves all rights not expressly granted to Client by this Agreement.
>
> **(Check applicable provision.)**
>
> ☐ The rights granted above are exclusive to Client.
>
> OR:
>
> ☐ The rights granted above are nonexclusive.
>
> ☐ ALTERNATIVE B
>
> Contractor assigns to Client all patent, copyright, and trade secret rights in anything created or developed by Contractor for Client under this Agreement. This assignment is conditioned upon full payment of the compensation due Contractor under this Agreement.
>
> Contractor shall help prepare any documents Client considers necessary to secure any copyright, patent, or other intellectual property rights at no charge to Client. However, Client shall reimburse Contractor for reasonable out-of-pocket expenses.

This is one of the most important parts of your agreement. There are many options regarding ownership of intellectual property created by independent contractors. Typically, your client will want to own all the intellectual property rights in your work, but you don't have to agree to this. For example, you could grant the client only some of your exclusive copyright rights and keep the others. The advantage to you is that you can resell work to which you retain your rights.

Before we explain the different ways you can keep and give away your intellectual property rights, you'll need some basic information on copyright law.

A copyright is made up of a bundle of rights, including:

- reproduction rights—the right to make copies of a protected work
- distribution rights—the right to sell or otherwise distribute copies to the public
- rights to create adaptations (or derivative works)—the right to prepare new works based on the protected work, and
- performance and display rights—the right to perform or display in public a protected literary, musical, dramatic, choreographic, pantomime, motion picture, or audiovisual work, whether in person or by means of a device like a television.

Each of these rights can be subdivided and sold separately or together with others. A copyright is infinitely divisible—for example, any of your rights can be divided and sold by:

- time period—a few weeks, months, years, or forever
- geographic area—for example, North America or worldwide
- category of use—for example, advertising or educational uses
- medium of use—for example, magazines, books, television, movies, public performances, and
- language—for example, English or other languages.

Any of these rights can be transferred on an exclusive or nonexclusive basis. An exclusive transfer gives the client the sole right to exercise the rights transferred. You can't transfer the same rights to anybody else because you don't own them anymore. By contrast, a nonexclusive transfer gives the client the right to exercise one or more of your copyright rights, but it does not prevent you from giving others permission to exercise the same right or rights at the same time. A nonexclusive transfer is the most limited form of rights transfer you can grant a client.

Why would you want the client to take less than all your rights? Because the client may not need all of them, and you might be able to make use of those rights you keep.

> **Example:** Schooldaze Publishing needs 25 spot illustrations for use in a school textbook. It contracts with Sally, an IC illustrator, to create them. Schooldaze wants just the illustrations for its textbooks and doesn't care if Sally makes other use of them. So Schooldaze obtains an assignment of Sally's exclusive rights to copy and reproduce the illustrations for use in school textbooks in North America for the life of the copyright. Sally retains all her other rights in the illustrations. For example, she can sell them to a magazine or use them in a book other than a textbook.

Two ownership options are provided here, but these are not the only options available. The only limit on how you deal with ownership of your work is your own imagination.

a. You retain ownership

Choose Alternative A if you want to grant the client a license to some of your copyright rights while you keep the others. You'll need to define clearly what rights you're giving up.

This clause also provides that you reserve any rights not expressly transferred to the client. This may include, for example, electronic rights such as the right to place your work on the Internet or in electronic databases such as Lexis. If the client wants additional use rights, an additional fee must be negotiated.

b. You transfer ownership to client

Choose Alternative B if you want to transfer to the client all your intellectual property rights. This clause makes this transfer contingent upon the client paying all of your compensation for the project. You also agree to help prepare any documents necessary to help the client obtain any copyright, patent, or other intellectual property rights at no charge to the client. This would probably amount to no more than signing a patent or copyright registration application. However, the client is required to reimburse you for these expenses.

8. Clause 8. Reusable Materials (Optional)

(Check if applicable.)

☐ Contractor owns or holds a license to use and sublicense various materials in existence before the start date of this Agreement ("Contractor's Materials"). Contractor's Materials include, but are not limited to, those items identified in Exhibit ____, attached to and made part of this Agreement. Contractor may, at its option, include Contractor's Materials in the work performed under this Agreement. Contractor retains all right, title, and interest, including all copyrights, patent rights, and trade secret rights, in Contractor's Materials. Contractor grants Client a royalty-free nonexclusive license to use any Contractor's Materials incorporated into the work performed by Contractor under this Agreement. The license shall have a perpetual term and may not be transferred by Client.

Many independent contractors who create intellectual property for clients have certain materials they use over and over again. You may lose the legal right to reuse such materials if you transfer all your ownership rights in your work to the client. To avoid this, include this clause in your agreement. It provides that you retain ownership of such materials and gives the client only a nonexclusive license to use them.

The license is royalty-free—meaning that the sole payment you receive for it is what the client paid you for your services. The client will make no additional payments for the license. The license also has a perpetual term, meaning it will last as long as your copyright, patent, or other intellectual property rights do.

Choose this optional clause if you want to preserve your right to reuse certain material in connection with the intellectual property you created. If you know what materials you'll use in advance, it's a good idea to list them in an exhibit attached to the agreement, but this isn't absolutely necessary.

9. Clause 9. Releases

(Choose Alternative A or B.)

☐ ALTERNATIVE A

Client shall obtain all necessary copyright permissions and privacy releases for materials included in the Designs at Client's request. Client shall indemnify Contractor against all claims and expenses, including reasonable attorney fees, due to Client's failure to obtain such permissions or releases.

☐ ALTERNATIVE B

Contractor shall obtain all necessary copyright permissions and privacy releases for materials included in the Designs at Client's request.

Ordinarily, a publicity/privacy release is necessary when you use a person's name, likeness, or voice for commercial purposes—for example, in a TV commercial or magazine ad. By signing the release, people agree to let you use these elements of their persona. If you don't obtain a release, people could sue you (and the client) for violating their right to privacy or publicity. Usually, obtaining a release is not difficult—the subject just has to sign a simple form.

 You'll find a release form on the CD-ROM.

You'll find a tear-out release form in Appendix III.

Choose Alternative A to require the client to obtain any necessary releases or copyright permissions. The client is required to indemnify—that is, repay—you if you are sued over a breach of any third party's copyright, privacy, publicity, or other rights resulting from the client's failure to obtain such permissions or releases.

Choose Alternative B if you're going to obtain any necessary releases. You'll often be in the best position to obtain publicity/privacy releases, so it's not unreasonable for the client to require you to do so. Indeed, obtaining such releases is standard practice for many creative ICs—for example, commercial photographers customarily obtain releases from their models. Freelance writers should also obtain releases from interview subjects.

10.   Clause 10. Copyright Notice and Credit Line

A copyright notice and credit line in Contractor's name shall accompany any reproduction of the Designs in the following form: _____

_____

_____

Your finished work doesn't have to be accompanied by a copyright notice—the familiar © symbol followed by the publication date and copyright owner's name—to be protected by copyright law. Using copyright notices is optional. If the work is published and someone copies it and uses it without your permission, you can always sue for copyright infringement.

However, including the symbol gives you the opportunity to argue for higher damages in a lawsuit. If your published work has a copyright notice, infringers can hardly claim they didn't know it was copyrighted. Such in-your-face thievery is often punished more harshly by judges and juries than less deliberate misuses.

The copyright notice and credit line clause requires the client to include a copyright notice and credit line when the client reproduces your work. Insert the form of name that should appear in the notice.

11.   Clause 11. Term of Agreement

See Chapter 5, Section B7, for a discussion of this clause.

12.   Clause 12. Terminating the Agreement

See Chapter 5, Section B8, for a discussion of this clause.

13.   Clause 13. Independent Contractor Status

See Chapter 5, Section B9, for a discussion of this clause.

14.   Clause 14. Local, State, and Federal Taxes

See Chapter 5, Section B10, for a discussion of this clause.

15.   Clause 15. Exclusive Agreement

See Chapter 5, Section B11, for a discussion of this clause.

16.   Clause 16. Modifying the Agreement (Optional)

See Chapter 5, Section B12, for a discussion of this clause.

17.   Clause 17. Resolving Disputes

See Chapter 5, Section B13, for a discussion of this clause.

18.   Clause 18. Limiting Your Liability to the Client (Optional)

See Chapter 5, Section B14, for a discussion of this clause.

19. Clause 19. Notices

See Chapter 5, Section B15, for a discussion of this clause.

20. Clause 20. No Partnership

See Chapter 5, Section B16, for a discussion of this clause.

21. Clause 21. Applicable Law

See Chapter 5, Section B17, for a discussion of this clause.

22. Clause 22. Assignment and Delegation (Optional)

> **(Check if applicable.)**
>
> ☐ Either Contractor or Client may assign rights or may delegate duties under this Agreement.

The general rules governing assigning rights and delegating duties under a contract are discussed in Chapter 5, Section A22. Generally, if an IC has agreed to provide personal services, the IC cannot delegate that work to someone else without the client's consent. In other words, you must perform the services personally or have them performed by employees under your supervision.

A creative IC's services would definitely be considered personal services. You're being hired for your particular creative abilities. This means you ordinarily cannot delegate your duties under the agreement without the client's consent. This is true even if your agreement does not expressly bar delegation.

However, you may want the flexibility to delegate your duties or assign your rights under the agreement. Luckily, your duties can be made delegable if the client agrees. That's what this clause is designed to accomplish. It allows an unrestricted right of assignment and delegation. If you include this clause in the agreement, the normal restrictions on assignment and delegation will not apply. If a client balks at including this clause, point out that it can help the client if a government agency challenges your status as an IC. It is strong evidence of an IC relationship because it shows the client is only concerned with results, not who achieves them. It also demonstrates lack of control over how you do the work.

If the client refuses to agree to this, delete the clause and make no mention of the subject. The normal assignment and delegation rules will apply. This means you'll still be able to assign your benefits under the agreement, but you will be unable to delegate your duties without the client's consent.

23. Signatures

See Chapter 5, Section B19, for an explanation of how to sign your agreement. ■

Agreements for Construction Contractors

The agreements in this chapter are for construction services. Use the agreement in Section A if you're hiring a construction contractor. Use the agreement in Section B if you're working as a construction contractor.

 Do not use these forms if you are working as or hiring a subcontractor.

A. Agreement for Use by Hiring Firm

Use this agreement if you are hiring a construction contractor. It can be used for all types of construction projects.

 The entire text of a construction contractor agreement for use by the hiring firm is on the CD-ROM. You can modify individual clauses to tailor the text to your specific needs.

A tear-out construction contractor agreement for use by the hiring firm is in Appendix II. Use this form if you don't have access to a computer. It contains blank spaces for you to fill in by hand or typewriter.

Although we provide a tear-out form for those of you who either don't have computers or who would prefer not to use the CD-ROM, don't just tear out the form and use it as your IC agreement if you can avoid doing so. A fill-in-the-blank IC agreement is the least persuasive type of contract when you are trying to convince a government agency that the worker is an IC and not an employee. The best way to create your agreement is to use the CD-ROM version and tailor it to your needs. If you can't use the CD-ROM version, then retype the version at the back of the book and tailor it to your needs. Use the fill-in-the-blank form only as a last resort.

Call up the agreement on your computer or tear out the form in Appendix II, and read it along with the following instructions and explanations.

Title of agreement. You don't have to have a title for the agreement, but if you want one you should call it "Independent Contractor Agreement." Do not use "Employment Agreement" as a title.

Names of IC and hiring firm. For guidance on what names to use, see Chapter 5, Section A.

1. Clause 1. Services to Be Performed

> Contractor shall furnish all labor and materials to construct and complete the project shown on the contract documents contained in Exhibit A, which is attached to and made part of this Agreement.

The agreement should describe in as much detail as possible what the contractor is expected to do. Word the description carefully to show only the results the IC is expected to achieve. Don't tell the IC how to achieve those results—this indicates that you have the right to control how the IC performs the agreed-upon services. Such a right of control is the hallmark of an employment relationship. (See Chapter 2 for a discussion of why this is important.)

It's perfectly all right for you to establish very detailed specifications for the IC's finished work product and to require the IC to comply with all applicable building code requirements. But the specs should describe only the end results the IC must achieve, not how to obtain those results.

Construction projects usually involve lengthy specifications or project descriptions. You should attach these to the agreement and label them as Exhibit A. This way, they become a part of the agreement.

2. Clause 2. Payment

(Choose Alternative A or B.)

☐ ALTERNATIVE A

Owner shall pay Contractor for all labor and materials the sum of $_____.

☐ ALTERNATIVE B

Owner shall pay Contractor $_____ for labor. Materials shall be paid for by Owner upon delivery to the worksite or as follows: _____

_____

Two options for payment are provided in this agreement. You can either:

- pay the IC for all labor and materials, or
- pay for materials directly and pay the IC just for labor.

Paying for materials yourself upon delivery to the worksite avoids the possibility that the IC will fail to pay for them and the materials provider will file a lien for payment against your property. (See Section A8, below, for more about this issue.) However, it requires additional time and effort on your part. Use whichever clause applies and delete the other.

3. Clause 3. Terms of Payment

(Choose Alternative A, B, or C.)

☐ ALTERNATIVE A

Upon completing Contractor's services under this Agreement, Contractor shall submit an invoice. Owner shall pay Contractor within _____ days from the date of Contractor's invoice.

☐ ALTERNATIVE B

Contractor shall be paid $_____ upon signing this Agreement and the remaining amount due when Contractor completes the services and submits an invoice. Owner shall pay Contractor within _____ days from the date of Contractor's invoice.

☐ ALTERNATIVE C

Contractor shall be paid according to the Schedule of Payments set forth in Exhibit __, attached to and made part of this agreement.

This clause refers to how and when you'll pay the IC. Three payment options are provided. Choose whichever one applies and delete the others.

a. Payment upon completing services

For brief projects, you may pay the IC in full after the work is completed. Choose Alternative A for this form of payment.

b. Down payment required

Choose Alternative B if you'll provide the IC with a down payment before starting work. Many ICs will insist upon a down payment. Obviously, it's in your interest to pay as little as possible in advance. This is a matter for negotiation.

c. Payment according to schedule

For longer projects, typically a payment schedule is used—for example, you pay the IC a specified amount when the agreement is signed, progress payments while the project is under construction, and the rest when the project is completed. Choose Alternative C to use this type of arrangement and attach the payment schedule to the agreement as an exhibit.

4. Clause 4. Time of Completion

The work to be performed under this Agreement shall commence on _____ and be substantially completed on or before _____. Time is of the essence.

State when the work will begin and end in the spaces provided. Be sure to give the IC adequate time to complete the project.

5. Clause 5. What Constitutes Completion

> The work specified in Clause 1 shall be considered completed upon approval by Owner; however, Owner's approval shall not be unreasonably withheld.

This clause makes clear to the contractor that the work is not completed until you say it is. However, you can't be unreasonable in withholding your approval of the work. If the contractor believes you're being unreasonable, the contractor can invoke the dispute resolution remedies set forth in Section A21, below.

6. Clause 6. Permits and Approvals

> (Choose Alternative A or B.)
> ☐ ALTERNATIVE A
> Owner shall be responsible for determining which state and local permits are necessary for performing the specified work and for obtaining and paying for the permits.
> ☐ ALTERNATIVE B
> Contractor shall be responsible for determining which state and local permits are necessary for performing the specified work and for obtaining and paying for the permits.

Building permits or other government approvals are typically required for construction projects. It's up to you and the IC to decide who will be responsible for obtaining and paying for any required permits.

Use one of the two clauses provided. The first clause requires you to obtain and pay for all permits and approvals. The second clause makes this the IC's responsibility.

7. Clause 7. Warranty

> Contractor warrants that all work shall be completed in a good workmanlike manner and in compliance with all building codes and other applicable laws. Contractor agrees to correct any defective work at no cost to Owner. This warranty shall be in effect for one year from the date of completion of the work.

A warranty is a promise or statement regarding the quality, quantity, performance, or legal title of something being sold. If the IC makes and fails to live up to a warranty, you can sue for breach of warranty and obtain damages.

This agreement includes a warranty that the work will be completed in a good workmanlike manner in compliance with all building codes. Moreover, the contractor promises to fix any defects you discover free of charge. This warranty lasts for one year after the work is completed. Such a warranty is fairly standard in the construction industry.

Even if your agreement doesn't include this type of warranty, the contractor would still have a legal duty to complete the work in a good and workmanlike manner. This does not mean that the construction work must be perfect. Rather, it means that the contractor must use appropriate and reasonable skill and care in the construction. Defects that a reasonably skilled contractor would not leave in the finished work must be fixed.

Because the law implies this warranty in all construction contracts, this warranty will protect you even if it is not written explicitly in the contract. The purpose of including it is simply to remind a contractor.

Most states also recognize an implied warranty on the part of a builder of a new home or other residence that the structure will be suitable for human habitation. This means that a residential structure must be a fit place for a person to live. It should comply with local building code requirements for residential housing and have, for example, adequate ventilation, light, heating, and sanitary facilities.

8.   Clause 8. Liens and Lien Waivers

Contractor represents and warrants that there will be no liens for labor or materials or appliances against the work covered by this Agreement, and agrees to protect and hold Owner free and harmless from and against any and all liens and claims for labor, materials, services, or appliances furnished or used in connection with the work.

To protect Owner against liens being filed by Contractor, subcontractors, and materials providers, Contractor agrees that final payment to Contractor under Clause 3 shall be withheld by Owner until Contractor presents Owner with lien waivers, lien releases, or acknowledgment of full payment from each subcontractor and materials supplier.

While doing the job, the contractor will probably order materials from suppliers and may hire subcontractors (other ICs) to do some of the work. These people all have to be paid. Sometimes, contractors fail to pay subcontractors or materials suppliers. If this happens, the supplier or subcontractor can make a claim against your property, even though you weren't the one who failed to pay.

An unpaid supplier or subcontractor can place a lien on your property, called a mechanic's lien or materials lien. A lien is a claim for payment made against your property that is recorded with the county recorder's office. If you want to sell your property, you can't pass good title to the buyer without clearing all such liens. To do so, you might have to pay off the unpaid subcontractors or materials suppliers yourself, even though you've already paid the contractor for the work or supplies.

This clause is designed to protect you from such liens. First, the contractor warrants (prom-

ises) that no such liens will be recorded. In other words, the contractor promises timely payment to suppliers and subcontractors.

But what if the contractor breaks the promise? Here's where the second part of this clause comes into play. It tells the contractor that you won't make a final payment until you learn in writing from each supplier and subcontractor that they've been satisfactorily paid. In construction law lingo, the last payment is contingent on the contractor giving you the following things in writing:

- lien waivers or releases, in which subcontractors and suppliers promise that they will not record a lien on your property, or

- acknowledgments of full payment, in which subcontractors and suppliers state that they have been paid by the contractor and so have no grounds for filing a lien against you.

Either document assures you won't have any problems with liens. Of course, it's up to you to keep tabs on which suppliers and subs are working on the job and to make sure you get these releases or acknowledgments from each.

9.   Clause 9. Site Maintenance

Contractor agrees to be bound by the following conditions when performing the specified work:
- Contractor shall remove all debris and leave the premises in a broom-clean condition.
- Contractor shall perform the specified work during the following hours: _____ .
- Contractor agrees that disruptively loud activities shall be performed only at the following times: _____ .
- At the end of each day's work, Contractor's equipment shall be stored in the following location: _____ .

You may wish to require the construction work to take place at specific times, that equipment be stored in a specific location, or that clean up be daily and thorough. Use the provisions in this clause that apply.

10. Clause 10. Subcontractors

> Contractor may at its discretion engage subcontractors to perform services under this Agreement, but Contractor shall remain responsible for proper completion of this Agreement.

Few contractors have the expertise necessary to perform all of the many jobs required in a construction project. For this reason, it's a common practice in the construction industry for contractors to subcontract out some or most of the work to other contractors, called subcontractors. For example, the contractor may hire an IC electrician to install electrical wiring. This clause provides that the IC may engage such subcontractors to help perform the services upon your written consent, but the contractor remains ultimately responsible for making sure that all of the work is completed properly.

11. Clause 11. Independent Contractor Status

> Contractor is an independent contractor, not Owner's employee. Contractor's employees or subcontractors are not Owner's employees. Contractor and Owner agree to the following rights consistent with an independent contractor relationship (check all that apply):
> - ☐ Contractor has the right to perform services for others during the term of this Agreement.
> - ☐ Contractor has the sole right to control and direct the means, manner and method by which the services required by this Agreement will be performed.

> - ☐ The Contractor or Contractor's employees or subcontractors shall perform the services required by this Agreement; Owner shall not hire, supervise, or pay any assistants to help Contractor.
> - ☐ Owner shall not require Contractor or Contractor's employees or subcontractors to devote full time to performing the services required by this Agreement.

One of the most important functions of an independent contractor agreement is to help establish that the worker is an IC, not your employee. The key to doing this is to make clear that the IC, not the hiring firm, has the right to control how the work will be performed.

The language in this clause addresses most of the factors the IRS and other agencies consider in determining whether an IC controls how the work is done. (See Chapter 2 for a discussion of why this is important.)

Check all that apply. When you draft your own agreement, include only those provisions that apply to your particular situation. The more that apply, the more likely the worker will be viewed as an IC.

12. Clause 12. Business Permits, Certificates, and Licenses

> Contractor represents and warrants that Contractor has complied with all federal, state, and local laws requiring business permits, certificates, and licenses required to carry out the services to be performed under this Agreement.
>
> Contractor's license or registration is for the following type of work and carries the following number: _____.

The IC should have all business permits, certificates, and licenses needed to perform the work. For example, the IC should have a contractor's license if your state's law requires one. A worker who lacks such licenses and permits looks like an employee, not an IC running an independent business. The IC should obtain such licenses and permits; you should not pay for them. This provision requires that the IC have all necessary licenses and permits. It also includes a space where you can insert the contractor's license number.

⚠ **It's particularly important for contractors to be licensed in California.** Under California law, if you hire an unlicensed worker to perform work requiring a contractor's license, you're automatically deemed the worker's employer for all state payroll tax purposes, including unemployment compensation and state income tax (Cal. Unemployment Ins. Code § 621.5, Cal. Labor Code § 2750.5). Neither the IRS nor virtually any other state has a rule similar to California's. If you're not sure whether the work you want done requires a contractor's license, contact the California State License Board in Sacramento. If a contractor's license is required, ask to see one before hiring the IC.

13. Clause 13. State and Federal Taxes

See Chapter 5, Section A8, for a discussion of this clause.

14. Clause 14. Fringe Benefits

See Chapter 5, Section A9, for a discussion of this clause.

15. Clause 15. Workers' Compensation

See Chapter 5, Section A10, for a discussion of this clause.

16. Clause 16. Unemployment Compensation

See Chapter 5, Section A11, for a discussion of this clause.

17. Clause 17. Insurance

See Chapter 5, Section A12, for a discussion of this clause.

18. Clause 18. Terminating the Agreement

See Chapter 5, Section A14, for a discussion of this clause.

19. Clause 19. Exclusive Agreement

See Chapter 5, Section A15, for a discussion of this clause.

20. Clause 20. Modifying the Agreement (Optional)

See Chapter 5, Section A16, for a discussion of this clause.

21. Clause 21. Resolving Disputes

See Chapter 5, Section A18, for a discussion of this clause.

22. Clause 22. Applicable Law

See Chapter 5, Section A19, for a discussion of this clause.

23. Clause 23. Notices

See Chapter 5, Section A20, for a discussion of this clause.

24. Clause 24. No Partnership

See Chapter 5, Section A21, for a discussion of this clause.

25. Signatures

See Chapter 5, Section A23, for a discussion of this clause.

B. Agreement for Use by Contractor

Use this agreement if you are a construction contractor. It can be used for all types of construction projects.

A construction contractor agreement for use by the contractor is on the CD-ROM. You can modify individual clauses to tailor the text to your specific needs.

A tear-out agreement for use by the contractor is in Appendix III. Use this form if you don't have access to a computer. It contains blank spaces for you to fill in by hand or typewriter.

Although we provide a tear-out form for those of you who either don't have computers or who would prefer not to use the CD-ROM, don't just tear out the form and use it as your IC agreement if you can avoid doing so. A fill-in-the-blank IC agreement is the least persuasive type of contract when you are trying to convince a government agency that you are an IC and not an employee of the hiring firm. The best way to create your agreement is to use the CD-ROM version and tailor it to your needs. If you can't use the CD-ROM version, then retype the version at the back of the book and tailor it to your needs. Use the fill-in-the-blank form only as a last resort.

Call up the agreement on your computer or tear out the form in Appendix III, and read it along with the following instructions and explanations.

Title of agreement. You don't need a title for an independent contractor agreement, but if you want one, call it "Independent Contractor Agreement" or "Agreement for Construction Services." Since you are not your client's employee, do not use "Employment Agreement" as a title.

Names of independent contractor and hiring firm. See Chapter 5, Section A, for guidance on what names to use.

1. Clause 1. Services to Be Performed

> Contractor shall furnish all labor and materials to construct and complete the project shown on the contract documents contained in Exhibit A, which is attached to and made part of this Agreement.

The single most important part of the agreement is the description of the services you'll perform for the client. This description will serve as the yardstick to measure whether your performance was satisfactory.

Describe in as much detail as possible what you're expected to accomplish. However, word the description carefully to emphasize only the results you're expected to achieve. Don't describe the method by which you will achieve the results. As an IC, it should be up to you to decide how to do the work. The client's control should be limited to accepting or rejecting your final results.

It's perfectly okay for the agreement to establish very detailed specifications for your finished work product. But the specs should only describe the end results you must achieve, not how to obtain those results.

Construction projects usually involve lengthy specifications or project descriptions. You should attach these to the agreement and label them as Exhibit A. This way, they become a part of the agreement.

2. Clause 2. Payment

> (Choose Alternative A or B.)
>
> ☐ ALTERNATIVE A
>
> Owner shall pay Contractor for all labor and materials the sum of $ _____.
>
> ☐ ALTERNATIVE B
>
> Owner shall pay Contractor $_____ for labor. Materials shall be paid for by Owner upon delivery to the worksite or as follows: _____
>
> _____
>
> _____

Two options for payment are provided in this agreement. The owner can either:

- pay you for all labor and materials, or
- pay for materials directly and pay you just for labor.

Use whichever clause you and the client choose.

A client who wants to pay for materials directly is probably interested in making sure that the materials suppliers actually do get paid and have no reason to file a lien against the property. (Liens are explained above, in Subsection A8.) However, making the client your bill payer at every step of a complex project can be an unwieldy arrangement, since clients will not have accounts at your favorite suppliers, nor are they necessarily on site whenever you need a check. If possible, arrange to pay for materials yourself. Assure the client that you will, indeed, promptly pay all suppliers and subs, and that every charge for supplies and outside labor that appears on your bill will be accompanied by a signed receipt.

3. Clause 3. Terms of Payment

(Choose Alternative A, B, or C.)

☐ ALTERNATIVE A

Upon completing Contractor's services under this Agreement, Contractor shall submit an invoice. Owner shall pay Contractor within _____ days from the date of Contractor's invoice.

☐ ALTERNATIVE B

Contractor shall be paid $_____ upon signing this Agreement and the remaining amount due when Contractor completes the services and submits an invoice. Owner shall pay Contractor within _____ days from the date of Contractor's invoice.

☐ ALTERNATIVE C

Contractor shall be paid according to the Schedule of Payments set forth in Exhibit __, attached to and made part of this agreement.

The terms of payment describe how you will bill the client and be paid. Before you are paid for your work, you should submit an invoice to the client setting out the amount due. An invoice doesn't have to be fancy. It should include:

- an invoice number
- the dates covered by the invoice
- the hours expended (if you're being paid by the hour), and
- a summary of the work performed.

Three payment methods are provided.

a. Payment upon completing services

For brief projects, you may be paid in full by the client after you complete the work. Choose Alternative A for this form of payment.

b. Down payment required

Choose Alternative B to require the client to provide you with a down payment before you start work.

c. Payment according to schedule

For longer projects, a payment schedule is typically used—for example, you may be paid a specified amount when the agreement is signed, progress payments while the project is under construction, and the rest when the project is completed. Choose Alternative C to use this type of arrangement and attach the payment schedule to the agreement as an exhibit.

4. Clause 4. Late Fees (Optional)

See Chapter 5, Section B4, for a discussion of this clause.

5. Clause 5. Time of Completion

> The work to be performed under this Agreement shall commence on _____ and be substantially completed on or before _____
>
> _____.

State when the work will begin and end in the spaces provided. Be sure to give yourself adequate time to complete the project.

6. Clause 6. Permits and Approvals

> (Choose Alternative A or B.)
>
> ☐ ALTERNATIVE A
>
> Owner shall be responsible for determining which state and local permits are necessary for performing the specified work and for obtaining and paying for the permits.
>
> ☐ ALTERNATIVE B
>
> Contractor shall be responsible for determining which state and local permits are necessary for performing the specified work and for obtaining and paying for the permits.

Typically, building permits or other government approvals are required for construction projects. It's up to you and your client to decide who will be responsible for obtaining and paying for any required permits.

7. Clause 7. Warranty

> Contractor warrants that all work shall be completed in a good workmanlike manner and in compliance with all building codes and other applicable laws.

A warranty is a promise or statement regarding the quality, quantity, performance, or legal title of something being sold. When you make an actual promise about these things, you are making an express, or stated, warranty.

A warranty can be created by using words such as *guarantee*, *affirm*, or *warrant* in a contract. However, no magic words are necessary to create a warranty. Promises you make in proposals and contractual specifications can also serve as express warranties. This can be true even where you did not intend to create a warranty. Depending on the circumstances, express warranties can last for any period of time, ranging from a few months to many years.

If your work fails to live up to your warranties, the client can sue you in court for breach of warranty and obtain damages. The client doesn't have to prove that you were negligent—that is, that you failed to do your work properly. All it has to show is that you didn't perform the way you said you would. This makes it much easier for the client to obtain damages.

It's up to you to decide if you want to make any warranties in your agreement. This agreement includes a limited warranty that the work will be completed in a good workmanlike manner in compliance with all building codes. Such a warranty is fairly standard in the construction industry.

Even in the absence of such a warranty in your agreement, you would still have a legal duty to complete the work in a "good and workmanlike manner." This does not mean that the construction work must be perfect. Rather, it means that you must use appropriate and reasonable skill and care in the construction. Defects that a reasonably skilled contractor would not leave in the finished work must be fixed.

Because the law implies this warranty in all construction contracts, this warranty will obligate you even if it is not written explicitly in the contract.

Most states also recognize an implied warranty on the part of a builder of a new home or other residence that the structure will be suitable for human habitation. This means that a residential structure must be a fit place for a person to live. It should comply with local building code requirements for residential housing and have, for example, adequate ventilation, light, heating, and sanitary facilities.

Because this warranty is implied, it will obligate you, even though it is not written explicitly in your contract.

8. Clause 8. Site Maintenance

Contractor agrees to be bound by the following conditions when performing the specified work:

- Contractor shall remove all debris and leave the premises in a broom clean condition.
- Contractor shall perform the specified work during the following hours:

  _____ .

- Contractor agrees that disruptively loud activities shall be performed only at the following times: _____ .
- At the end of each day's work, Contractor's equipment shall be stored in the following location: _____ .

Some clients may require that the construction work occur at specific times, that equipment be stored in a specific location, or that work be commenced or finished at a particular time each day. Use the provisions in this clause that apply.

9. Clause 9. Subcontractors

Contractor may at its discretion engage subcontractors to perform services under this Agreement, but Contractor shall remain responsible for proper completion of this Agreement.

The agreement provides that you may engage subcontractors to help you perform your services, but you remain ultimately responsible for making sure that the work is completed properly.

10. Clause 10. Independent Contractor Status

Contractor is an independent contractor, not Owner's employee. Contractor's employees or subcontractors are not Owner's employees. Contractor and Owner agree to the following rights consistent with an independent contractor relationship (check all that apply):

- ☐ Contractor has the right to perform services for others during the term of this Agreement.
- ☐ Contractor has the sole right to control and direct the means, manner, and method by which the services required by this Agreement will be performed.
- ☐ The Contractor or Contractor's employees or subcontractors shall perform the services required by this Agreement; Owner shall not hire, supervise, or pay any assistants to help Contractor.
- ☐ Owner shall not require Contractor or Contractor's employees or subcontractors to devote full time to performing the services required by this Agreement.
- ☐ Neither Contractor nor Contractor's employees or subcontractors are eligible to participate in any employee pension, health, vacation pay, sick pay, or other fringe benefit plan of Owner.

One of the most important functions of an independent contractor agreement is to help establish that you are an independent contractor, not your client's employee. The key to doing this is to make clear that you, not the client, have the right to control how the work will be performed.

You will need to emphasize the factors the IRS and other agencies consider in determining whether a client controls how the work is done. Of course, if you merely recite these factors but fail to live up to them, agency auditors won't be fooled. Think of this clause as a reminder to you and your client about how to conduct your business relationship.

11. Clause 11. Local, State, and Federal Taxes

Contractor shall pay all income taxes and FICA (Social Security and Medicare taxes) incurred while performing services under this Agreement. Owner will not:

- withhold FICA from Contractor's payments or make FICA payments on Contractor's behalf
- make state or federal unemployment compensation contributions on Contractor's behalf, or
- withhold state or federal income tax from Contractor's payments.

The charges included here do not include taxes. If Contractor is required to pay any federal, state, or local sales, use, property, or value added taxes based on the services provided under this Agreement, the taxes shall be billed separately to Owner. Owner shall be responsible for paying any interest or penalties incurred due to late payment or nonpayment of any taxes by Owner.

The agreement should address federal and state income taxes, Social Security taxes, and sales taxes.

a. Income taxes

Your client should not pay or withhold any income or Social Security taxes on your behalf. Doing so is a very strong indicator that you are an employee, not an independent contractor. Indeed, some courts have classified workers as employees based upon this factor alone. Keep in mind that one of the best things about being an independent contractor is that you don't have taxes withheld from your paychecks.

This straightforward provision tells the client that you'll pay all applicable taxes due on your compensation and that, therefore, the client should not withhold taxes from your payments. You need not insert any information here.

b. Sales taxes

A few states require independent contractors to pay sales taxes, even if they provide only services to their clients. These states include Hawaii, New Mexico, and South Dakota. Many other states require sales taxes to be paid for certain specified services.

Whether or not you're required to collect sales taxes, include this provision in your agreement to make clear that the client will have to pay these and similar taxes. States change sales tax laws constantly and more and more are beginning to look at services as a good source of sales tax revenue. So this provision could come in handy in the future even if you don't really need it now.

12. Clause 12. Insurance

The agreement requires you to obtain adequate liability insurance to provide coverage for injuries to your employees or others. This is a standard provision.

13. Clause 13. Terminating the Agreement

See Chapter 5, Section B8, for a discussion of this clause.

14. Clause 14. Exclusive Agreement

See Chapter 5, Section B11, for a discussion of this clause.

15. Clause 15. Modifying the Agreement (Optional)

See Chapter 5, Section B12, for a discussion of this clause.

16. Clause 16. Resolving Disputes

See Chapter 5, Section B13, for a discussion of this clause.

17. Clause 17. Notices

See Chapter 5, Section B15, for a discussion of this clause.

18. Clause 18. No Partnership

See Chapter 5, Section B16, for a discussion of this clause.

19. Clause 19. Applicable Law

See Chapter 5, Section B17, for a discussion of this clause.

20. Signatures

See Chapter 5, Section B19, for a discussion of this clause. ■

Agreements for Messengers, Couriers, and Delivery People

The agreements in this chapter are for use by ICs who work as messengers, couriers, or delivery people. Use the agreement in Section A if you are a hiring firm. Use the agreement in Section B if you are an IC.

A. Agreement for Hiring Firm to Use

Use this agreement if you're hiring an IC messenger, courier, or delivery person. The agreement is designed to protect your interests as much as possible without being unduly one-sided or unfair to the IC. It also helps establish that the worker is an IC, not your employee. (See Chapter 2 for information on why this is important.)

 The entire text of the independent contractor agreement for use by the hiring firm is on the CD-ROM. You can modify individual clauses to tailor the text to your specific needs. However, be sure to read the following discussion before making any major changes.

The agreement is also reproduced as a tear-out form in Appendix II. Use this form if you don't have access to a computer. It contains blank spaces for you to fill in by hand or typewriter.

Although we provide a tear-out form for those of you who either don't have computers or who would prefer not to use the CD-ROM, don't just tear out the form and use it as your IC agreement if you can avoid doing so. A fill-in-the-blank IC agreement is the least persuasive type of contract when you are trying to convince a government agency that the worker is an IC and not an employee. The best way to create your agreement is to use the CD-ROM version and tailor it to your needs. If you can't use the CD-ROM version, then retype the version at the back of the book and tailor it to your needs. Use the fill-in-the-blank form only as a last resort.

Title of agreement. You don't have to title the IC agreement, but if you want to, you should call it "Independent Contractor Agreement." Never use "Employment Agreement" as a title, because the worker is not your employee.

Names of independent contractor and hiring firm. Refer to Chapter 5, Section A, for guidance on what names to use in your agreement.

1. Clause 1. Services to Be Performed

> (Check and complete applicable provision.)
> ☐ ALTERNATIVE A
> Contractor agrees to perform the following services:
> _____ .
>
> ☐ ALTERNATIVE B
> Contractor agrees to perform the services described in Exhibit A, which is attached to this Agreement.

The agreement should describe in detail what you expect the contractor to do. Word the description carefully to show only the results you expect the IC to achieve. Don't tell the IC how to achieve those results—this would indicate that you have the right to control how the IC performs the agreed-upon services. Having a right of control is the hallmark of an employment relationship. (See Chapter 2 for a discussion of why this is important.)

You can include the description in the main body of the agreement or, if it's lengthy, on a separate attachment. These options are identified as Alternative A and Alternative B, respectively.

Tailor your work description to your particular needs. Here's an example of a work description for courier services:

> Contractor will provide courier service five days per week between Client's main office at 701 Main Street, Anytown, IA, and the Felton Distribution Facility at 1223 Highway 2, Felton, IA. This requires Contractor to do the following:
> • Pick up from the Main Office at 7:30 a.m., Monday through Friday, with delivery to the Felton facility by 8:00 a.m.

- Pick up from the Felton facility at 5:00 p.m., Monday through Friday, hold overnight, and deliver to the Main Office by 7:30 a.m. the next morning.
- If a holiday intervenes, pick up and/or delivery shall occur the next business day.

2. Clause 2. Payment

> In consideration for the services to be performed by Contractor, Client agrees to pay Contractor at the following rates: _____ .
>
> Contractor shall be paid within a reasonable time after Contractor submits an invoice to Client. The invoice should include the following: an invoice number, the dates covered by the invoice, and a summary of the work performed.

You can pay the IC a fixed fee, by unit of time, or any other way. Paying a fixed fee instead of an hourly or daily rate will help you establish the worker's IC status in the eyes of the government. A fixed fee could be based on the number of trips made, the number of items delivered, the distance traveled, or some other means. It's up to you and the IC to determine the method of payment. Whatever formula you use, it should be explained in full in the IC Agreement.

Pay the Contractor only after you receive an invoice. This agreement requires you to pay the Contractor within a reasonable time after you receive the invoice. It does not require you to pay a fee or any other penalty if your payments are late. But, of course, paying late is not a good way to keep up good relations with a valuable IC. And if you don't pay at all, the IC can sue you for the amount due.

3. Clause 3. Expenses

> (Choose Alternative A or B.)
> ☐ ALTERNATIVE A
> Contractor shall be responsible for all expenses incurred while performing services under this Agreement. This includes automobile, truck, and other travel expenses; vehicle maintenance and repair costs; vehicle and other license fees and permits; insurance premiums; road, fuel, and other taxes; fines; radio, pager, or cell phone expenses; meals; and all salary, expenses, and other compensation paid to employees or contract personnel the Contractor hires to complete the work under this Agreement.
> ☐ ALTERNATIVE B
> Client shall reimburse Contractor for the following expenses that are attributable directly to work performed under this Agreement: _____
> _____ .
>
> Contractor shall submit an itemized statement of Contractor's expenses. Client shall pay Contractor within 30 days after receipt of each statement.

Estimating expenses and factoring in these costs when bidding on a job is a strong indicator that a person is in business and not an employee. For this reason, you should not usually reimburse an IC for expenses. Make this clear to the IC. Choose Alternative A if the IC will pay all expenses.

However, if you must reimburse the IC for expenses, choose Alternative B and be sure to list them.

4. Clause 4. Vehicles and Equipment

> Contractor will furnish all vehicles, equipment, tools, and materials used to provide the services required by this Agreement. Client will not require Contractor to rent or purchase any equipment, product, or service as a condition of entering into this Agreement.

Providing workers with tools and equipment is an indicator of employee status. Therefore, it's not advisable to provide an IC with materials, tools, or equipment. For example, if you're hiring the IC to make deliveries with a delivery van, don't provide the van. The IC should buy or lease the van elsewhere.

If you must break this rule and provide the IC with equipment, such as a vehicle, it's best for the agreement to be silent on the matter. Delete this clause entirely and renumber the remaining clauses. If you lease equipment to the IC, have the IC sign a separate written lease agreement for the equipment. Charge the IC the same amount you'd charge anyone else. If you sell the IC equipment—for example, a radio—use a written bill of sale. You can find forms for selling or leasing equipment in *Legal Forms for Starting & Running a Small Business*, by Fred Steingold (Nolo).

5. Clause 5. Independent Contractor Status

Contractor is an independent contractor, and neither Contractor nor Contractor's employees or contract personnel are, or shall be deemed, Client's employees. In its capacity as an independent contractor, Contractor agrees and represents, and Client agrees, as follow (check all that apply):

☐ Contractor has the right to perform services for others during the term of this Agreement.

☐ Contractor has the sole right to control and direct the means, manner, and method by which the services required by this Agreement will be performed. Contractor shall select the routes taken, starting and quitting times, days of work, and order the work is performed.

☐ Contractor has the right to hire assistants as subcontractors or to use employees to provide the services required by this Agreement.

☐ Neither Contractor nor Contractor's employees or contract personnel shall be required to wear any uniforms provided by Client.

☐ The services required by this Agreement shall be performed by Contractor, Contractor's employees, or contract personnel, and Client shall not hire, supervise, or pay any assistants to help Contractor.

☐ Neither Contractor nor Contractor's employees or contract personnel shall receive any training from Client in the professional skills necessary to perform the services required by this Agreement.

☐ Neither Contractor nor Contractor's employees or contract personnel shall be required by Client to devote full time to the performance of the services required by this Agreement.

One of the most important functions of an independent contractor agreement is to help establish that the worker is an IC, not your employee. The key to doing this is to make clear that the IC, not the hiring firm, has the right to control how the work will be performed.

The language in this clause addresses most of the factors the IRS and other government agencies consider in determining whether an IC controls how the work is done. (You can find a complete explanation of these factors in Chapter 2.)

When you draft your own agreement, include only those provisions in Clause 6 that apply to your particular situation. The more that apply, the more likely the worker will be viewed as an IC.

6. Clause 6. Business Licenses, Permits, and Certificates

Contractor represents and warrants that Contractor and Contractor's employees and contract personnel will comply with all federal, state, and local laws requiring driver's and other licenses, business permits, and certificates required to carry out the services required by this Agreement.

The contractor should have all licenses, permits, and certificates needed to perform the services. This would likely include a driver's license and a business license in some cities and states. You should not be involved in either the application or the certification procedure, nor should you pay for them.

It's a good idea to obtain copies of the IC's licenses and keep them in your file. They will help establish that the worker is an IC if you're audited.

Many states have nonprofit or government-sponsored agencies devoted to small businesses that can provide you with information on whether your state requires specific types of ICs to have a license. You can find a state-by-state list of small business resources on the Small Business Administration's website at www.sba.gov.

Also, many state agencies now have websites that contain information about licensing requirements. A good place to start an Internet search for such information is the State Web Locator at www.infoctr.edu/swl/.

7. Clause 7. State and Federal Taxes

Client will not:

- withhold FICA (Social Security and Medicare taxes) from Contractor's payments or make FICA payments on Contractor's behalf

- make state or federal unemployment compensation contributions on Contractor's behalf, or

- withhold state or federal income tax from Contractor's payments.

Contractor shall pay all taxes incurred while performing services under this Agreement—including all applicable income taxes and, if Contractor is not a corporation, self-employment (Social Security) taxes. Upon demand, Contractor shall provide Client with proof that such payments have been made.

You should never pay or withhold any taxes on an IC's behalf. Doing so is a very strong indicator that the worker is an employee. Indeed, some courts have held that workers were employees based upon this factor alone.

8. Clause 8. Fringe Benefits

Contractor understands that neither Contractor nor Contractor's employees or contract personnel are eligible to participate in any employee pension, health, vacation pay, sick pay, or other fringe benefit plan of Client.

Do not provide ICs or their employees with fringe benefits that you extend to your own employees, such as health insurance, pension benefits, child-care allowances, or even the right to use employee facilities like an exercise room.

9. Clause 9. Unemployment Compensation

See Chapter 5, Section A10, for a discussion of this clause.

10. Clause 10. Workers' Compensation

See Chapter 5, Section A10, for a discussion of this clause.

11. Clause 11. Insurance

Client shall not provide insurance coverage kind for Contractor or Contractor's employees or contract personnel. Contractor shall obtain the following insurance coverage and maintain it during the entire term of this Agreement

☐ Automobile liability insurance for each vehicle used in the performance of this Agreement—including owned, non-owned (for example, owned by Contractor's employees), leased, or hired vehicles—in the minimum amount of $_____ combined single limit per occurrence for bodily injury and property damage.

☐ Comprehensive or commercial general liability insurance coverage in the minimum amount of $_____ combined single limit, including coverage for bodily injury, personal injury, broad form property damage, contractual liability, and cross-liability.

Before commencing any work, Contractor shall provide Client with proof of this insurance and with proof that Client has been made an additional insured under the policies.

ICs should have their own insurance coverage just like any other business; you need not provide it. This clause requires the IC to obtain two policies: an automobile liability policy and a general liability policy. The automobile policy covers claims arising from auto or other motor vehicle accidents. The general liability policy insures the IC against personal injury or property damage claims arising for other reasons—for example, if the IC accidentally injures a bystander or one of your employees while making a delivery inside your office. The IC's liability policy will pay the costs of defending a lawsuit and money damages up to the limits of the policy coverage. If the IC will not be using a motor vehicle, delete the paragraph regarding automotive insurance.

You need to state how much coverage is required. $500,000 to $1 million is usually adequate. If you're not sure how much coverage to ask for, ask your own insurance broker or agent for advice.

The insurance clause also requires the IC to add you to the policies as an additional insured. This means that you will have all of the benefits of the policy's protections. If you are sued along with the IC in a lawsuit by someone injured by the IC's negligence, the IC's policy will provide defense and pay any claim or verdict up to the limits of the policy.

> ### WATCH YOUR WORDS: "NAMED" AND "ADDITIONAL" AREN'T THE SAME
>
> Do not confuse "named insured" with "additional insured." If you are added to a policy as a named insured, the protections of the policy extend beyond you to your partners, officers, employees, agents, and affiliates. Because the coverage is so broad, it may cost as much as 50% of the original premium—a cost few ICs would be willing to incur. And there's no need to. Adding you as an additional insured is inexpensive, and it provides you with sufficient protection.

12. Clause 12. Indemnification

> Contractor shall indemnify and hold Client harmless from any loss or liability arising from performing services under this Agreement.

This clause requires the IC to indemnify you—that is, to personally repay you—if somebody injured by the IC decides to sue you. This is an important protection even if you are an additional insured under the IC's policy (as discussed above). There may be circumstances under which you're unable to collect from the IC's insurer—for example, the insurer may deny coverage for some reason. Or the claim may exceed the policy limits of the IC's insurance. In this event, the indemnity clause requires the IC to repay you personally. This gives you additional protection but, of course, will be useful only if the IC has the money to repay you.

13. Clause 13. Term of Agreement

See Chapter 5, Section A13, for a discussion of this clause.

14. Clause 14. Terminating the Agreement

See Chapter 5, Section A14, for a discussion of this clause.

15. Clause 15. Exclusive Agreement

See Chapter 5, Section A15, for a discussion of this clause.

16. Clause 16. Modifying the Agreement (Optional)

See Chapter 5, Section A16, for a discussion of this clause.

17. Clause 17. Resolving Disputes

See Chapter 5, Section A18, for a discussion of this clause.

18. Clause 18. Confidentiality (Optional)

See Chapter 5, Section A17, for a discussion of this clause.

19. Clause 19. Applicable Law

See Chapter 5, Section A19, for a discussion of this clause.

20. Clause 20. No Partnership

See Chapter 5, Section A21, for a discussion of this clause.

21. Clause 21. Assignment and Delegation

See Chapter 5, Section A22, for a discussion of this clause.

22. Signatures

See Chapter 5, Section A23, for guidance on signing your agreement.

B. Agreement for Use by IC

Use this agreement if you're an IC. It contains a number of provisions favorable to you that a hiring firm wouldn't ordinarily include in an agreement it prepared—for example, requiring the hiring firm to pay a late fee if it doesn't pay you on time.

 The entire text of the agreement is on the CD-ROM. You can modify individual clauses to tailor the text to your needs. However, be sure to read the following discussion before you make any major changes.

The agreement is also reproduced as a tear-out form in Appendix III. Use this form if you don't have access to a computer. It contains blank spaces for you to fill in by hand or typewriter.

Although we provide a tear-out form for those of you who either don't have computers or who would prefer not to use the CD-ROM, don't just tear out the form and use it as your IC agreement if you can avoid doing so. A fill-in-the-blank IC agreement is the least persuasive type of contract when you are trying to convince a government agency that you are an IC and not an employee of the hiring firm. The best way to create your agreement is to use the CD-ROM version and tailor it to your needs. If you can't use the CD-ROM version, then retype the version at the back of the book and tailor it to your needs. Use the fill-in-the-blank form only as a last resort.

Title of agreement. You don't need a title for an independent contractor agreement, but if you want one, call it "Independent Contractor Agreement" or "Agreement for Professional Services." Because you are not your client's employee, do not use "Employment Agreement" as a title.

Names of independent contractor and hiring firm. See Chapter 5, Section B, for guidance on what names to use in your agreement.

1. Clause 1. Services to Be Performed

See Chapter 5, Section B1, for a discussion of this clause.

2. Clause 2. Payment

> In consideration for the services to be performed by Contractor, Client agrees to pay Contractor at the following rates: _____.

There are a number of ways that the hiring firm can pay you for your service, including by fixed fee or by a unit of time. Being paid a fixed fee instead of an hourly or daily rate will help establish that you are an IC. Such a fee could be based on the number of trips made, the number of items delivered, or the distance traveled.

3. Clause 3. Terms of Payment

> Contractor shall send Client an invoice [CHOOSE ONE: upon completion of the services required by this Agreement OR _____ (choose time period—for example, monthly)]. Client shall pay Contractor within ____ days from the date of each invoice.

As an IC, you should submit an invoice to get paid. An invoice doesn't have to be fancy or filled with legalese. It need include only an invoice number, the dates covered by the invoice, the hours expended if you're being paid by the hour, and a summary of the work performed. There are several computer accounting programs, such as QuickBooks, that can generate invoices for you.

 An invoice you can use with your clients is on the CD-ROM.

The timing of your invoice depends on the work that you are doing. If you're performing only a single job for a client, send the invoice when you complete the work. If you're perform-ing services on an ongoing basis, submit an invoice each month or other time period.

Specify in the contract how long the client has to pay the invoice. Thirty days is a common period, but you can choose a shorter period if the client will agree. Note that the time period begins to run from the date of your invoice, not the date the client receives the invoice, which may be several days later.

4. Clause 4. Late Fees (Optional)

See Chapter 5, Section B4, for a discussion of this clause.

5. Clause 5. Expenses

> (Choose Alternative A or B.)
> ☐ ALTERNATIVE A
> Contractor shall be responsible for all expenses incurred while performing services under this Agreement.
> ☐ ALTERNATIVE B
> Client shall reimburse Contractor for the following expenses that are attributable directly to work performed under this Agreement: _____.

Expenses are the costs you incur that you can attribute directly to your work for a client. Setting your compensation at a level high enough to cover your expenses is simple and safe. On the practical side, it frees you from having to keep records of your expenses. Covering your own expenses will also help bolster your status as an IC. In the recent past, the IRS viewed the payment of a worker's expenses by a client as a sign of employee status. Although the agency now downgrades the importance of this factor, it's not entirely gone. You'll be on more solid ground if you don't ask the client to reimburse you for expenses.

And, even though the IRS has changed its stance, other government agencies may still consider payment of a worker's business or traveling expenses to be a strong indication of employment relationship. For this reason, it is usually best that you don't bill separately for your expenses.

6. Clause 6. Equipment and Materials (Optional)

> **(Check if applicable.)**
>
> ☐ Contractor will furnish all equipment and materials used to provide the services required by this Agreement.

Generally, an independent contractor should provide all the equipment and materials necessary to do the work. For example, a bicycle messenger should provide the bicycle. This helps bolster IC status.

If the client is providing you with a vehicle or other equipment for free, delete this clause. In this situation, it's best for the agreement not to mention equipment.

If you're buying or leasing a vehicle or other equipment from the client, you should enter into a separate purchase or lease agreement. You can find forms for selling or leasing equipment in *Legal Forms for Starting & Running a Small Business,* by Fred Steingold (Nolo).

7. Clause 7. Term of Agreement

See Chapter 5, Section B7, for a discussion of this clause.

8. Clause 8. Terminating the Agreement

See Chapter 5, Section B8, for a discussion of this clause.

9. Clause 9. Independent Contractor Status

> Contractor is an independent contractor, and neither Contractor nor Contractor's employees or contract personnel are, or shall be deemed, Client's employees. In its capacity as an independent contractor, Contractor agrees and represents, and Client agrees, as follows (check all that apply):
>
> ☐ Contractor has the right to perform services for others during the term of this Agreement.
>
> ☐ Contractor has the sole right to control and direct the means, manner, and method by which the services required by this Agreement will be performed. Contractor shall select the routes taken, starting and quitting times, days of work, and order the work is performed.
>
> ☐ Contractor has the right to hire assistants as subcontractors or to use employees to provide the services required by this Agreement.
>
> ☐ Neither Contractor nor Contractor's employees or contract personnel shall be required to wear any uniforms provided by Client.
>
> ☐ The services required by this Agreement shall be performed by Contractor, Contractor's employees, or contract personnel. Client shall not hire, supervise, or pay any assistants to help Contractor.
>
> ☐ Neither Contractor nor Contractor's employees or contract personnel shall receive any training from Client in the professional skills necessary to perform the services required by this Agreement.
>
> ☐ Neither Contractor nor Contractor's employees or contract personnel shall be required by Client to devote full time to the performance of the services required by this Agreement.

One of the most important functions of an independent contractor agreement is to help establish that you are an IC, not an employee. The key to doing this is to make clear that you, not the hiring firm, have the right to control how the work will be performed.

The language in this clause addresses most of the factors the IRS and other agencies consider in determining whether an IC controls how the work is done. (You can find a complete explanation of these factors in Chapter 2.)

When you draft your own agreement, include only those provisions in Clause 6 that apply to your particular situation. The more that apply, the more likely you will be viewed as an IC.

10. Clause 10. Local, State, and Federal Taxes

> Contractor shall pay all income taxes and FICA (Social Security and Medicare taxes) incurred while performing services under this Agreement. Client will not:
>
> - withhold FICA from Contractor's payments or make FICA payments on Contractor's behalf
> - make state or federal unemployment compensation contributions on Contractor's behalf, or
> - withhold state or federal income tax from Contractor's payments.
>
> Contractor shall not be required to pay any federal, state, or local sales taxes based on Contractor's compensation for the services provided under this Agreement. If client is charged for such sales taxes, the taxes shall be separately billed to Client. Client shall be responsible for any interest or penalties incurred due to late payment or nonpayment of any taxes by Client.

The agreement should address the issue of who pays federal and state income taxes, Social Security taxes, and sales taxes.

a.   Income taxes

Your client should not pay or withhold any income or Social Security taxes on your behalf. Doing so is a very strong indicator that you are an employee, not an independent contractor. Indeed, some courts have classified workers as employees based upon this factor alone. Keep in mind that

one of the best things about being an independent contractor is that you don't have taxes withheld from your paychecks.

This straightforward provision in your agreement makes clear that you'll pay all applicable taxes due on your compensation and, therefore, that the client should not withhold taxes from your payments. You need not insert any information here.

b.   Sales taxes

A few states require independent contractors to pay sales taxes, even if they provide their clients with services only and don't provide goods. These states include Hawaii, New Mexico, and South Dakota. Many other states require sales taxes to be paid for certain types of services.

Whether or not your state collects sales taxes for your services, include this provision in your agreement making it clear that the client will have to pay these and similar taxes. States change sales tax laws constantly, and more and more are beginning to look at services as a good source of tax revenue. So this provision could come in handy in the future, even if you don't really need it now.

Note that you are responsible for paying other types of sales taxes—for example, taxes on the fuel or supplies you use to perform your services.

11. Clause 11. Exclusive Agreement

See Chapter 5, Section B11, for a discussion of this clause.

12. Clause 12. Modifying the Agreement (Optional)

See Chapter 5, Section B12, for a discussion of this clause.

13. Clause 13. Resolving Disputes

See Chapter 5, Section B13, for a discussion of this clause.

14. Clause 14. Limiting Liability (Optional)

See Chapter 5, Section B14, for a discussion of this clause.

15. Clause 15. Notices

See Chapter 5, Section B15, for a discussion of this clause.

16. Clause 16. No Partnership

See Chapter 5, Section B16, for a discussion of this clause.

17. Clause 17. Applicable Law

See Chapter 5, Section B17, for a discussion of this clause.

18. Clause 18. Assignment and Delegation

See Chapter 5, Section B18, for a discussion of this clause.

19. Signatures

See Chapter 5, Section B19, for guidance on signing your agreement. ■

How to Use the CD-ROM

The tear-out forms in Appendix II and III are included on a CD-ROM in the back of the book. This CD-ROM, which can be used with Windows computers, installs files that can be opened, printed, and edited using a word processor or other software. It is *not* a stand-alone software program. Please read this Appendix and the README.TXT file included on the CD-ROM for instructions on using the Forms CD.

Note to Mac users: This CD-ROM and its files should also work on Macintosh computers. Please note, however, that Nolo cannot provide technical support for non-Windows users.

HOW TO VIEW THE README FILE

If you do not know how to view the file README.TXT, insert the Forms CD-ROM into your computer's CD-ROM drive and follow these instructions:

- Windows 9x, 2000, Me, and XP: (1) On your PC's desktop, double click the My Computer icon; (2) double click the icon for the CD-ROM drive into which the Forms CD-ROM was inserted; (3) double click the file README.TXT.

- Macintosh: (1) On your Mac desktop, double click the icon for the CD-ROM that you inserted; (2) double click on the file README.TXT.

While the README file is open, print it out by using the Print command in the File menu.

A. Installing the Form Files Onto Your Computer

Word processing forms that you can open, complete, print, and save with your word processing program (see Section B, below) are contained on the CD-ROM. Before you can do anything with the files on the CD-ROM, you need to install them onto your hard disk. In accordance with U.S. copyright laws, remember that copies of the CD-ROM and its files are for your personal use only.

Insert the Forms CD and do the following:

1. Windows 9x, 2000, Me, and XP Users

Follow the instructions that appear on the screen. (If nothing happens when you insert the Forms CD-ROM, then (1) double click the My Computer icon; (2) double click the icon for the CD-ROM drive into which the Forms CD-ROM was inserted; and (3) double click the file WELCOME.EXE.)

By default, all the files are installed to the \IC Forms folder in the \Program Files folder of your computer. A folder called "IC Forms" is added to the "Programs" folder of the Start menu.

2. Macintosh Users

Step 1: If the "IC Forms CD" window is not open, open it by double clicking the "IC Forms CD" icon.

Step 2: Select the "IC Forms" folder icon.

Step 3: Drag and drop the folder icon onto the icon of your hard disk.

B. Using the Word-Processing Files to Create Documents

This section concerns the files for forms that can be opened and edited with your word processing program.

All word processing forms come in rich text format. These files have the extension ".RTF." For example, the form for the Contract Amendment Form discussed in Chapter 5 is on the file Amendment.rtf. All forms, their file names, and file formats are listed in Appendix II and III.

RTF files can be read by most recent word processing programs, including all versions of MS

Sample Agreements for Use by Hiring Firm

Chapter	File Name	Form Name
5 Section A	General1.rtf	Independent Contractor Agreement
6 Section A	Consultant1.rtf	Independent Contractor Agreement for Consultant
7 Section A	Household1.rtf	Independent Contractor Agreement for Household Worker
8 Section B	Seller1.rtf	Independent Contractor Agreement for Direct Seller
9 Section A	Accountant1.rtf	Independent Contractor Agreement for Accountant & Bookkeeper
10 Section A	Software1.rtf	Independent Contractor Agreement for Software Consultant
11 Section A	Creative1.rtf	Independent Contractor Agreement for Creative Contractor
12 Section A	Construction1.rtf	Independent Contractor Agreement for Construction Contractor
13 Section A	Courier1.rtf	Independent Contractor Agreement for Couriers, Messengers, and Delivery People
	Amendment1.rtf	Contract Amendment Form
	Release1.rtf	Publicity/Privacy Release
	Letter1.rtf	Letter
	Question1.rtf	Independent Contractor Questionnaire
	Doccheck.rtf	Document Checklist

Word for Windows and Macintosh, WordPad for Windows, and recent versions of WordPerfect for Windows and Macintosh.

To use a form from the CD to create your documents you must: (1) open a file in your word processor or text editor; (2) edit the form by filling in the required information; (3) print it out; (4) rename and save your revised file.

The following are general instructions on how to do this. However, each word processor uses different commands to open, format, save, and print documents. Please read your word processor's manual for specific instructions on performing these tasks.

Do not call Nolo's technical support if you have questions on how to use your word processor.

Step 1: Opening a File

There are three ways to open the word processing files included on the CD-ROM after you have installed them onto your computer.

- Windows users can open a file by selecting its "shortcut" as follows: (1) Click the Windows "Start" button; (2) open the "Programs" folder; (3) open the "IC Forms" subfolder; and (4) open the "RTF" subfolder; and (5) click on the shortcut to the form you want to work with.
- Both Windows and Macintosh users can open a file directly by double clicking on it. Use My Computer or Windows Explorer (Windows 9x, 2000, Me or XP) or the Finder (Macintosh) to go to the folder you installed or copied the CD-ROM's files to. Then, double click on the specific file you want to open.
- You can also open a file from within your word processor. To do this, you must first start your word processor. Then, go to the File menu and choose the Open command. This opens a dialog box where you will tell

the program (1) the type of file you want to open (*.RTF); and (2) the location and name of the file (you will need to navigate through the directory tree to get to the folder on your hard disk where the CD's files have been installed). If these directions are unclear you will need to look through the manual for your word processing program—Nolo's technical support department will *not* be able to help you with the use of your word processing program.

WHERE ARE THE FILES INSTALLED?

Windows Users

- RTF files are installed by default to a folder named \IC Forms\RTF in the \Program Files folder of your computer.

Macintosh Users

- RTF files are located in the "RTF" folder within the "IC Forms" folder.

Step 2: Editing Your Document

Fill in the appropriate information according to the instructions and sample agreements in the book. Underlines are used to indicate where you need to enter your information, frequently followed by instructions in brackets. *Be sure to delete the underlines and instructions from your edited document.* If you do not know how to use your word processor to edit a document, you will need to look through the manual for your word processing program—Nolo's technical support department will *not* be able to help you with the use of your word processing program.

EDITING FORMS THAT HAVE OPTIONAL OR ALTERNATIVE TEXT

Some of the forms have check boxes before text. The check boxes indicate:

- Optional text, where you choose whether to include or exclude the given text.
- Alternative text, where you select one alternative to include and exclude the other alternatives.

If you are using the tear-out forms in Appendix II and III, you simply mark the appropriate box to make your choice.

If you are using the Forms CD, however, we recommend that instead of marking the check boxes, you do the following:

Optional text

If you don't want to include optional text, just delete it from your document.

If you do want to include optional text, just leave it in your document.

In either case, delete the check box itself as well as the italicized instructions that the text is optional.

Alternative text

First delete all the alternatives that you do not want to include.

Then delete the remaining check boxes, as well as the italicized instructions that you need to select one of the alternatives provided.

Step 3: Printing Out the Document

Use your word processor's or text editor's "Print" command to print out your document. If you do not know how to use your word processor to print a document, you will need to look through the manual for your word processing program—Nolo's technical support department will *not* be able to help you with the use of your word processing program.

Step 4: Saving Your Document

After filling in the form, use the "Save As" command to save and rename the file. Because all the files are "read-only," you will not be able to use the "Save" command. This is for your protection. *If you save the file without renaming it, the underlines that indicate where you need to enter your information will be lost, and you will not be able to create a new document with this file without recopying the original file from the CD-ROM.*

If you do not know how to use your word processor to save a document, you will need to look through the manual for your word processing program—Nolo's technical support department will *not* be able to help you with the use of your word processing program. ■

INDEPENDENT CONTRACTOR AGREEMENT

This Agreement is made between _____ ("Client")

with a principal place of business at _____

and _____ ("Contractor")

with a principal place of business at _____ .

1. Services to be Performed

(Choose Alternative A or B.)

☐ ALTERNATIVE A

Contractor agrees to perform the following services: _____

_____

_____

_____

☐ ALTERNATIVE B

Contractor agrees to perform the services described in Exhibit A attached to this Agreement.

2. Payment

(Choose Alternative A or B and optional clause if desired.)

☐ ALTERNATIVE A

In consideration for the services to be performed by Contractor, Client agrees to pay Contractor $_____ according to the terms of payment set forth below.

☐ ALTERNATIVE B

In consideration for the services to be performed by Contractor, Client agrees to pay Contractor at the rate of $_____ per hour according to the terms of payment set forth below.

(Optional: Check if applicable.)

☐ Unless otherwise agreed upon in writing by Client, Client's maximum liability for all services performed during the term of this Agreement shall not exceed $_____ .

3. Terms of Payment

(Choose Alternative A, B, C, or D.)

☐ ALTERNATIVE A

Upon completion of Contractor's services under this Agreement, Contractor shall submit an invoice. Client shall pay Contractor the compensation described within _____ [15, 30, 45, 60] days after receiving Contractor's invoice.

☐ ALTERNATIVE B

Contractor shall be paid $_____ upon the signing of this Agreement and the remainder of the compensation described above upon completion of Contractor's services and submission of an invoice.

☐ ALTERNATIVE C

Client shall pay Contractor according to the schedule of payments set forth in Exhibit _____ attached to this Agreement.

☐ ALTERNATIVE D

Contractor shall invoice Client on a monthly basis for all hours worked pursuant to this Agreement during the preceding month. Invoices shall be submitted on Contractor's letterhead specifying an invoice number, the dates covered in the invoice, the hours expended, and the work performed (in summary) during the invoice period. Client shall pay Contractor's fee within _____ days after receiving Contractor's invoice.

4. Expenses

(Choose Alternative A or B.)

☐ ALTERNATIVE A

Contractor shall be responsible for all expenses incurred while performing services under this Agreement. This includes license fees, memberships, and dues; automobile and other travel expenses; meals and entertainment; insurance premiums; and all salary, expenses, and other compensation paid to employees or contract personnel the Contractor hires to complete the work under this Agreement.

☐ ALTERNATIVE B

Client shall reimburse Contractor for the following expenses that are attributable directly to work performed under this Agreement: _____.

Contractor shall submit an itemized statement of Contractor's expenses. Client shall pay Contractor within 30 days after receipt of each statement.

5. Materials, Equipment, and Office Space

Contractor will furnish all materials, tools, and equipment used to provide the services required by this Agreement.

(Optional: Check and complete if applicable.)

☐ Client agrees to rent to Contractor the following office space and facilities:

_____

The rental will begin on _____ [date] and will end on the earlier of (1) the date Contractor completes the services required by this Agreement; (2) _____ [date]; or (3) the date a party terminates this Agreement. The rental amount is $_____ per ☐ day, ☐ week, ☐ month.

6. Independent Contractor Status

Contractor is an independent contractor, and neither Contractor nor Contractor's employees or contract personnel are, or shall be deemed, Client's employees. In its capacity as an independent contractor, Contractor agrees and represents, and Client agrees, as follows (check all that apply):

    ☐ Contractor has the right to perform services for others during the term of this Agreement.

    ☐ Contractor has the sole right to control and direct the means, manner, and method by which the services required by this Agreement will be performed.

    ☐ Contractor has the right to perform the services required by this Agreement at any place or location and at such times as Contractor may determine.

    ☐ Contractor has the right to hire assistants as subcontractors or to use employees to provide the services required by this Agreement.

    ☐ The services required by this Agreement shall be performed by Contractor, Contractor's employees, or contract personnel, and Client shall not hire, supervise, or pay any assistants to help Contractor.

☐ Neither Contractor nor Contractor's employees or contract personnel shall receive any training from Client in the professional skills necessary to perform the services required by this Agreement.

☐ Neither Contractor nor Contractor's employees or contract personnel shall be required by Client to devote full time to the performance of the services required by this Agreement.

7. Business Permits, Certificates, and Licenses

Contractor has complied with all federal, state, and local laws requiring business permits, certificates, and licenses required to carry out the services to be performed under this Agreement.

8. State and Federal Taxes

Client will not:

- withhold FICA (Social Security and Medicare taxes) from Contractor's payments or make FICA payments on Contractor's behalf
- make state or federal unemployment compensation contributions on Contractor's behalf, or
- withhold state or federal income tax from Contractor's payments.

Contractor shall pay all taxes incurred while performing services under this Agreement—including all applicable income taxes and, if Contractor is not a corporation, self-employment (Social Security) taxes. Upon demand, Contractor shall provide Client with proof that such payments have been made.

9. Fringe Benefits

Contractor understands that neither Contractor nor Contractor's employees or contract personnel are eligible to participate in any employee pension, health, vacation pay, sick pay, or other fringe benefit plan of Client.

10. Workers' Compensation

Client shall not obtain workers' compensation insurance on behalf of Contractor or Contractor's employees. If Contractor hires employees to perform any work under this Agreement, Contractor will cover them with workers' compensation insurance to the extent required by law and provide Client with a certificate of workers' compensation insurance before the employees begin the work.

(Optional: Check if applicable.)

☐ Contractor shall obtain workers' compensation insurance coverage for Contractor. Contractor shall provide Client with proof that such coverage has been obtained before starting work.

11. Unemployment Compensation

Client shall make no state or federal unemployment compensation payments on behalf of Contractor or Contractor's employees or contract personnel. Contractor will not be entitled to these benefits in connection with work performed under this Agreement.

12. Insurance

Client shall not provide any insurance coverage of any kind for Contractor or Contractor's employees or contract personnel. Contractor shall obtain and maintain a broad form commercial general liability insurance policy providing for coverage of at least $_____ for each occurrence. Before commencing any work, Contractor shall provide Client with proof of this insurance and with proof that Client has been made an additional insured under the policy.

Contractor shall indemnify and hold Client harmless from any loss or liability arising from performing services under this Agreement.

13. Term of Agreement

This agreement will become effective when signed by both parties and will terminate on the earlier of:

- the date Contractor completes the services required by this Agreement
- _____ [date], or
- the date a party terminates the Agreement as provided below.

14. Terminating the Agreement

(Choose Alternative A or B.)

☐ ALTERNATIVE A

With reasonable cause, either Client or Contractor may terminate this Agreement, effective immediately upon giving written notice.

Reasonable cause includes:

- a material violation of this Agreement, or
- any act exposing the other party to liability to others for personal injury or property damage.

☐ ALTERNATIVE B

Either party may terminate this Agreement at any time by giving _____ days' written notice to the other party of the intent to terminate.

15. Exclusive Agreement

This is the entire Agreement between Contractor and Client.

16. Modifying the Agreement (Optional)

(Check box if applicable.)

☐ This Agreement may be modified only by a writing signed by both parties.

17. Confidentiality (Optional)

(Check box and complete if applicable.)

☐ Contractor acknowledges that it will be necessary for Client to disclose certain confidential and proprietary information to Contractor in order for Contractor to perform duties under this Agreement. Contractor acknowledges that any disclosure to any third party or any misuse of this proprietary or confidential information would irreparably harm Client. Accordingly, Contractor will not disclose or use, either during or after the term of this Agreement, any proprietary or confidential information of Client without Client's prior written permission except to the extent necessary to perform services on Client's behalf.

Proprietary or confidential information includes:

- the written, printed, graphic, or electronically recorded materials furnished by Client for Contractor to use
- any written or tangible information stamped "confidential," "proprietary," or with a similar legend or any information that Client makes reasonable efforts to maintain the secrecy of
- business or marketing plans or strategies, customer lists, operating procedures, trade secrets, design formulas, know-how and processes, computer programs and inventories, discoveries and improvements of any kind, sales projections, and pricing information
- information belonging to customers and suppliers of Client about whom Contractor gained knowledge as a result of Contractor's services to Client, and

• other: _____

Contractor shall not be restricted in using any material that is publicly available, already in Contractor's possession, known to Contractor without restriction, or rightfully obtained by Contractor from sources other than Client.

Upon termination of Contractor's services to Client, or at Client's request, Contractor shall deliver to Client all materials in Contractor's possession relating to Client's business.

Contractor acknowledges that any breach or threatened breach of this clause will result in irreparable harm to Client for which damages would be an inadequate remedy. Therefore, Client shall be entitled to equitable relief, including an injunction, in the event of such breach or threatened breach of this clause. Such equitable relief shall be in addition to Client's rights and remedies otherwise available at law.

18. Resolving Disputes

(Choose Alternative A, B, or C and any desired optional clauses.)

☐ ALTERNATIVE A

If a dispute arises under this Agreement, any party may take the matter to court.

(Optional: Check if applicable.)

☐ If any court action is necessary to enforce this Agreement, the prevailing party shall be entitled to reasonable attorney fees, costs, and expenses in addition to any other relief to which the party may be entitled.

☐ ALTERNATIVE B

If a dispute arises under this Agreement, the parties agree to first try to resolve the dispute with the help of a mutually agreed-upon mediator in _____ [list city or county where mediation will occur]. Any costs and fees other than attorney fees associated with the mediation shall be shared equally by the parties. If the dispute is not resolved within 30 days after it is referred to the mediator, any party may take the matter to court.

(Optional: Check if applicable.)

☐ If any court action is necessary to enforce this Agreement, the prevailing party shall be entitled to reasonable attorney fees, costs, and expenses in addition to any other relief to which the party may be entitled.

☐ ALTERNATIVE C

If a dispute arises under this Agreement, the parties agree to first try to resolve the dispute with the help of a mutually agreed-upon mediator in _____ [list city or county where mediation will occur]. Any costs and fees other than attorney fees associated with the mediation shall be shared equally by the parties. If it proves impossible to arrive at a mutually satisfactory solution through mediation, the parties agree to submit the dispute to a mutually agreed-upon arbitrator in _____ _____ [list city or county where arbitration will occur]. Judgment upon the award rendered by the arbitrator may be entered in any court having jurisdiction to do so. Costs of arbitration, including attorney fees, will be allocated by the arbitrator.

19. Applicable Law

This Agreement will be governed by the laws of the state of _____ .

20. Notices

All notices and other communications in connection with this Agreement shall be in writing and shall be considered given as follows:

- when delivered personally to the recipient's address as stated on this Agreement
- three days after being deposited in the United States mail, with postage prepaid to the recipient's address as stated on this Agreement, or
- when sent by fax or electronic mail, such notice is effective upon receipt provided that a duplicate copy of the notice is promptly given by first class mail, or the recipient delivers a written confirmation of receipt.

21. No Partnership

This Agreement does not create a partnership relationship. Contractor does not have authority to enter into contracts on Client's behalf.

22. Assignment and Delegation (Optional)

(Choose Alternative A or B, if applicable.)

☐ ALTERNATIVE A

Either Contractor or Client may assign its rights and may delegate its duties under this Agreement.

☐ ALTERNATIVE B

Contractor may not assign or subcontract any rights or delegate any of its duties under this Agreement without Client's prior written approval.

Signatures

Client: _____
Name of Client

By: _____
Signature

_____
Typed or Printed Name

Title: _____

Date: _____

Contractor: _____
Name of Contractor

By: _____
Signature

_____
Typed or Printed Name

Title: _____

Taxpayer ID Number: _____

Date: _____

INDEPENDENT CONTRACTOR AGREEMENT
FOR CONSULTANT

This Agreement is made between _____ ("Client")

with a principal place of business at _____

and _____ ("Consultant")

with a principal place of business at _____ .

1. Services to be Performed

(Choose Alternative A or B.)

☐ ALTERNATIVE A

Consultant agrees to perform the following consulting services on Client's behalf:

_____

_____

_____

☐ ALTERNATIVE B

Consultant agrees to perform the consulting services described in Exhibit A attached to this Agreement.

2. Payment

(Choose Alternative A or B and optional provision, if desired.)

☐ ALTERNATIVE A

In consideration for the services to be performed by Consultant, Client agrees to pay Consultant

$ _____ according to the terms of payment set forth below.

☐ ALTERNATIVE B

In consideration for the services to be performed by Consultant, Client agrees to pay Consultant at the rate of

$_____ per hour according to the terms of payment set forth below.

(Optional: Check if applicable.)

☐ Unless otherwise agreed upon in writing by Client, Client's maximum liability for all services performed
during the term of this Agreement shall not exceed $_____ .

3. Terms of Payment

(Choose Alternative A, B, C, or D.)

☐ ALTERNATIVE A

Upon completion of Consultant's services under this Agreement, Consultant shall submit an invoice. Client shall
pay Consultant the compensation described within _____ days after receiving Consultant's invoice.

☐ ALTERNATIVE B

Consultant shall be paid $_____ upon the signing of this Agreement and the remainder of the
compensation described above upon completion of Consultant's services and submission of an invoice.

☐ ALTERNATIVE C

Client shall pay Consultant according to the schedule of payments set forth in Exhibit _____ attached to this
Agreement.

☐ ALTERNATIVE D

Consultant shall invoice Client on a monthly basis for all hours worked pursuant to this Agreement during the preceding month. Invoices shall be submitted on Consultant's letterhead specifying an invoice number, the dates covered in the invoice, the hours expended, and the work performed (in summary) during the invoice period. Client shall pay Consultant's fee within _____ days after receiving Consultant's invoice.

4. Expenses

(Choose Alternative A or B.)

☐ ALTERNATIVE A

Consultant shall be responsible for all expenses incurred while performing services under this Agreement. This includes license fees, memberships and dues; automobile and other travel expenses; meals and entertainment; insurance premiums; and all salary, expenses, and other compensation paid to employees or contract personnel the Consultant hires to complete the work under this Agreement.

☐ ALTERNATIVE B

Client shall reimburse Consultant for the following expenses that are directly attributable to work performed under this Agreement: _____

_____

_____

Consultant shall submit an itemized statement of Consultant's expenses. Client shall pay Consultant within 30 days after receipt of each statement

5. Materials

Consultant will furnish all materials, equipment, and supplies used to provide the services required by this Agreement.

6. Independent Contractor Status

Consultant is an independent contractor, and neither Consultant nor Consultant's employees or contract personnel are, or shall be deemed, Client's employees. In its capacity as an independent contractor, Consultant agrees and represents, and Client agrees, as follows (check all that apply):

☐ Consultant has the right to perform services for others during the term of this Agreement.

☐ Consultant has the sole right to control and direct the means, manner and method by which the services required by this Agreement will be performed.

☐ Consultant has the right to perform the services required by this Agreement at any place or location and at such times as Consultant may determine.

☐ Consultant has the right to hire assistants as subcontractors or to use employees to provide the services required by this Agreement.

☐ The services required by this Agreement shall be performed by Consultant, Consultant's employees, or contract personnel, and Client shall not hire, supervise, or pay any assistants to help Consultant.

☐ Neither Consultant nor Consultant's employees or contract personnel shall receive any training from Client in the professional skills necessary to perform the services required by this Agreement.

☐ Neither Consultant nor Consultant's employees or contract personnel shall be required by Client to devote full time to the performance of the services required by this Agreement.

7. Business Permits, Certificates, and Licenses

Consultant has complied with all federal, state, and local laws requiring business permits, certificates, and licenses required to carry out the services to be performed under this Agreement.

8. State and Federal Taxes

Client will not:

- withhold FICA (Social Security and Medicare taxes) from Consultant's payments or make FICA payments on Consultant's behalf
- make state or federal unemployment compensation contributions on Consultant's behalf, or
- withhold state or federal income tax from Consultant's payments.

Consultant shall pay all taxes incurred while performing services under this Agreement—including all applicable income taxes and, if Consultant is not a corporation, self-employment (Social Security) taxes. Upon demand, Consultant shall provide Client with proof that such payments have been made.

9. Fringe Benefits

Consultant understands that neither Consultant nor Consultant's employees or contract personnel are eligible to participate in any employee pension, health, vacation pay, sick pay, or other fringe benefit plan of Client.

10. Workers' Compensation

Client shall not obtain workers' compensation insurance on behalf of Consultant or Consultant's employees. If Consultant hires employees to perform any work under this Agreement, Consultant will cover them with workers' compensation insurance and provide Client with a certificate of workers' compensation insurance before the employees begin the work.

(Optional: Check if applicable.)

☐ If not operating as a corporation, Consultant shall obtain workers' compensation insurance coverage for Consultant. Consultant shall provide Client with proof that such coverage has been obtained before starting work.

11. Unemployment Compensation

Client shall make no state or federal unemployment compensation payments on behalf of Consultant or Consultant's employees or contract personnel. Consultant will not be entitled to these benefits in connection with work performed under this Agreement.

12. Insurance

Client shall not provide any insurance coverage of any kind for Consultant or Consultant's employees or contract personnel. Consultant shall obtain and maintain a broad form Commercial General Liability Insurance policy providing for coverage of at least $_____ for each occurrence. Before commencing any work, Consultant shall provide Client with proof of this insurance and with proof that Client has been made an additional insured under the policy.

Consultant shall indemnify and hold Client harmless from any loss or liability arising from performing services under this Agreement.

☐ Consultant shall obtain professional liability insurance coverage for malpractice or errors or omissions committed by Consultant or Consultant's employees during the term of this Agreement. The policy shall provide for coverage of at least $_____ for each occurrence. Before commencing any work, Consultant shall provide Client with proof of this insurance.

13. Term of Agreement

This agreement will become effective when signed by both parties and will terminate on the earlier of:

- the date Consultant completes the services required by this Agreement
- _____ [date], or
- the date a party terminates the Agreement as provided below.

14. Terminating the Agreement

(Choose Alternative A or B.)

☐ ALTERNATIVE A

With reasonable cause, either Client or Consultant may terminate this Agreement, effective immediately upon giving written notice.

Reasonable cause includes:

- a material violation of this Agreement, or
- any act exposing the other party to liability to others for personal injury or property damage.

☐ ALTERNATIVE B

Either party may terminate this Agreement any time by giving _____ days' written notice to the other party of the intent to terminate.

15. Exclusive Agreement

This is the entire Agreement between Consultant and Client.

16. Modifying the Agreement (Optional)

(Check if applicable.)

☐ This Agreement may be modified only by a writing signed by both parties.

17. Intellectual Property Ownership (Optional)

(Check if applicable.)

☐ Consultant assigns to Client all patent, copyright, trademark, and trade secret rights in anything created or developed by Consultant for Client under this Agreement. Consultant shall help prepare any papers that Client considers necessary to secure any patents, copyrights, trademarks, or other proprietary rights at no charge to Client. However, Client shall reimburse Consultant for reasonable out-of-pocket expenses incurred.

Consultant must obtain written assurances from Consultant's employees and contract personnel that they agree with this assignment.

Consultant agrees not to use any of the intellectual property mentioned above for the benefit of any other party without Client's prior written permission.

18. Confidentiality (Optional)

(Check if applicable.)

☐ Consultant acknowledges that it will be necessary for Client to disclose certain confidential and proprietary information to Consultant in order for Consultant to perform duties under this Agreement. Consultant acknowledges that any disclosure to any third party or any misuse of this proprietary or confidential information would irreparably harm Client. Accordingly, Consultant will not disclose or use, either during or after the term of this Agreement, any proprietary or confidential information of Client without Client's prior written permission except to the extent necessary to perform services on Client's behalf.

Proprietary or confidential information includes:

- the written, printed, graphic, or electronically recorded materials furnished by Client for Consultant to use
- business or marketing plans or strategies, customer lists, operating procedures, trade secrets, design formulas, know-how and processes, computer programs and inventories, discoveries and improvements of any kind, sales projections, and pricing information
- information belonging to customers and suppliers of Client about whom Consultant gained knowledge as a result of Consultant's services to Client
- any written or tangible information stamped "confidential," "proprietary," or with a similar legend, and
- any information that Client makes reasonable efforts to maintain the secrecy of.

Consultant shall not be restricted in using any material which is publicly available, already in Consultant's possession or known to Consultant without restriction, or which is rightfully obtained by Consultant from sources other than Client.

Upon termination of Consultant's services to Client, or at Client's request, Consultant shall deliver to Client all materials in Consultant's possession relating to Client's business.

19. Resolving Disputes

(Choose Alternative A, B, or C and any desired optional clauses.)

☐ ALTERNATIVE A

If a dispute arises under this Agreement, any party may take the matter to court.

(Optional: Check if applicable.)

☐ If any court action is necessary to enforce this Agreement, the prevailing party shall be entitled to reasonable attorney fees, costs, and expenses in addition to any other relief to whichthe party may be entitled.

☐ ALTERNATIVE B

If a dispute arises under this Agreement, the parties agree to first try to resolve the dispute with the help of a mutually agreed-upon mediator in _____ [list city or county where mediation will occur]. Any costs and fees other than attorney fees associated with the mediation shall be shared equally by the parties. If the dispute is not resolved within 30 days after it is referred to the mediator, any party may take the matter to court.

(Optional: Check if applicable.)

☐ If any court action is necessary to enforce this Agreement, the prevailing party shall be entitled to reasonable attorney fees, costs, and expenses in addition to any other relief to which he or she may be entitled.

☐ ALTERNATIVE C

If a dispute arises under this Agreement, the parties agree to first try to resolve the dispute with the help of a mutually agreed-upon mediator in _____ [list city or county where mediation will occur].

Any costs and fees other than attorney fees associated with the mediation shall be shared equally by the parties. If it proves impossible to arrive at a mutually satisfactory solution through mediation, the parties agree to submit the dispute to a mutually agreed-upon arbitrator in _____ [list city or county where arbitration will occur]. Judgment upon the award rendered by the arbitrator may be entered in any court having jurisdiction to do so. Costs of arbitration, including attorney fees, will be allocated by the arbitrator.

20. Applicable Law

This Agreement will be governed by the laws of the state of _____ .

21. Notices

All notices and other communications in connection with this Agreement shall be in writing and shall be considered given as follows:

- when delivered personally to the recipient's address as stated on this Agreement
- three days after being deposited in the United States mail, with postage prepaid to the recipient's address as stated on this Agreement, or
- when sent by fax or electronic mail, such notice is effective upon receipt provided that a duplicate copy of the notice is promptly given by first class mail, or the recipient delivers a written confirmation of receipt.

22. No Partnership

This Agreement does not create a partnership relationship. Consultant does not have authority to enter into contracts on Client's behalf.

23. Assignment and Delegation (Optional)

(Check if applicable.)

☐ Consultant may not assign or subcontract any rights or obligations under this Agreement without Client's prior written approval.

Signatures

Client: _____
Name of Client

By: _____
Signature

_____
Typed or Printed Name

Title: _____

Date: _____

Consultant: _____
Name of Consultant

By: _____
Signature

_____
Typed or Printed Name

Title: _____

Taxpayer ID Number: _____

Date: _____

INDEPENDENT CONTRACTOR AGREEMENT FOR HOUSEHOLD WORKER

This Agreement is made between _____ ("Client")

whose address is _____

and _____ ("Contractor")

with a principal place of business at _____ .

1. Services to be Performed

(Check and complete applicable provisions.)

Contractor agrees to perform the following services:

a. Cleaning Interior

☐ Contractor will clean the following rooms and areas: _____

_____

_____

b. Cleaning Exterior

Contractor will clean the following:

☐ Front porch or deck: _____

☐ Back porch or deck: _____

☐ Garage: _____

☐ Pool, hot tub, or sauna:

_____

☐ Other exterior areas: _____

c. Gardening

☐ Contractor will perform the following gardening services: _____

_____

_____

d. Other Responsibilities

☐ Cooking: _____

☐ Laundry: _____

☐ Ironing: _____

☐ Shopping and errands: _____

☐ Other: _____

2. Payment

(Choose Alternative A or B.)

☐ ALTERNATIVE A

In consideration for the services to be performed by Contractor, Client agrees to pay Contractor $_____ .

☐ ALTERNATIVE B

In consideration for the services to be performed by Contractor, Client agrees to pay Contractor at the rate of $_____ per ☐ hour, ☐ day, ☐ week, ☐ month according to the terms of payment set forth below.

3. Terms of Payment

(Choose Alternative A or B.)

☐ ALTERNATIVE A

Upon completion of Contractor's services under this Agreement, Contractor shall submit an invoice. Client shall pay Contractor the compensation described within _____ [15, 30, 45, 60] days after receiving Contractor's invoice.

☐ ALTERNATIVE B

Contractor shall invoice Client on a monthly basis for all hours worked pursuant to this Agreement during the preceding month. Invoices shall specify an invoice number, the dates covered in the invoice, the hours expended, and the work performed (in summary) during the invoice period. Client shall pay Contractor's fee within _____ [15, 30, 45, 60] days after receiving Contractor's invoice.

4. Expenses

(Choose Alternative A or B.)

☐ ALTERNATIVE A

Contractor shall be responsible for all expenses incurred while performing services under this Agreement. This includes license fees, memberships, and dues; automobile and other travel expenses; meals and entertainment; insurance premiums; and all salary, expenses, and other compensation paid to employees or contract personnel the Contractor hires to complete the work under this Agreement.

☐ ALTERNATIVE B

Client shall reimburse Contractor for the following expenses that are attributable directly to work performed under this Agreement: _____

_____

Contractor shall submit an itemized statement of Contractor's expenses. Client shall pay Contractor within 30 days after receipt of each statement.

5. Materials (Optional)

(Check if applicable.)

☐ Contractor will furnish all equipment, tools, materials, and supplies needed to perform the services required by this Agreement.

6. Independent Contractor Status

Contractor is an independent contractor, not Client's employee. Contractor's employees or contract personnel are not Client's employees. Contractor and Client agree to the following rights consistent with an independent contractor relationship (check all that apply):

☐ Contractor has the right to perform services for others during the term of this Agreement.

☐ Contractor has the sole right to control and direct the means, manner, and method by which the services required by this Agreement will be performed.

☐ Contractor has the right to perform the services required by this Agreement at any place, location, or time.

☐ Contractor has the right to hire assistants as subcontractors or to use employees to provide the services required by this Agreement.

☐ The Contractor or Contractor's employees or contract personnel shall perform the services required by this Agreement; Client shall not hire, supervise, or pay any assistants to help Contractor.

☐ Neither Contractor nor Contractor's employees or contract personnel shall receive any training from Client in the skills necessary to perform the services required by this Agreement.

☐ Client shall not require Contractor or Contractor's employees or contract personnel to devote full time to performing the services required by this Agreement.

7. Work Schedule

Contractor shall perform the services at _____

during reasonable hours on a schedule to be mutually agreed upon by Client and Contractor based upon Client's needs and Contractor's availability to perform such services.

8. Business Permits, Certificates, and Licenses

Contractor has complied with all federal, state, and local laws requiring business permits, certificates, and licenses required to carry out the services to be performed under this Agreement.

9. State and Federal Taxes

Client will not:

- withhold FICA (Social Security and Medicare taxes) from Contractor's payments or make FICA payments on Contractor's behalf
- make state or federal unemployment compensation contributions on Contractor's behalf, or
- withhold state or federal income tax from Contractor's payments.

Contractor shall pay all taxes incurred while performing services under this Agreement—including all applicable income taxes and, if Contractor is not a corporation, self-employment (Social Security) taxes. Upon demand, Contractor shall provide Client with proof that such payments have been made.

10. Workers' Compensation

Client shall not obtain workers' compensation insurance on behalf of Contractor or Contractor's employees. If Contractor hires employees to perform any work under this Agreement, Contractor will cover them with workers' compensation insurance to the extent required by law and provide Client with a certificate of workers' compensation insurance before the employees begin the work.

11. Unemployment Compensation

Client shall make no state or federal unemployment compensation payments on behalf of Contractor or Contractor's employees or contract personnel. Contractor will not be entitled to these benefits in connection with work performed under this Agreement.

12. Term of Agreement

This agreement will become effective when signed by both parties and will terminate on the earlier of:

- the date Contractor completes the services required by this Agreement
- _____ [date], or
- the date a party terminates the Agreement as provided below.

13. Terminating the Agreement

(Choose Alternative A or B.)

☐ ALTERNATIVE A

With reasonable cause, either Client or Contractor may terminate this Agreement, effective immediately upon giving written notice.

Reasonable cause includes:

- a material violation of this Agreement, or
- any act exposing the other party to liability to others for personal injury or property damage.

☐ ALTERNATIVE B

Either party may terminate this Agreement any time by giving 30 days' written notice to the other party of the intent to terminate.

Signatures

Client: _____
Name of Client

By: _____
Signature

_____
Typed or Printed Name

Title: _____

Date: _____

Contractor: _____
Name of Contractor

By: _____
Signature

_____
Typed or Printed Name

Title: _____

Taxpayer ID Number: _____

Date: _____

INDEPENDENT CONTRACTOR AGREEMENT
FOR DIRECT SELLER

This Agreement is made between _____ ("Client")

with a principal place of business at _____

and _____ ("Contractor")

with a principal place of business at _____ .

1. Services to be Performed

Contractor agrees to sell the following product or merchandise for Client:

_____

(Optional: Check if applicable.)

☐ Contractor shall seek sales of the product in the homes of various individuals.

2. Compensation

In consideration for the services to be performed by Contractor, Client agrees to pay Contractor a commission on completed sales as follows:

_____

_____

_____

Contractor acknowledges that no other compensation is payable by Client and that all of Contractor's compensation will depend on sales made by Contractor. None of Contractor's compensation shall be based on the number of hours worked by Contractor.

3. Expenses

(Choose Alternative A or B.)

☐ ALTERNATIVE A

Contractor shall be responsible for all expenses incurred while performing services under this Agreement. This includes license fees, memberships and dues; automobile and other travel expenses; meals and entertainment; insurance premiums; and all salary, expenses, and other compensation paid to employees or contract personnel the Contractor hires to complete the work under this Agreement.

☐ ALTERNATIVE B

Client shall reimburse Contractor for the following expenses that are directly attributable to work performed under this Agreement: _____

_____

Contractor shall submit an itemized statement of Contractor's expenses. Client shall pay Contractor within 30 days after receipt of each statement.

4. Materials

☐ Contractor will furnish all materials, tools, and equipment used to provide the services required by this Agreement.

5. Independent Contractor Status

Contractor is an independent contractor, not Client's employee. Contractor's employees or contract personnel are not Client's employees. Contractor and Client agree to the following rights consistent with an independent contractor relationship (check all that apply):

☐ Contractor has the right to perform services for others during the term of this Agreement.

☐ Contractor shall have no obligation to perform any services other than the sale of the product described here.

☐ Contractor has the sole right to control and direct the means, manner, and method by which the services required by this Agreement will be performed. Consistent with this freedom from Client's control, Contractor:

  ☐ does not have to pursue or report on leads furnished by Client

  ☐ is not required to attend sales meetings organized by Client

  ☐ does not have to obtain Client's pre-approval for orders, and

  ☐ shall adopt and carry out Contractor's own sales strategy.

☐ Subject to any restrictions on Contractor's sales territory contained in this Agreement, Contractor has the right to perform the services required by this Agreement at any location or time.

☐ Contractor has the right to hire assistants as subcontractors or to use employees to provide the services required by this Agreement, except that Client may supply Contractor with sales forms.

☐ The Contractor or Contractor's employees or contract personnel shall perform the services required by this Agreement; Client shall not hire, supervise, or pay any assistants to help Contractor.

☐ Neither Contractor nor Contractor's employees or contract personnel shall receive any training from Client in the skills necessary to perform the services required by this Agreement.

☐ Client shall not require Contractor or Contractor's employees or contract personnel to devote full time to performing the services required by this Agreement.

6. Business Permits, Certificates, and Licenses

Contractor has complied with all federal, state, and local laws requiring business permits, certificates, and licenses required to carry out the services to be performed under this Agreement.

7. State and Federal Taxes

Client will not:

- withhold FICA (Social Security and Medicare taxes) from Contractor's payments or make FICA payments on Contractor's behalf

- make state or federal unemployment compensation contributions on Contractor's behalf, or

- withhold state or federal income tax from Contractor's payments.

Contractor shall pay all taxes incurred while performing services under this Agreement—including all applicable income taxes and, if Contractor is not a corporation, self-employment (Social Security) taxes. Upon demand, Contractor shall provide Client with proof that such payments have been made.

8. Fringe Benefits

Contractor understands that neither Contractor nor Contractor's employees or contract personnel are eligible to participate in any employee pension, health, vacation pay, sick pay, or other fringe benefit plan of Client.

9. Workers' Compensation

Client shall not obtain workers' compensation insurance on behalf of Contractor or Contractor's employees. If Contractor hires employees to perform any work under this Agreement, Contractor will cover them with workers' compensation insurance to the extent required by law and provide Client with a certificate of workers' compensation insurance before the employees begin the work.

(Optional: Check if applicable.)

☐ Contractor shall obtain workers' compensation insurance coverage for Contractor. Contractor shall provide Client with proof that such coverage has been obtained before starting work.

10. Unemployment Compensation

Client shall make no state or federal unemployment compensation payments on behalf of Contractor or Contractor's employees or contract personnel. Contractor will not be entitled to these benefits in connection with work performed under this Agreement.

11. Insurance

Client shall not provide any insurance coverage of any kind for Contractor or Contractor's employees or contract personnel.

Contractor shall obtain and maintain a broad form Commercial General Liability Insurance policy providing for coverage of at least $_____ for each occurrence and naming Client as an additional insured under the policy.

Contractor shall maintain automobile liability insurance for injuries to person and property, including coverage for all non-owned and rented automotive equipment, providing for coverage of at least $_____ for each person, $_____ for each accident, and $_____ for property damage.

Before commencing any work under this Agreement, Contractor shall provide Client with proof of this insurance.

Contractor shall indemnify and hold Client harmless from any loss or liability arising from performing services under this Agreement, including any claim for injuries or damages caused by Contractor while traveling in Contractor's automobile and performing services under this Agreement.

12. Confidentiality

Contractor acknowledges that it will be necessary for Client to disclose certain confidential and proprietary information to Contractor in order for Contractor to perform duties under this Agreement. Contractor acknowledges that any disclosure to any third party or any misuse of this proprietary or confidential information would irreparably harm Client. Accordingly, Contractor will not disclose or use, either during or after the term of this Agreement, any proprietary or confidential information of Client without Client's prior written permission except to the extent necessary to perform services on Client's behalf.

Proprietary or confidential information includes:

- the written, printed, graphic, or electronically recorded materials furnished by Client for Contractor to use

- business or marketing plans or strategies, customer lists, operating procedures, trade secrets, design formulas, know-how and processes, computer programs and inventories, discoveries and improvements of any kind, sales projections, and pricing information

- information belonging to customers and suppliers of Client about whom Contractor gained knowledge as a result of Contractor's services to Client

- any written or tangible information stamped "confidential," "proprietary," or with a similar legend, and

- any information that Client makes reasonable efforts to maintain the secrecy of.

Contractor shall not be restricted in using any material that is publicly available, already in Contractor's possession, known to Contractor without restriction, or rightfully obtained by Contractor from sources other than Client.

Upon termination of Contractor's services to Client, or at Client's request, Contractor shall deliver to Client all materials in Contractor's possession relating to Client's business.

13. Term of Agreement

This agreement will become effective when signed by both parties and will terminate on the earlier of:

- the date Contractor completes the services required by this Agreement

- _____ [date], or

- the date a party terminates the Agreement as provided below.

14. Terminating the Agreement

(Choose Alternative A or B.)

☐ ALTERNATIVE A

With reasonable cause, either Client or Contractor may terminate this Agreement, effective immediately upon giving written notice.

Reasonable cause includes:

- a material violation of this Agreement, or

- any act exposing the other party to liability to others for personal injury or property damage.

☐ ALTERNATIVE B

Either party may terminate this Agreement any time by giving thirty days' written notice to the other party of the intent to terminate.

15. Exclusive Agreement

This is the entire Agreement between Contractor and Client.

16. Resolving Disputes

(Choose Alternative A, B, or C and any desired optional clauses.)

☐ ALTERNATIVE A

If a dispute arises under this Agreement, any party may take the matter to court.

(Optional: Check if applicable.)

☐ If any court action is necessary to enforce this Agreement, the prevailing party shall be entitled to reasonable attorney fees, costs, and expenses in addition to any other relief to which the party may be entitled.

☐ ALTERNATIVE B

If a dispute arises under this Agreement, the parties agree to first try to resolve the dispute with the help of a mutually agreed-upon mediator in _____ [list city or county where mediation will occur]. Any costs and fees other than attorney fees associated with the mediation shall be shared equally by the parties. If the dispute is not resolved within 30 days after it is referred to the mediator, any party may take the matter to court.

(Optional: Check if applicable.)

☐ If any court action is necessary to enforce this Agreement, the prevailing party shall be entitled to reasonable attorney fees, costs, and expenses in addition to any other relief to which the party may be entitled.

☐ ALTERNATIVE C

If a dispute arises under this Agreement, the parties agree to first try to resolve the dispute with the help of a mutually agreed-upon mediator in _____ [list city or county where mediation will occur]. Any costs and fees other than attorney fees associated with the mediation shall be shared equally by the parties. If it proves impossible to arrive at a mutually satisfactory solution through mediation, the parties agree to submit the dispute to a mutually agreed-upon arbitrator in _____
[list city or county where arbitration will occur]. Judgment upon the award rendered by the arbitrator may be entered in any court having jurisdiction to do so. Costs of arbitration, including attorney fees, will be allocated by the arbitrator.

17. Applicable Law

This Agreement will be governed by the laws of the state of _____.

18. Notices

All notices and other communications in connection with this Agreement shall be in writing and shall be considered given as follows:

- when delivered personally to the recipient's address as stated on this Agreement
- three days after being deposited in the United States mail, with postage prepaid to the recipient's address as stated on this Agreement, or
- when sent by fax or electronic mail, such notice is effective upon receipt provided that a duplicate copy of the notice is promptly given by first class mail, or the recipient delivers a written confirmation of receipt.

19. No Partnership

This Agreement does not create a partnership relationship. Contractor does not have authority to enter into contracts on Client's behalf.

20. Assignment and Delegation (Optional)

(Check if applicable.)

☐ Either Contractor or Client may assign its rights and may delegate its duties under this Agreement.

Signatures

Client: _____

Name of Client

By: _____

Signature

_____

Typed or Printed Name

Title: _____

Date: _____

Contractor: _____

Name of Contractor

By: _____

Signature

_____

Typed or Printed Name

Title: _____

Taxpayer ID Number: _____

Date: _____

INDEPENDENT CONTRACTOR AGREEMENT FOR ACCOUNTANT & BOOKKEEPER

This Agreement is made between _____ ("Client")

with a principal place of business at _____

and _____ ("Contractor")

with a principal place of business at _____ .

1. Services to be Performed

(Choose Alternative A or B.)

☐ ALTERNATIVE A

Contractor agrees to perform the following services:

_____

_____

_____

☐ ALTERNATIVE B

Contractor agrees to perform the services described in Exhibit A attached to this Agreement.

2. Payment

(Choose Alternative A or B and optional clause, if desired.)

☐ ALTERNATIVE A

In consideration for the services to be performed by Contractor, Client agrees to pay Contractor

$_____ according to the terms of payment set forth below.

☐ ALTERNATIVE B

In consideration for the services to be performed by Contractor, Client agrees to pay Contractor at the rate of

$_____ per hour according to the terms of payment set forth below.

(Optional: Check and complete if applicable.)

☐ Unless otherwise agreed upon in writing by Client, Client's maximum liability for all services performed during the term of this Agreement shall not exceed $_____.

3. Terms of Payment

(Choose Alternative A, B, or C.)

☐ ALTERNATIVE A

Upon completion of Contractor's services under this Agreement, Contractor shall submit an invoice. Client shall pay Contractor the compensation described within _____ [15, 30, 45, 60] days after receiving Contractor's invoice.

☐ ALTERNATIVE B

Contractor shall be paid $_____ upon the signing of this Agreement and the remainder of the compensation described above upon completion of Contractor's services and submissilon of an invoice.

☐ ALTERNATIVE C

Contractor shall invoice Client on a monthly basis for all hours worked pursuant to this Agreement during the preceding month. Invoices shall be submitted on Contractor's letterhead specifying an invoice number, the dates covered in the invoice, the hours expended, and the work performed (in summary) during the invoice period. Client shall pay Contractor's fee within _____ days after receiving Contractor's invoice.

4. Expenses

Client shall reimburse Contractor for the following expenses that are attributable directly to work performed under this Agreement:

- travel expenses other than normal commuting, including airfares, rental vehicles, and highway mileage in company or personal vehicles at _____ cents per mile
- telephone, fax, online, and telegraph charges
- postage and courier services
- printing and reproduction
- computer services, and
- other expenses resulting from the work performed under this Agreement.

Contractor shall submit an itemized statement of Contractor's expenses. Client shall pay Contractor within 30 days after receipt of each statement.

5. Materials

Client shall make available to Contractor, at Client's expense, the following materials:

These items shall be provided to Contractor by _____

6. Independent Contractor Status

Contractor is an independent contractor, and neither Contractor nor Contractor's employees or contract personnel are, or shall be deemed, Client's employees. In its capacity as an independent contractor, contractor agrees and represents, and Client agrees, as follows (check all that apply):

☐ Contractor has the right to perform services for others during the term of this Agreement.

☐ Contractor has the sole right to control and direct the means, manner, and method by which the services required by this Agreement will be performed.

☐ Contractor has the right to perform the services required by this Agreement at any place or location and at such times as Contractor may determine.

☐ Contractor has the right to hire assistants as subcontractors or to use employees to provide the services required by this Agreement.

☐ The services required by this Agreement shall be performed by Contractor, Contractor's employees, or contract personnnel, and Client shall not hire, supervise, or pay any assistants to help Contractor.

☐ Neither Contractor nor Contractor's employees or contract personnel shall receive any training from Client in the professional skills necessary to perform the services required by this Agreement.

☐ Neither Contractor nor Contractor's employees or contract personnel shall be required by Client to devote full time to the performance of the services required by this Agreement.

7. Business Permits, Certificates, and Licenses

Contractor has complied with all federal, state, and local laws requiring business permits, certificates, and licenses required to carry out the services to be performed under this Agreement.

8. Professional Obligations

Contractor shall perform all services under this Agreement in accordance with generally accepted accounting practices and principles. This Agreement is subject to the laws, rules, and regulations governing the accounting profession imposed by government authorities or professional associations of which Contractor is a member.

9. Insurance

Client shall not provide any insurance coverage of any kind for Contractor or Contractor's employees or contract personnel. Contractor shall maintain a broad form Commercial General Liability Insurance policy providing for coverage of at least $_____ for each occurrence. Before commencing any work, Contractor shall provide Client with proof of this insurance and proof that Client has been made an additional insured under the policy.

(Optional: Check and complete if applicable.)

☐ Contractor shall maintain an errors and omission Insurance policy providing for coverage of at least $_____ for each occurrence. Before commencing any work, Contractor shall provide Client with proof of this insurance.

10. Term of Agreement

This Agreement will become effective on _____ and will end no later than _____ .

11. Terminating the Agreement

(Choose Alternative A or B.)

☐ ALTERNATIVE A

With reasonable cause, either Client or Contractor may terminate this Agreement, effective immediately upon giving written notice.

Reasonable cause includes:

- a material violation of this Agreement, or
- any act exposing the other party to liability to others for personal injury or property damage.

☐ ALTERNATIVE B

Either party may terminate this Agreement at any time by giving thirty days' written notice to the other party of the intent to terminate.

12. Exclusive Agreement

This is the entire Agreement between Contractor and Client.

13. Resolving Disputes

(Choose Alternative A, B, or C and any desired optional clauses.)

☐ ALTERNATIVE A

If a dispute arises under this Agreement, any party may take the matter to court.

(Optional: Check if applicable.)

☐ If any court action is necessary to enforce this Agreement, the prevailing party shall be entitled to reasonable attorney fees, costs, and expenses in addition to any other relief to which the party may be entitled.

☐ ALTERNATIVE B

If a dispute arises under this Agreement, the parties agree to first try to resolve the dispute with the help of a mutually agreed-upon mediator in _____ [list city or county where mediation will occur]. Any costs and fees other than attorney fees associated with the mediation shall be shared equally by the parties. If the dispute is not resolved within 30 days after it is referred to the mediator, any party may take the matter to court.

(Optional: Check if applicable.)

☐ If any court action is necessary to enforce this Agreement, the prevailing party shall be entitled to reasonable attorney fees, costs, and expenses in addition to any other relief to which the party may be entitled.

☐ ALTERNATIVE C

If a dispute arises under this Agreement, the parties agree to first try to resolve the dispute with the help of a mutually agreed-upon mediator in _____ [list city or county where mediation will occur]. Any costs and fees other than attorney fees associated with the mediation shall be shared equally by the parties. If it proves impossible to arrive at a mutually satisfactory solution through mediation, the parties agree to submit the dispute to a mutually agreed-upon arbitrator in _____ [list city or county where arbitration will occur]. Judgment upon the award rendered by the arbitrator may be entered in any court having jurisdiction to do so. Costs of arbitration, including attorney fees, will be allocated by the arbitrator.

14. Applicable Law

This Agreement will be governed by the laws of the state of _____.

15. Notices

All notices and other communications in connection with this Agreement shall be in writing and shall be considered goven as follows:

- when delivered personally to the recipient's address as stated on this Agreement
- three days after being deposited in the United States mail, with postage prepaid to the recipient's address as stated on this Agreement, or
- when sent by fax or electronic mail, such notice is effective upon receipt provided that a duplicate copy of the notice is promptly given by first class mail, or the recipient delivers a written confirmation of receipt.

16. No Partnership

This Agreement does not create a partnership relationship. Contractor does not have authority to enter into contracts on Client's behalf.

17. Assignment and Delegation (Optional)

(Check if applicable.)

(Choose Alternative A or B.)

☐ ALTERNATIVE A

Either Contractor or Client may assign rights and may delegate duties under this Agreement.

☐ ALTERNATIVE B

Contractor may not assign or subcontract any rights or delegate any duties under this Agreement without Client's prior written approval.

Signatures

Client: _____

Name of Client

By: _____

Signature

_____

Typed or Printed Name

Title: _____

Date: _____

Contractor: _____

Name of Contractor

By: _____

Signature

_____

Typed or Printed Name

Title: _____

Taxpayer ID Number: _____

Date: _____

INDEPENDENT CONTRACTOR AGREEMENT
FOR SOFTWARE CONSULTANT

This Agreement is made between _____ ("Client")

with a principal place of business at _____

and _____ ("Consultant")

with a principal place of business at _____ .

1. Services to Be Performed:

(Choose Alternative A or B.)

☐ ALTERNATIVE A

Consultant agrees to perform the following services for Client:

_____

_____

_____

☐ ALTERNATIVE B

Consultant agrees to perform the services described in Exhibit A, which is attached to and made part of this Agreement.

2. Payment

(Choose Alternative A, B, or C and optional clause, if desired.)

☐ ALTERNATIVE A

Consultant shall be paid $_____ upon completion of the work as detailed in Clause 1.

☐ ALTERNATIVE B

Client shall pay Consultant a fixed fee of $_____, in _____ installments as follows:

   (a) $_____ upon completion of the following services: _____

      _____

   (b) $_____ upon completion of the following services: _____

      _____

   (c) $_____ upon completion of all the remaining work to be performed and the services to be rendered in accordance with the schedule set forth in Clause 1, above, and written acceptance by Client.

☐ ALTERNATIVE C

Consultant shall be compensated at the rate of $_____ per _____ ☐ hour, ☐ day, ☐ week, ☐ month.

(Optional: Check and complete if applicable.)

☐ Unless otherwise agreed upon in writing by Client, Client's maximum liability for all services performed during the term of this Agreement shall not exceed $_____ .

3. Expenses

(Choose Alternative A or B.)

☐ ALTERNATIVE A

Consultant shall be responsible for all expenses incurred while performing services under this Agreement.

☐ ALTERNATIVE B

Consultant will not be reimbursed for any expenses incurred in connection with the performance of services under this Agreement, unless those expenses are approved in advance in writing by Client.

4. Invoices

Consultant shall submit invoices for all services rendered. Client shall pay Consultant within _____ days after receipt of each invoice.

5. Independent Contractor Status

Consultant is an independent contractor, and neither Consultant nor Consultant's staff is, or shall be deemed, Client's employees. In its capacity as an independent contractor, Consultant agrees and represents, and Client agrees, as follows (check all that apply):

☐ Consultant has the right to perform services for others during the term of this Agreement subject to noncompetition provisions set out in this Agreement, if any.

☐ Consultant has the sole right to control and direct the means, manner, and method by which the services required by this Agreement will be performed.

☐ Consultant has the right to perform the services required by this Agreement at any place or location and at such times as Consultant may determine.

☐ Consultant will furnish all equipment and materials used to provide the services required by this Agreement, except to the extent that Consultant's work must be performed on or with Client's computer or existing software.

☐ The services required by this Agreement shall be performed by Consultant or Consultant's staff, and Client shall not be required to hire, supervise, or pay any assistants to help Consultant.

☐ Consultant is responsible for paying all ordinary and necessary expenses of its staff.

☐ Neither Consultant nor Consultant's staff shall receive any training from Client in the professional skills necessary to perform the services required by this Agreement.

☐ Neither Consultant nor Consultant's staff shall be required to devote full time to the performance of the services required by this Agreement.

☐ Client shall not withhold from Consultant's compensation any amount that would normally be withheld from an employee's pay.

6. Intellectual Property Ownership

(Choose Alternative A or B.)

☐ ALTERNATIVE A

Work Product includes, but is not limited to, the programs and documentation, including all ideas, routines, object and source codes, specifications, flow charts, and other materials, in whatever form, developed solely for Client under this Agreement.

Consultant hereby assigns to Client its entire right, title, and interest, including all patent, copyright, trade secret, trademark, and other proprietary rights, in the Work Product.

Consultant shall, at no charge to Client, execute and aid in the preparation of any papers that Client may consider necessary or helpful to obtain or maintain—at Client's expense—any patents, copyrights, trademarks, or other proprietary rights. Client shall reimburse Consultant for reasonable out-of-pocket expenses incurred under this provision.

☐ ALTERNATIVE B

Work Product includes, but is not limited to, the programs and documentation, including all ideas, routines, object and source codes, specifications, flow charts, and other materials, in whatever form, developed solely for Client under this Agreement.

Client agrees that Consultant shall retain any and all rights Consultant may have in the Work Product.

Consultant hereby grants Client an unrestricted, nonexclusive, perpetual, fully paid-up, worldwide license to use and sublicense the use of the Work Product for the purpose of developing and marketing its products, but not for the purpose of marketing Work Product separate from its products.

7. Ownership of Consultant's Materials

Consultant's Materials means all programs and documentation, including routines, object and source codes, tools, utilities, and other copyrightable materials, that:

- do not constitute Work Product
- are incorporated into the Work Product, and
- are owned solely by Consultant or licensed to Consultant with a right to sublicense.

Consultant's Materials include, but are not limited to, the following: _____

_____

_____

Consultant shall retain any and all rights Consultant may have in Consultant's Materials. Consultant hereby grants Client an unrestricted, nonexclusive, perpetual, fully paid-up, worldwide license to use and sublicense the use of Consultant's Materials for the purpose of developing and marketing its products.

8. Confidential Information (Optional)

(Check if applicable.)

☐ Consultant agrees that the Work Product is Client's sole and exclusive property. Consultant shall treat the Work Product on a confidential basis and not disclose it to any third party without Client's written consent, except when reasonably necessary to perform the services under this Agreement.

Consultant acknowledges that it will be necessary for Client to disclose certain confidential and proprietary information to Consultant in order for Consultant to perform duties under this Agreement. Consultant acknowledges that any disclosure to any third party or any misuse of this proprietary or confidential information would irreparably harm Client. Accordingly, Consultant will not use or disclose to others without Client's written consent Client's confidential information, except when reasonably necessary to perform the services under this Agreement. Confidential Information includes, but is not limited to:

- the written, printed, graphic, or electronically recorded materials furnished by Client for use by Contractor
- Client's business or marketing plans or strategies, customer lists, operating procedures, trade secrets, design formulas, know-how and processes, computer programs and inventories, discoveries, and improvements of any kind
- any written or tangible information stamped "confidential," "proprietary," or with a similar legend, and
- any written or tangible information not marked with a confidentiality legend, or information disclosed orally to Consultant, that is treated as confidential when disclosed and later summarized sufficiently for identification purposes in a written memorandum marked "confidential" and delivered to Consultant within 30 days after the disclosure.

Consultant shall not be restricted in the use of any material that is publicly available, already in Consultant's possession, known to Consultant without restriction, or rightfully obtained by Consultant from sources other than Client.

Consultant's obligations regarding proprietary or confidential information extend to information belonging to customers and suppliers of Client about whom Consultant may have gained knowledge as a result of Consultant's services to Client.

Consultant will not disclose to Client information or material that is a trade secret of any third party.

The provisions of this clause shall survive any termination of this Agreement.

9. Term of Agreement

This agreement will become effective when signed by both parties and will terminate on the earlier of:

- the date Consultant completes the services required by this Agreement
- _____ [date], or
- the date a party terminates the Agreement as provided below.

10. Termination of Agreement

- Each party has the right to terminate this Agreement if the other party has materially breached any obligation herein and such breach remains uncured for a period of 30 days after notice thereof is sent to the other party.
- If at any time after commencement of the services required by this Agreement, Client shall, in its sole reasonable judgment, determine that such services are inadequate, unsatisfactory, no longer needed, or substantially not conforming to the descriptions, warranties, or representations contained in this Agreement, Client may terminate this Agreement upon _____ days' written notice to Consultant.

11. Return of Materials

Upon termination of this Agreement, each party shall promptly return to the other all data, materials, and other property of the other held by it.

12. Warranties and Representations

Consultant warrants and represents that:

- Consultant has the authority to enter into this Agreement and to perform all obligations hereunder.
- The Work Product and Consultant's Materials are and shall be free and clear of all encumbrances including security interests, licenses, liens, or other restrictions except as follows: _____
- The use, reproduction, distribution, or modification of the Work Product and Consultant's Materials does not and will not violate the copyright, patent, trade secret, or other property right of any former client, employer, or third party.
- For a period of _____ days following acceptance of the Work Product, the Work Product will be:
  - free from reproducible programming errors and defects in workmanship and materials under normal use, and
  - substantially in conformance with the product specifications.
- The Work Product shall be created solely by Consultant, Consultant's employees during the course of their employment, or independent contractors who assigned all right, title, and interest in the work to Consultant.

13. Indemnification

Consultant agrees to indemnify and hold harmless Client against any claims, actions, or demands, including without limitation reasonable attorney and accounting fees, alleging or resulting from the breach of the warranties contained in this Agreement. Client shall provide notice to Consultant promptly of any such claim, suit, or proceeding and shall assist Consultant, at Consultant's expense, in defending any such claim, suit, or proceeding.

14. Assignment and Delegation (Optional)

(Check if applicable.)

☐ Consultant may not assign or subcontract any rights or obligations under this Agreement without Client's prior written approval.

15. Insurance

Client shall not provide any insurance coverage of any kind for Consultant or Consultant's employees or contract personnel. Consultant shall obtain and maintain a broad form Commercial General Liability Insurance policy providing for coverage of at least $_____ for each occurrence. Before commencing any work, Consultant shall provide Client with proof of this insurance and with proof that Client has been made an additional insured under the policy.

(Optional: Check and complete if applicable.)

☐ Consultant shall obtain professional liability insurance coverage for malpractice or errors or omissions committed by Consultant or Consultant's employees during the term of this Agreement. The policy shall provide for coverage of at least $_____ for each occurrence. Before commencing any work, Consultant shall provide Client with proof of this insurance.

16. Resolving Disputes

(Choose Alternative A, B, or C and any desired optional clauses.)

☐ ALTERNATIVE A

If a dispute arises under this Agreement, any party may take the matter to court.

(Optional: Check if applicable.)

☐ If any court action is necessary to enforce this Agreement, the prevailing party shall be entitled to reasonable attorney fees, costs, and expenses in addition to any other relief to which the party may be entitled.

☐ ALTERNATIVE B

If a dispute arises under this Agreement, the parties agree to first try to resolve the dispute with the help of a mutually agreed-upon mediator in _____ [list city or county where mediation will occur]. Any costs and fees other than attorney fees associated with the mediation shall be shared equally by the parties. If the dispute is not resolved within 30 days after it is referred to the mediator, any party may take the matter to court.

(Optional: Check if applicable.)

☐ If any court action is necessary to enforce this Agreement, the prevailing party shall be entitled to reasonable attorney fees, costs, and expenses in addition to any other relief to which the party may be entitled.

☐ ALTERNATIVE C

If a dispute arises under this Agreement, the parties agree to first try to resolve the dispute with the help of a mutually agreed-upon mediator in _____ [list city or county where mediation will occur]. Any costs and fees other than attorney fees associated with the mediation shall be shared equally by the parties. If it proves impossible to arrive at a mutually satisfactory solution through mediation, the parties agree to submit the dispute to a mutually agreed-upon arbitrator in _____ [list city or county where arbitration will occur]. Judgment upon the award rendered by the arbitrator may be entered in any court having jurisdiction to do so. Costs of arbitration, including attorney fees, will be allocated by the arbitrator.

17. Exclusive Agreement

This is the entire Agreement between Consultant and Client.

18. Applicable Law

This Agreement will be governed by the laws of the state of _____ .

19. Notices

All notices and other communications in connection with this Agreement shall be in writing and shall be considered given as follows:

- when delivered personally to the recipient's address as stated on this Agreement
- three days after being deposited in the United States mail, with postage prepaid to the recipient's address as stated on this Agreement, or
- when sent by fax or electronic mail, such notice is effective upon receipt provided that a duplicate copy of the notice is promptly given by first class mail, or the recipient delivers a written confirmation of receipt.

20. No Partnership

This Agreement does not create a partnership relationship. Consultant does not have authority to enter into contracts on Client's behalf.

Signatures

Client: _____
Name of Client

By: _____
Signature

_____
Typed or Printed Name

Title:_____

Date: _____

Consultant:_____
Name of Consultant

By: _____
Signature

_____
Typed or Printed Name

Title:_____

Taxpayer ID Number: _____

Date: _____

INDEPENDENT CONTRACTOR AGREEMENT FOR CREATIVE CONTRACTOR

This Agreement is made between _____ ("Client")

with a principal place of business at _____

and _____ ("Contractor")

with a principal place of business at _____ .

1. Services to be Performed

(Choose Alternative A or B.)

☐ ALTERNATIVE A

Contractor agrees to perform the following services:

_____

_____

_____

☐ ALTERNATIVE B

Contractor agrees to perform the services described in Exhibit A attached to and made part of this Agreement.

2. Payment

(Choose Alternative A or B and optional clause, if applicable.)

☐ ALTERNATIVE A

In consideration for the services to be performed by Contractor, Client agrees to pay Contractor

$_____ according to the terms of payment set forth below.

☐ ALTERNATIVE B

In consideration for the services to be performed by Contractor, Client agrees to pay Contractor at the rate of

$_____ per ☐ hour, ☐ day, ☐ week, ☐ month according to the terms of payment set forth below.

(Optional: Check and complete if applicable.)

☐ Unless otherwise agreed upon in writing by Client, Client's maximum liability for all services performed during the term of this Agreement shall not exceed $_____.

3. Terms of Payment

(Choose Alternative A, B, C, or D.)

☐ ALTERNATIVE A

Upon completion of Contractor's services under this Agreement, Contractor shall submit an invoice. Client shall pay Contractor the compensation described within _____ days after receiving Contractor's invoice.

☐ ALTERNATIVE B

Contractor shall be paid $_____ upon the signing of this Agreement and the remainder of the compensation described above upon completion of Contractor's services and submission of an invoice.

☐ ALTERNATIVE C

Client shall pay Contractor according to the schedule of payments set forth in Exhibit _____ attached to this Agreement.

☐ ALTERNATIVE D

Contractor shall invoice Client on a monthly basis for all hours worked pursuant to this Agreement during the preceding month. Invoices shall be submitted on Contractor's letterhead specifying an invoice number, the dates covered in the invoice, the hours expended, and the work performed (in summary) during the invoice period. Client shall pay Contractor's fee within _____ [15, 30, 45, 60] days after receiving Contractor's invoice.

4. Expenses

(Choose Alternative A or B.)

☐ ALTERNATIVE A

Contractor shall be responsible for all expenses incurred while performing services under this Agreement.

☐ ALTERNATIVE B

Client shall reimburse Contractor for the following expenses that are directly attributable to work performed under this Agreement: _____

_____

_____

Contractor shall submit an itemized statement of Contractor's expenses. Client shall pay Contractor within 30 days after receipt of each statement.

5. Materials

Contractor will furnish all materials, tools, and equipment used to provide the services required by this Agreement.

6. Intellectual Property Ownership

(Choose Alternative A, B, or C.)

☐ ALTERNATIVE A

Contractor assigns to Client Contractor's entire right, title, and interest in anything created or developed by Contractor for Client under this Agreement. Contractor shall help prepare any papers that Client considers necessary to secure any copyrights, trademarks, or other proprietary rights at no charge to Client. However, Client shall reimburse Contractor for reasonable out-of-pocket expenses incurred.

Contractor must obtain written assurances from Contractor's employees and contract personnel that they agree with this assignment.

Contractor agrees not to use any of the intellectual property mentioned above for the benefit of any other party without Client's prior written permission.

☐ ALTERNATIVE B

To the extent that the work performed by Contractor under this Agreement ("Contractor's Work") includes any work of Authorship entitled to protection under the copyright laws, the parties agree to the following provisions.

- Contractor's Work has been specially ordered and commissioned by Client as a contribution to a collective work, a supplementary work, or other category of work eligible to be treated as a work made for hire under the United States Copyright Act.

- Contractor's Work shall be deemed a commissioned work and a work made for hire to the greatest extent permitted by law.
- Client shall be the sole author of Contractor's Work and any work embodying the Contractor's Work according to the United States Copyright Act.
- To the extent that Contractor's Work is not properly characterized as a work made for hire, Contractor grants to Client all right, title, and interest in Contractor's Work, including all copyright rights, in perpetuity and throughout the world.
- Contractor shall help prepare any papers Client considers necessary to secure any copyrights, patents, trademarks, or intellectual property rights at no charge to Client. However, Client shall reimburse Contractor for reasonable out-of-pocket expenses incurred.

Contractor agrees not to use any of the intellectual property mentioned above for the benefit of any other party without Client's prior written permission.

☐ ALTERNATIVE C

Contractor assigns to Client the following intellectual property rights in the work created or developed by Contractor under this Agreement: _____

_____

_____

_____

_____

(Check applicable provision.)

    ☐ The rights granted above are exclusive to Client.

OR:

    ☐ The rights granted above are nonexclusive.

7. Releases (Optional)

(Check if applicable.)

☐ Contractor shall obtain all necessary copyright permissions and privacy releases for materials included in the Images at Client's request. Contractor shall indemnify Client against all claims and expenses, including reasonable attorney fees, due to Contractor's failure to obtain such permissions or releases.

8. Moral Rights Waiver for Works of Fine Art (Optional)

(Check if applicable.)

☐ Contractor waives any and all moral rights or any similar rights in the work created or developed by Contractor under this Agreement ("Work Product") and agrees not to institute, support, maintain, or permit any action or lawsuit on the grounds that Client's use of the Work Product:
- constitutes an infringement of any moral right or any similar right
- is in any way a defamation or mutilation of the Work Product
- damages Contractor's reputation, or
- contains unauthorized variations, alterations, changes, or translations of the Work Product.

9. Independent Contractor Status

Contractor is an independent contractor, and neither Contractor nor Contractor's employees or contract personnel are, or shall be deemed, Client's employees. In its capacity as an independent contractor, Contractor agrees and represents, and Client agrees, as follows (check all that apply):

☐ Contractor has the right to perform services for others during the term of this Agreement.

☐ Contractor has the sole right to control and direct the means, manner, and method by which the services required by this Agreement will be performed.

☐ Contractor has the right to perform the services required by this Agreement at any place or location and at such times as Contractor may determine.

☐ Contractor has the right to hire assistants as subcontractors or to use employees to provide the services required by this Agreement.

☐ The services required by this Agreement shall be performed by Contractor, Contractor's employees, or contract personnel, and Client shall not hire, supervise, or pay any assistants to help Contractor.

☐ Neither Contractor nor Contractor's employees or contract personnel shall receive any training from Client in the professional skills necessary to perform the services required by this Agreement.

☐ Neither Contractor nor Contractor's employees or contract personnel shall be required by Client to devote full time to the performance of the services required by this Agreement.

10. Business Permits, Certificates, and Licenses

Contractor has complied with all federal, state, and local laws requiring business permits, certificates, and licenses required to carry out the services to be performed under this Agreement.

11. State and Federal Taxes

Client will not:

- withhold FICA (Social Security and Medicare taxes) from Contractor's payments or make FICA payments on Contractor's behalf
- make state or federal unemployment compensation contributions on Contractor's behalf, or
- withhold state or federal income tax from Contractor's payments.

Contractor shall pay all taxes incurred while performing services under this Agreement—including all applicable income taxes and, if Contractor is not a corporation, self-employment (Social Security) taxes. Upon demand, Contractor shall provide Client with proof that such payments have been made.

12. Fringe Benefits

Contractor understands that neither Contractor nor Contractor's employees or contract personnel are eligible to participate in any employee pension, health, vacation pay, sick pay, or other fringe benefit plan of Client.

13. Workers' Compensation

Client shall not obtain workers' compensation insurance on behalf of Contractor or Contractor's employees. If Contractor hires employees to perform any work under this Agreement, Contractor will cover them with workers' compensation insurance to the extent required by law and will provide Client with a certificate of workers' compensation insurance before the employees begin the work.

☐ If not operating as a corporation, Contractor shall obtain workers' compensation insurance coverage for Contractor. Contractor shall provide Client with proof that such coverage has been obtained before starting work.

14. Unemployment Compensation

Client shall make no state or federal unemployment compensation payments on behalf of Contractor or Contractor's employees or contract personnel. Contractor will not be entitled to these benefits in connection with work performed under this Agreement.

15. Contractor's Insurance

Client shall not provide any insurance coverage of any kind for Contractor or Contractor's employees or contract personnel. Contractor shall obtain and maintain a broad form Commercial General Liability Insurance policy providing for coverage of at least $_____ for each occurrence. Before commencing any work, Contractor shall provide Client with proof of this insurance and with proof that Client has been made an additional insured under the policy.

Contractor shall indemnify and hold Client harmless from any loss or liability arising from performing services under this Agreement.

16. Client's Liability

Client shall exercise the same care with regard to any materials or work product belonging to Contractor that are in Client's possession as it would for its own property. However, if any property received from Contractor is lost or damaged, Client's liability shall be limited to the coverage available under its business liability policy in force at the time this Agreement is signed.

17. Warranties and Representations

Contractor warrants and represents that:

- Contractor is free to enter into this Agreement
- the work created or developed by Contractor under this Agreement ("Work Product") shall be original or all necessary permissions and releases obtained and paid for, and
- Contractor's Work Product shall not infringe upon any copyright or other proprietary right of any other person or entity.

Contractor agrees to indemnify Client for loss, liability, or expense resulting from actual breach of these Warranties.

18. Term of Agreement

This agreement will become effective when signed by both parties and will terminate on the earlier of:

- the date Contractor completes the services required by this Agreement
- _____ [date], or
- the date a party terminates the Agreement as provided below.

19. Terminating the Agreement

(Choose Alternative A or B.)

☐ ALTERNATIVE A

With reasonable cause, either Client or Contractor may terminate this Agreement, effective immediately upon giving written notice.

Reasonable cause includes:

• a material violation of this Agreement, or

• any act exposing the other party to liability to others for personal injury or property damage.

☐ ALTERNATIVE B

Either party may terminate this Agreement any time by giving thirty days' written notice to the other party of the intent to terminate.

20. Exclusive Agreement

This is the entire Agreement between Contractor and Client.

(Optional: Check if applicable.)

21. Modifying the Agreement

☐ This Agreement may be modified only by a writing signed by both parties.

(Optional: Check if applicable.)

22. Confidentiality

☐ Contractor acknowledges that it will be necessary for Client to disclose certain confidential and proprietary information to Contractor in order for Contractor to perform duties under this Agreement. Contractor acknowledges that any disclosure to any third party or any misuse of this proprietary or confidential information would irreparably harm Client. Accordingly, Contractor will not disclose or use, either during or after the term of this Agreement, any proprietary or confidential information of Client without Client's prior written permission except to the extent necessary to perform services on Client's behalf.

Proprietary or confidential information includes:

• the written, printed, graphic, or electronically recorded materials furnished by Client for Contractor to use

• business or marketing plans or strategies, customer lists, operating procedures, trade secrets, design formulas, know-how and processes, computer programs and inventories, discoveries and improvements of any kind, sales projections, and pricing information

• information belonging to customers and suppliers of Client about whom Contractor gained knowledge as a result of Contractor's services to Client, and

• any written or tangible information stamped "confidential," "proprietary," or with a similar legend, or any information that Client makes reasonable efforts to maintain the secrecy of.

Contractor shall not be restricted in using any material that is publicly available, already in Contractor's possession or known to Contractor without restriction, or that is rightfully obtained by Contractor from sources other than Client.

Upon termination of Contractor's services to Client, or at Client's request, Contractor shall deliver to Client all materials in Contractor's possession relating to Client's business.

23. Resolving Disputes

(Choose Alternative A, B, or C and any desired optional clauses.)

☐ ALTERNATIVE A

If a dispute arises under this Agreement, any party may take the matter to court.

(Optional: Check if applicable.)

☐ If any court action is necessary to enforce this Agreement, the prevailing party shall be entitled to reasonable attorney fees, costs, and expenses in addition to any other relief to which the party may be entitled.

☐ ALTERNATIVE B

If a dispute arises under this Agreement, the parties agree to first try to resolve the dispute with the help of a mutually agreed-upon mediator in _____ [list city or county where mediation will occur]. Any costs and fees other than attorney fees associated with the mediation shall be shared equally by the parties. If the dispute is not resolved within 30 days after it is referred to the mediator, any party may take the matter to court.

(Optional: Check if applicable.)

☐ If any court action is necessary to enforce this Agreement, the prevailing party shall be entitled to reasonable attorney fees, costs, and expenses in addition to any other relief to which the party may be entitled.

☐ ALTERNATIVE C

If a dispute arises under this Agreement, the parties agree to first try to resolve the dispute with the help of a mutually agreed-upon mediator in _____ [list city or county where mediation will occur]. Any costs and fees other than attorney fees associated with the mediation shall be shared equally by the parties. If it proves impossible to arrive at a mutually satisfactory solution through mediation, the parties agree to submit the dispute to a mutually agreed-upon arbitrator in _____ _____ [list city or county where arbitration will occur]. Judgment upon the award rendered by the arbitrator may be entered in any court having jurisdiction to do so. Costs of arbitration, including attorney fees, will be allocated by the arbitrator.

24. Applicable Law

This Agreement will be governed by the laws of the state of _____ .

25. Notices

All notices and other communications in connection with this Agreement shall be in writing and shall be considered given as follows:

- when delivered personally to the recipient's address as stated on this Agreement
- three days after being deposited in the United States mail, with postage prepaid to the recipient's address as stated on this Agreement, or
- when sent by fax or electronic mail, such notice is effective upon receipt provided that a duplicate copy of the notice is promptly given by first class mail, or the recipient delivers a written confirmation of receipt.

26. No Partnership

This Agreement does not create a partnership relationship. Contractor does not have authority to enter into contracts on Client's behalf.

27. Assignment and Delegation (Optional)

(Check if applicable.)

☐ Contractor may not assign or subcontract any rights or obligations under this Agreement without Client's prior written approval.

Signatures

Client: _____
Name of Client

By: _____
Signature

_____
Typed or Printed Name

Title: _____

Date: _____

Contractor: _____
Name of Contractor

By: _____
Signature

_____
Typed or Printed Name

Title: _____

Taxpayer ID Number: _____

Date: _____

INDEPENDENT CONTRACTOR AGREEMENT FOR CONSTRUCTION CONTRACTOR

This Agreement is made between _____ ("Client")

with a principal place of business at _____

and _____ ("Contractor")

with a principal place of business at _____ .

1. Services to Be Performed

Contractor shall furnish all labor and materials to construct and complete the project shown on the contract documents contained in Exhibit A, which is attached to and made part of this Agreement.

2. Payment

(Choose Alternative A or B.)

☐ ALTERNATIVE A

Owner shall pay Contractor for all labor and materials the sum of $_____ .

☐ ALTERNATIVE B

Owner shall pay Contractor $_____ for labor. Materials shall be paid for by Owner upon delivery to the worksite or as follows: _____

_____

3. Terms of Payment

(Choose Alternative A, B, or C.)

☐ ALTERNATIVE A

Upon completing Contractor's services under this Agreement, Contractor shall submit an invoice. Owner shall pay Contractor within _____ days from the date of Contractor's invoice.

☐ ALTERNATIVE B

Contractor shall be paid $_____ upon signing this Agreement and the remaining amount due when Contractor completes the services and submits an invoice. Owner shall pay Contractor within _____ days from the date of Contractor's invoice.

☐ ALTERNATIVE C

Contractor shall be paid according to the Schedule of Payments set forth in Exhibit ___ , attached to and made part of this agreement.

4. Time of Completion

The work to be performed under this Agreement shall commence on _____ and be substantially completed on or before _____ . Time is of the essence.

5. What Constitutes Completion

The work specified in Clause 1 shall be considered completed upon approval by Owner; however, Owner's approval shall not be unreasonably withheld.

6. Permits and Approvals

(Choose Alternative A or B.)

☐ ALTERNATIVE A

Owner shall be responsible for determining which state and local permits are necessary for performing the specified work and for obtaining and paying for the permits.

☐ ALTERNATIVE B

Contractor shall be responsible for determining which state and local permits are necessary for performing the specified work and for obtaining and paying for the permits.

7. Warranty

Contractor warrants that all work shall be completed in a good workmanlike manner and in compliance with all building codes and other applicable laws. Contractor agrees to correct any defective work at no cost to Owner. This warranty shall be in effect for one year from the date of completion of the work.

8. Liens and Lien Waivers

Contractor represents and warrants that there will be no liens for labor or materials or appliances against the work covered by this Agreement, and agrees to protect and hold Owner free and harmless from and against any and all liens and claims for labor, materials, services, or appliances furnished or used in connection with the work.

To protect Owner against liens being filed by Contractor, subcontractors, and materials providers, Contractor agrees that final payment to Contractor under Clause 3 shall be withheld by Owner until Contractor presents Owner with lien waivers, lien releases, or acknowledgment of full payment from each subcontractor and materials supplier.

9. Site Maintenance

Contractor agrees to be bound by the following conditions when performing the specified work:

- Contractor shall remove all debris and leave the premises in a broom-clean condition.
- Contractor shall perform the specified work during the following hours: _____ .
- Contractor agrees that disruptively loud activities shall be performed only at the following times:

  _____ .

- At the end of each day's work, Contractor's equipment shall be stored in the following location:

  _____ .

10. Subcontractors

Contractor may at its discretion engage subcontractors to perform services under this Agreement, but Contractor shall remain responsible for proper completion of this Agreement.

11. Independent Contractor Status

Contractor is an independent contractor, not Owner's employee. Contractor's employees or subcontractors are not Owner's employees. Contractor and Owner agree to the following rights consistent with an independent contractor relationship (check all that apply):

☐ Contractor has the right to perform services for others during the term of this Agreement.

☐ Contractor has the sole right to control and direct the means, manner, and method by which the services required by this Agreement will be performed.

☐ The Contractor or Contractor's employees or subcontractors shall perform the services required by this Agreement; Owner shall not hire, supervise, or pay any assistants to help Contractor.

☐ Owner shall not require Contractor or Contractor's employees or subcontractors to devote full time to performing the services required by this Agreement.

12. Business Permits, Certificates, and Licenses

Contractor represents and warrants that Contractor has complied with all federal, state, and local laws requiring business permits, certificates, and licenses required to carry out the services to be performed under this Agreement.

Contractor's license or registration is for the following type of work and carries the following number:

_____ .

13. State and Federal Taxes

Client will not:

- withhold FICA (Social Security and Medicare taxes) from Contractor's payments or make FICA payments on Contractor's behalf
- make state or federal unemployment compensation contributions on Contractor's behalf, or
- withhold state or federal income tax from Contractor's payments.

Contractor shall pay all taxes incurred while performing services under this Agreement—including all applicable income taxes and, if Contractor is not a corporation, self-employment (Social Security) taxes. Upon demand, Contractor shall provide Client with proof that such payments have been made.

14. Fringe Benefits

Contractor understands that neither Contractor nor Contractor's employees or contract personnel are eligible to participate in any employee pension, health, vacation pay, sick pay, or other fringe benefit plan of Client.

15. Workers' Compensation

Client shall not obtain workers' compensation insurance on behalf of Contractor or Contractor's employees. If Contractor hires employees to perform any work under this Agreement, Contractor will cover them with workers' compensation insurance to the extent required by law and provide Client with a certificate of workers' compensation insurance before the employees begin the work.

(Optional: Check if applicable.)

☐ Contractor shall obtain workers' compensation insurance coverage for Contractor. Contractor shall provide Client with proof that such coverage has been obtained before starting work.

16. Unemployment Compensation

Client shall make no state or federal unemployment compensation payments on behalf of Contractor or Contractor's employees or contract personnel. Contractor will not be entitled to these benefits in connection with work performed under this Agreement.

17. Insurance

Client shall not provide any insurance coverage of any kind for Contractor or Contractor's employees or contract personnel. Contractor shall obtain and maintain a broad form Commercial General Liability Insurance policy providing for coverage of at least $_____ for each occurrence. Before commencing any work, Contractor shall provide Client with proof of this insurance and with proof that Client has been made an additional insured under the policy.

Contractor shall indemnify and hold Client harmless from any loss or liability arising from performing services under this Agreement.

18. Terminating the Agreement

(Choose Alternative A or B.)

☐ ALTERNATIVE A

With reasonable cause, either Client or Contractor may terminate this Agreement, effective immediately upon giving written notice.

Reasonable cause includes:
- a material violation of this Agreement, or
- any act exposing the other party to liability to others for personal injury or property damage.

☐ ALTERNATIVE B

Either party may terminate this Agreement any time by giving _____ days' written notice to the other party of the intent to terminate.

19. Exclusive Agreement

This is the entire Agreement between Contractor and Client.

20. Modifying the Agreement (Optional)

(Check if applicable.)

☐ This Agreement may be modified only by a writing signed by both parties.

21. Resolving Disputes

(Choose Alternative A, B, or C and any desired clauses.)

☐ ALTERNATIVE A

If a dispute arises under this Agreement, any party may take the matter to court.

(Optional: Check if applicable.)

☐ If any court action is necessary to enforce this Agreement, the prevailing party shall be entitled to reasonable attorney fees, costs, and expenses in addition to any other relief to which the party may be entitled.

☐ ALTERNATIVE B

If a dispute arises under this Agreement, the parties agree to first try to resolve the dispute with the help of a mutually agreed-upon mediator in _____ [list city or county where mediation will occur]. Any costs and fees other than attorney fees associated with the mediation shall be shared equally by the parties. If the dispute is not resolved within 30 days after it is referred to the mediator, any party may take the matter to court.

(Optional: Check if applicable.)

☐ If any court action is necessary to enforce this Agreement, the prevailing party shall be entitled to reasonable attorney fees, costs, and expenses in addition to any other relief to which the party may be entitled.

☐ ALTERNATIVE C

If a dispute arises under this Agreement, the parties agree to first try to resolve the dispute with the help of a mutually agreed-upon mediator in _____ [list city or county where mediation will occur]. Any costs and fees other than attorney fees associated with the mediation shall be shared equally by the parties. If it proves impossible to arrive at a mutually satisfactory solution through mediation, the parties agree to submit the dispute to a mutually agreed-upon arbitrator in _____ [list city or county where arbitration will occur]. Judgment upon the award rendered by the arbitrator may be entered in any court having jurisdiction to do so. Costs of arbitration, including attorney fees, will be allocated by the arbitrator.

22. Applicable Law

This Agreement will be governed by the laws of the state of _____

23. Notices

All notices and other communications in connection with this Agreement shall be in writing and shall be considered given as follows:

- when delivered personally to the recipient's address as stated on this Agreement
- three days after being deposited in the United States mail, with postage prepaid to the recipient's address as stated on this Agreement, or
- when sent by fax or electronic mail, such notice is effective upon receipt provided that a duplicate copy of the notice is promptly given by first class mail, or the recipient delivers a written confirmation of receipt.

24. No Partnership

This Agreement does not create a partnership relationship. Neither party has authority to enter into contracts on the other's behalf.

Signatures

Client: _____
Name of Owner

By: _____
Signature

_____
Typed or Printed Name

Title: _____

Date: _____

Contractor: _____
Name of Contractor

By: _____
Signature

_____
Typed or Printed Name

Title: _____

Taxpayer ID Number: _____

Date: _____

INDEPENDENT CONTRACTOR AGREEMENT FOR COURIERS, MESSENGERS, AND DELIVERY PEOPLE

This Agreement is made between _____ ("Client")

with a principal place of business at _____

and _____ ("Contractor")

with a principal place of business at _____ .

1. Services to Be Performed

(Check and complete applicable provision.)

☐ ALTERNATIVE A

Contractor agrees to perform the following services: _____

☐ ALTERNATIVE B

Contractor agrees to perform the services described in Exhibit A, which is attached to this Agreement.

2. Payment

In consideration for the services to be performed by Contractor, Client agrees to pay Contractor at the following rates: _____ .

Contractor shall be paid within a reasonable time after Contractor submits an invoice to Client. The invoice should include the following: an invoice number, the dates covered by the invoice, and a summary of the work performed.

3. Expenses

(Choose Alternative A or B.)

☐ ALTERNATIVE A

Contractor shall be responsible for all expenses incurred while performing services under this Agreement. This includes automobile, truck, and other travel expenses; vehicle maintenance and repair costs; vehicle and other license fees and permits; insurance premiums; road, fuel, and other taxes; fines; radio, pager, or cell phone expenses; meals; and all salary, expenses, and other compensation paid to employees or contract personnel the Contractor hires to complete the work under this Agreement.

☐ ALTERNATIVE B

Client shall reimburse Contractor for the following expenses that are attributable directly to work performed under this Agreement: _____ .

Contractor shall submit an itemized statement of Contractor's expenses. Client shall pay Contractor within 30 days after receipt of each statement.

4. Vehicles and Equipment

Contractor will furnish all vehicles, equipment, tools, and materials used to provide the services required by this Agreement. Client will not require Contractor to rent or purchase any equipment, product, or service as a condition of entering into this Agreement.

5. Independent Contractor Status

Contractor is an independent contractor, and neither Contractor nor Contractor's employees or contract personnel are, or shall be deemed, Client's employees. In its capacity as an independent contractor, Contractor agrees and represents, and Client agrees, as follows (check all that apply):

☐ Contractor has the right to perform services for others during the term of this Agreement.

☐ Contractor has the sole right to control and direct the means, manner, and method by which the services required by this Agreement will be performed. Contractor shall select the routes taken, starting and quitting times, days of work, and order the work is performed.

☐ Contractor has the right to hire assistants as subcontractors or to use employees to provide the services required by this Agreement.

☐ Neither Contractor nor Contractor's employees or contract personnel shall be required to wear any uniforms provided by Client.

☐ The services required by this Agreement shall be performed by Contractor, Contractor's employees, or contract personnel, and Client shall not hire, supervise, or pay any assistants to help Contractor.

☐ Neither Contractor nor Contractor's employees or contract personnel shall receive any training from Client in the professional skills necessary to perform the services required by this Agreement.

☐ Neither Contractor nor Contractor's employees or contract personnel shall be required by Client to devote full time to the performance of the services required by this Agreement.

6. Business Licenses, Permits, and Certificates

Contractor represents and warrants that Contractor and Contractor's employees and contract personnel will comply with all federal, state, and local laws requiring driver's and other licenses, business permits, and certificates required to carry out the services to be performed under this Agreement.

7. State and Federal Taxes

Client will not:

- withhold FICA (Social Security and Medicare taxes) from Contractor's payments or make FICA payments on Contractor's behalf
- make state or federal unemployment compensation contributions on Contractor's behalf, or
- withhold state or federal income tax from Contractor's payments.

Contractor shall pay all taxes incurred while performing services under this Agreement—including all applicable income taxes and, if Contractor is not a corporation, self-employment (Social Security) taxes. Upon demand, Contractor shall provide Client with proof that such payments have been made.

8. Fringe Benefits

Contractor understands that neither Contractor nor Contractor's employees or contract personnel are eligible to participate in any employee pension, health, vacation pay, sick pay, or other fringe benefit plan of Client.

9. Unemployment Compensation

Client shall make no state or federal unemployment compensation payments on behalf of Contractor or Contractor's employees or contract personnel. Contractor will not be entitled to these benefits in connection with work performed under this Agreement.

10. Workers' Compensation

Client shall not obtain workers' compensation insurance on behalf of Contractor or Contractor's employees. If Contractor hires employees to perform any work under this Agreement, Contractor will cover them with workers' compensation insurance to the extent required by law and provide Client with a certificate of workers' compensation insurance before the employees begin the work.

(Optional: Check if applicable.)

☐ Contractor shall obtain workers' compensation insurance coverage for Contractor. Contractor shall provide Client with proof that such coverage has been obtained before starting work.

11. Insurance

Client shall not provide insurance coverage of any kind for Contractor or Contractor's employees or contract personnel. Contractor shall obtain the following insurance coverage and maintain it during the entire term of this Agreement (check all that apply):

☐ Automobile liability insurance for each vehicle used in the performance of this Agreement, including owned, non-owned (for example, owned by Contractor's employees), leased, or hired vehicles, in the minimum amount of $_____ combined single limit per occurrence for bodily injury and property damage.

☐ Comprehensive or commercial general liability insurance coverage in the minimum amount of $_____ combined single limit, including coverage for bodily injury, personal injury, broad form property damage, contractual liability, and cross-liability.

Before commencing any work, Contractor shall provide Client with proof of this insurance and with proof that Client has been made an additional insured under the policies.

12. Indemnification

Contractor shall indemnify and hold Client harmless from any loss or liability arising from performing services under this Agreement.

13. Term of Agreement

This agreement will become effective when signed by both parties and will terminate on the earlier of:

- the date Contractor completes the services required by this Agreement
- _____ [date], or
- the date a party terminates the Agreement as provided below.

14. Terminating the Agreement

(Choose Alternative A or B.)

☐ ALTERNATIVE A

With reasonable cause, either Client or Contractor may terminate this Agreement, effective immediately upon giving written notice.

Reasonable cause includes:

- a material violation of this Agreement, or
- any act exposing the other party to liability to others for personal injury or property damage.

☐ ALTERNATIVE B

Either party may terminate this Agreement at any time by giving _____ days' written notice to the other party of the intent to terminate.

15. Exclusive Agreement

This is the entire Agreement between Contractor and Client.

16. Modifying the Agreement (Optional)

(Check box if applicable.)

☐ This Agreement may be modified only by a writing signed by both parties.

17. Resolving Disputes

(Choose Alternative A, B, or C and any desired optional clauses.)

☐ ALTERNATIVE A

If a dispute arises under this Agreement, any party may take the matter to court.

(Optional: Check if applicable.)

☐ If any court action is necessary to enforce this Agreement, the prevailing party shall be entitled to reasonable attorney fees, costs, and expenses in addition to any other relief to which the party may be entitled.

☐ ALTERNATIVE B

If a dispute arises under this Agreement, the parties agree to first try to resolve the dispute with the help of a mutually agreed-upon mediator in _____ [list city or county where mediation will occur]. Any costs and fees other than attorney fees associated with the mediation shall be shared equally by the parties. If the dispute is not resolved within 30 days after it is referred to the mediator, any party may take the matter to court.

(Optional: Check box if applicable.)

☐ If any court action is necessary to enforce this Agreement, the prevailing party shall be entitled to reasonable attorney fees, costs, and expenses in addition to any other relief to which the party may be entitled.

☐ ALTERNATIVE C

If a dispute arises under this Agreement, the parties agree to first try to resolve the dispute with the help of a mutually agreed-upon mediator in _____ [list city or county where mediation will occur]. Any costs and fees other than attorney fees associated with the mediation shall be shared equally by the parties. If it proves impossible to arrive at a mutually satisfactory solution through mediation, the parties agree to submit the dispute to a mutually agreed-upon arbitrator in _____ [list city or county where arbitration will occur]. Judgment upon the award rendered by the arbitrator may be entered in any court having jurisdiction to do so. Costs of arbitration, including attorney fees, will be allocated by the arbitrator.

18. Confidentiality (Optional)

(Check box and complete if applicable.)

☐ Contractor acknowledges that it will be necessary for Client to disclose certain confidential and proprietary information to Contractor in order for Contractor to perform duties under this Agreement. Contractor acknowledges that disclosure to a third party or misuse of this proprietary or confidential information would irreparably harm Client. Accordingly, Contractor will not disclose or use, either during or after the term of this Agreement, any proprietary or confidential information of Client without Client's prior written permission except to the extent necessary to perform services on Client's behalf.

Proprietary or confidential information includes:

- the written, printed, graphic, or electronically recorded materials furnished by Client for Contractor to use
- any written or tangible information stamped "confidential," "proprietary," or with a similar legend, or any information that Client makes reasonable efforts to maintain the secrecy of
- business or marketing plans or strategies, customer lists, operating procedures, trade secrets, design formulas, know-how and processes, computer programs and inventories, discoveries, and improvements of any kind, sales projections, and pricing information
- information belonging to customers and suppliers of Client about whom Contractor gained knowledge as a result of Contractor's services to Client, and
- other: _____.

Upon termination of Contractor's services to Client, or at Client's request, Contractor shall deliver to Client all materials in Contractor's possession relating to Client's business.

Contractor acknowledges that any breach or threatened breach of Clause 18 of this Agreement will result in irreparable harm to Client for which damages would be an inadequate remedy. Therefore, Client shall be entitled to equitable relief, including an injunction, in the event of such breach or threatened breach of Clause 18 of this Agreement. Such equitable relief shall be in addition to Client's rights and remedies otherwise available at law.

19. Applicable Law

This Agreement will be governed by the laws of the state of _____.

20. No Partnership

This Agreement does not create a partnership relationship. Contractor does not have authority to enter into contracts on Client's behalf.

21. Assignment and Delegation

(Choose Alternative A or B, if applicable.)

☐ ALTERNATIVE A

Either Contractor or Client may assign rights and may delegate duties under this Agreement.

☐ ALTERNATIVE B

Contractor may not assign or subcontract any rights or delegate any of its duties under this Agreement without Client's prior written approval.

Signatures

Owner: _____
Name of Owner

By: _____
Signature

_____
Typed or Printed Name

Title: _____

Date: _____

Contractor: _____
Name of Contractor

By: _____
Signature

_____
Typed or Printed Name

Title: _____

Taxpayer ID Number: _____

Date: _____

CONTRACT AMENDMENT FORM

This Amendment is made between _____ and
_____ to amend the Original Agreement titled
_____, signed by them on _____.

   The Original Agreement is amended as follows: _____

_____

_____

_____

_____

_____

_____

   All provisions of the Original Agreement, except as modified by this Amendment, remain in full force and effect, and are reaffirmed. If there is any conflict between this Amendment and any provision of the Original Agreement, the provisions of this Amendment shall control.

Client: _____
                                    Name of Client

By: _____
                                      Signature

_____
                               Typed or Printed Name

Title: _____

Date: _____

Consultant/Contractor: _____
                            Name of Consultant/Contractor

By: _____
                                      Signature

_____
                               Typed or Printed Name

Title: _____

Taxpayer ID Number: _____

Date: _____

PUBLICITY/PRIVACY RELEASE

For good and valuable consideration, the receipt and sufficiency of which is hereby acknowledged, I hereby grant _____ permission to use, adapt, modify, reproduce, distribute, publicly perform, and display, in any form now known or later developed, the Materials specified in this release (as indicated by my initials) throughout the world, by incorporating them into one or more Works and/or advertising and promotional materials relating thereto.

This release is for the following Materials [initial appropriate lines]:

_____ Name

_____ Voice

_____ Visual likeness (on photographs, video, film, etc.)

_____ Photographs, graphics, or other artwork as specified: _____

_____

_____ Film, videotape, or other audiovisual materials as specified: _____

_____

_____ Music or sound recordings as specified: _____

_____

_____ Other:

_____

I warrant and represent that the Materials identified above are either owned by me, and/or are original to me and/or that I have full authority from the owner of the Materials to grant this release.

I release _____, its agents, employees, licensees, and assigns from any and all claims I may have now or in the future for invasion of privacy, right of publicity, copyright infringement, defamation, or any other cause of action arising out of the use, reproduction, adaptation, distribution, broadcast, performance, or display of the Works.

I waive any right to inspect or approve any Works that may be created containing the Materials.

I understand and agree that _____ is and shall be the exclusive owner of all right, title, and interest, including copyright, in the Works, and any advertising or promotional materials containing the Materials, except as to preexisting rights in any of the Materials released hereunder.

I am of full legal age and have read this release and am fully familiar with its contents.

_____
Signature

_____
Typed or Printed Name

_____
Date

LETTER

[*Date*]

[*Contractor name*]
[*Contractor address*]

Dear [*hiring firm*] :

Enclosed, please find the independent contractor agreement for your services. The agreement makes clear that you are an independent contractor (self-employed), and not an employee of [*hiring firm*] . Please read it carefully. If you have any questions about your work status, please do not hesitate to contact me.

 Because you are an independent contractor, [*hiring firm*] will not withhold any taxes from your pay. You must pay all your state and federal taxes yourself. Usually, you'll have to pay estimated taxes four times a year.

 In addition, [*hiring firm*] will not provide you with unemployment insurance coverage. When your services end, you will not be legally entitled to apply for unemployment benefits based on your term of service with [*hiring firm*] .

 Failure to preserve your independent contractor status could prove costly not only to [*hiring firm*] , but to you personally because it could result in your loss of valuable tax deductions. To help preserve your status, please do not identify yourself as a [*hiring firm*] employee, either orally or in writing on tax or other government forms, your business cards, letterhead, resume, marketing literature, or any other document. If you are asked what your status was while working with [*hiring firm*] , please state that you were a self-employed independent contractor.

 Please inform [*hiring firm*] immediately if the IRS or other government agency contacts you regarding your work status while performing services for us.

 Please sign the acknowledgement of this letter, below, and return it with a copy of the signed Independent Contractor Agreement.

Very truly yours,

_____

by: _____

I have read and agree to be bound by the terms of this letter and the Independent Contractor Agreement.

Name of Contractor_____

By: _____

Date:_____

INDEPENDENT CONTRACTOR QUESTIONNAIRE

Please provide the following background information:

Name: _____

Fictitious business name (if any):_____

Business address: _____

Business phone number: _____ Fax: _____

Employer Identification number or Social Security number: _____

Form of business entity (check one):

☐ Corporation   ☐ Partnership   ☐ Sole Proprietorship   ☐ Limited Liability Company

Please provide the name, address, and dates of service of all companies for which you have performed services as an independent contractor for the past two years. But please do not provide any information you have a duty to keep confidential.

_____

_____

_____

_____

Have you ever hired employees?   ☐ Yes   ☐ No

If yes, please provide the following information for each employee (you can use a separate sheet if necessary):

Name: _____

Address: _____

Title: _____

Salary: _____

Dates of employment: _____

Workers' compensation carrier and policy number: _____

Have you paid federal and state payroll taxes for your employees?   ☐ Yes ☐ No

Describe the training you have received in your specialty.

School attended: _____

Dates of attendance: _____ Degrees received: _____

School attended: _____

Dates of attendance: _____ Degrees received: _____

School attended: _____

Dates of attendance: _____ Degrees received: _____

Do you advertise your services?   ☐ Yes ☐ No

If you don't advertise, how do you market your services?

_____

_____

Describe the business expenses you have paid in the past two years, including office or workplace rental, materials and equipment expenses, telephone, and other expenses:

_____

Describe the business expenses you pay now:

_____

Describe the equipment and facilities you own:

_____

Please describe the tools and materials you will use to perform the services in this job:

_____

Please provide the names and salaries of all assistants whom you will use on the job: _____

_____

Please provide the names and addresses of other clients or customers for whom you have performed services during the previous two years, but don't provide any information you have a duty to keep confidential:

_____

Please give the name of your general liability insurance carrier: _____

Policy number: _____

Please give the name of your auto insurance carrier: _____

Policy number: _____

Have you ever worked for us before?   ☐ Yes   ☐ No

If yes, please complete the following:

Dates of employment: _____

Services performed: _____

Do you have an independent contractor agreement form that you generally use?   ☐ Yes   ☐ No

If so, please attach a copy.

If you're a sole proprietor, have you paid self-employment taxes on your income and filed a Schedule C with your federal tax return?   ☐ Yes   ☐ No

If so, please provide copies of your tax returns for the past two years.

DOCUMENT CHECKLIST

Please provide the following documents:

☐ Copies of your business license and any professional licenses you have

☐ Certificates showing that you have insurance, including general liability insurance and workers' compensation insurance if you have employees

☐ Copies of your business cards and stationery

☐ Copies of any advertising you've done, such as a Yellow Pages listing

☐ A copy of your White Pages business phone listing, if there is one

☐ If you have a Website marketing your services, please provide a printout of the home page

☐ If you're operating under an assumed name, a copy of the fictitious business name statement

☐ A copy of your invoice form to be used for billing purposes

☐ A copy of any office lease and proof that you've paid the rent, such as copies of canceled rental checks

☐ If you're a sole proprietor, copies of your tax returns for the previous two years showing that you have filed a Schedule C, Profit or Loss From a Business.

Sample Agreements for Use by IC

Chapter	File Name	Form Name
5 Section B	General2.rtf	Independent Contractor Agreement
6 Section B	Consultant2.rtf	Independent Contractor Agreement for Consultant
7 Section B	Household2.rtf	Independent Contractor Agreement for Household Worker
8 Section C	Seller2.rtf	Independent Contractor Agreement for Direct Seller
9 Section B	Accountant2.rtf	Independent Contractor Agreement for Accountant & Bookkeeper
10 Section B	Software2.rtf	Independent Contractor Agreement for Software Consultant
11 Section B	Creative2.rtf	Independent Contractor Agreement for Creative Contractor
12 Section B	Construction2.rtf	Independent Contractor Agreement for Construction Contractor
13 Section B	Courier2.rtf	Independent Contractor Agreement for Couriers, Messengers, and Delivery People
	Amendment2.rtf	Contract Amendment Form
	Release2.rtf	Publicity/Privacy Release
	Invoice.rtf	Invoice

INDEPENDENT CONTRACTOR AGREEMENT
FOR CONSULTANT

This Agreement is made between _____ ("Client")

with a principal place of business at _____

and _____ ("Consultant"),

with a principal place of business at _____ .

1. Services to be Performed

(Choose Alternative A or B.)

☐ ALTERNATIVE A

Consultant agrees to perform the following services on Client's behalf: _____

_____

_____

_____

☐ ALTERNATIVE B

Consultant agrees to perform the services described in Exhibit A, which is attached to this Agreement.

2. Payment

(Choose Alternative A or B and optional clause, if desired.)

☐ ALTERNATIVE A

In consideration for the services to be performed by Consultant, Client agrees to pay Consultant $_____.

☐ ALTERNATIVE B

In consideration for the services to be performed by Consultant, Client agrees to pay Consultant at the rate of

$_____ per ☐ hour, ☐ day, ☐ week, ☐ month.

(Optional: Check and complete if applicable.)

☐ Consultant's total compensation shall not exceed $_____ without Client's written consent.

3. Terms of Payment

(Choose Alternative A, B, C, or D.)

☐ ALTERNATIVE A

Upon completing Consultant's services under this Agreement, Consultant shall submit an invoice. Client shall pay

Consultant within _____ days from the date of Consultant's invoice.

☐ ALTERNATIVE B

Consultant shall be paid $_____ upon signing this Agreement and the remaining amount due when

Consultant completes the services and submits an invoice. Client shall pay Consultant within _____ days

from the date of Consultant's invoice.

☐ ALTERNATIVE C

Consultant shall be paid according to the Schedule of Payments set forth in Exhibit _____ attached to and

made part of this Agreement.

☐ ALTERNATIVE D

Consultant shall send Client an invoice monthly. Client shall pay Consultant within _____ days from the date of each invoice.

4. Late Fees (Optional)

(Check and complete if applicable.)

☐ Late payments by Client shall be subject to late penalty fees of _____% per month from the due date until the amount is paid.

5. Expenses

(Choose Alternative A or B and optional clause, if desired.)

☐ ALTERNATIVE A

Consultant shall be responsible for all expenses incurred while performing services under this Agreement.

(Optional: Check if applicable.)

☐ However, Client shall reimburse Consultant for all reasonable travel and living expenses necessarily incurred by Consultant while away from Consultant's regular place of business to perform services under this Agreement. Consultant shall submit an itemized statement of such expenses. Client shall pay Consultant within 30 days from the date of each statement.

☐ ALTERNATIVE B

Client shall reimburse Consultant for the following expenses that are directly attributable to work performed under this Agreement:

- travel expenses other than normal commuting, including airfares, rental vehicles, and highway mileage in company or personal vehicles at __ cents per mile
- telephone, fax, online, and telegraph charges
- postage and courier services
- printing and reproduction
- computer services, and
- other expenses resulting from the work performed under this Agreement.

Consultant shall submit an itemized statement of Consultant's expenses. Client shall pay Consultant within 30 days from the date of each statement.

6. Materials (Optional)

(Check if applicable.)

☐ Consultant will furnish all materials, equipment, and supplies used to provide the services required by this Agreement.

(Optional: Check if applicable.)

7. Intellectual Property Ownership

(Choose Alternative A or B.)

☐ ALTERNATIVE A

Consultant grants to Client a royalty-free nonexclusive license to use anything created or developed by Consultant for Client under this Agreement ("Contract Property"). The license shall have a perpetual term and Client may not transfer it. Consultant shall retain all copyrights, patent rights, and other intellectual property rights to the Contract Property.

☐ ALTERNATIVE B

Consultant assigns to Client all patent, copyright, and trade secret rights in anything created or developed by Consultant for Client under this Agreement. This assignment is conditioned upon full payment of the compensation due Consultant under this Agreement.

Consultant shall help prepare any documents Client considers necessary to secure any copyright, patent, or other intellectual property rights at no charge to Client. However, Client shall reimburse Consultant for reasonable out-of-pocket expenses.

8. Consultant's Reusable Materials (Optional)

(Check and complete if applicable.)

☐ Consultant owns or holds a license to use and sublicense various materials in existence before the start date of this Agreement ("Consultant's Materials"). Consultant's Materials include, but are not limited to, those items identified in Exhibit _____, attached to and made part of this Agreement. Consultant may, at its option, include Consultant's Materials in the work performed under this Agreement. Consultant retains all right, title, and interest, including all copyrights, patent rights, and trade secret rights, in Consultant's Materials. Consultant grants Client a royalty-free nonexclusive license to use any of Consultant's Materials incorporated into the work performed by Consultant under this Agreement. The license shall have a perpetual term and may not be transferred by Client.

9. Term of Agreement

This agreement will become effective when signed by both parties and will terminate on the earlier of:

- the date Consultant completes the services required by this Agreement
- _____[date], or
- the date a party terminates the Agreement as provided below.

10. Terminating the Agreement

(Choose Alternative A or B.)

☐ ALTERNATIVE A

With reasonable cause, either party may terminate this Agreement effective immediately by giving written notice of cause for termination. Reasonable cause includes:

- a material violation of this Agreement, or
- nonpayment of Consultant's compensation after 20 days' written demand for payment.

Consultant shall be entitled to full payment for services performed prior to the effective date of termination.

☐ ALTERNATIVE B

Either party may terminate this Agreement at any time by giving _____ days' written notice of termination. Consultant shall be entitled to full payment for services performed prior to the date of termination.

11. Independent Contractor Status

Consultant is an independent contractor, not Client's employee. Consultant's employees or subcontractors are not Client's employees. Consultant and Client agree to the following rights consistent with an independent contractor relationship (check all that apply):

☐ Consultant has the right to perform services for others during the term of this Agreement.

☐ Consultant has the sole right to control and direct the means, manner, and method by which the services required by this Agreement will be performed.

☐ Consultant has the right to hire assistants as subcontractors or to use employees to provide the services required by this Agreement.

☐ Consultant or Consultant's employees or subcontractors shall perform the services required by this Agreement; Client shall not hire, supervise, or pay any assistants to help Consultant.

☐ Neither Consultant nor Consultant's employees or subcontractors shall receive any training from Client in the skills necessary to perform the services required by this Agreement.

☐ Client shall not require Consultant or Consultant's employees or subcontractors to devote full time to performing the services required by this Agreement.

☐ Neither Consultant nor Consultant's employees or subcontractors are eligible to participate in any employee pension, health, vacation pay, sick pay, or other fringe benefit plan of Client.

12. Local, State, and Federal Taxes

Consultant shall pay all income taxes and FICA (Social Security and Medicare taxes) incurred while performing services under this Agreement. Client will not:

- withhold FICA from Consultant's payments or make FICA payments on Consultant's behalf
- make state or federal unemployment compensation contributions on Consultant's behalf, or
- withhold state or federal income tax from Consultant's payments.

The charges included here do not include taxes. If Consultant is required to pay any federal, state, or local sales, use, property, or value added taxes based on the services provided under this Agreement, the taxes shall be separately billed to Client. Client shall be responsible for paying any interest or penalties incurred due to late payment or nonpayment of any taxes by Client.

13. Exclusive Agreement

This is the entire Agreement between Consultant and Client.

14. Modifying the Agreement (Optional)

(Check if applicable.)

☐ Client and Consultant recognize that:

- Consultant's original cost and time estimates may be too low due to unforeseen events or to factors unknown to Consultant when this Agreement was made
- Client may desire a mid-project change in Consultant's services that would add time and cost to the project and possibly inconvenience Consultant, or
- other provisions of this Agreement may be difficult to carry out due to unforeseen circumstances.

If any intended changes or any other events beyond the parties' control require adjustments to this Agreement, the parties shall make a good faith effort to agree on all necessary particulars. Such agreements shall be put in writing, signed by the parties, and added to this Agreement.

15. Resolving Disputes

(Choose Alternative A, B, or C and any desired optional clauses.)

☐ ALTERNATIVE A

If a dispute arises under this Agreement, any party may take the matter to court.

☐ If any court action is necessary to enforce this Agreement, the prevailing party shall be entitled to reasonable attorney fees, costs, and expenses in addition to any other relief to which the party may be entitled.

☐ ALTERNATIVE B

If a dispute arises under this Agreement, the parties agree to first try to resolve the dispute with the help of a mutually agreed-upon mediator in _____ [list city or county where mediation will occur]. Any costs and fees other than attorney fees associated with the mediation shall be shared equally by the parties. If the dispute is not resolved within 30 days after it is referred to the mediator, any party may take the matter to court.

☐ If any court action is necessary to enforce this Agreement, the prevailing party shall be entitled to reasonable attorney fees, costs, and expenses in addition to any other relief to which the party may be entitled.

☐ ALTERNATIVE C

If a dispute arises under this Agreement, the parties agree to first try to resolve the dispute with the help of a mutually agreed-upon mediator in _____ [list city or county where mediation will occur]. Any costs and fees other than attorney fees associated with the mediation shall be shared equally by the parties. If it proves impossible to arrive at a mutually satisfactory solution through mediation, the parties agree to submit the dispute to a mutually agreed-upon arbitrator in _____
[list city or county where arbitration will occur]. Judgment upon the award rendered by the arbitrator may be entered in any court having jurisdiction to do so. Costs of arbitration, including attorney fees, will be allocated by the arbitrator.

16. Limited Liability (Optional)

(Check if applicable.)

☐ This provision allocates the risks under this Agreement between Consultant and Client. Consultant's pricing reflects the allocation of risk and limitation of liability specified below.

Consultant's total liability to Client under this Agreement for damages, costs, and expenses shall not exceed $_____ or the compensation received by Consultant under this Agreement, whichever is less. However, contractor shall remain liable for bodily injury or personal property damage resulting from grossly negligent or willful actions of Consultant or Consultant's employees or agents while on Client's premises to the extent such actions or omissions were not caused by Client.

NEITHER PARTY TO THIS AGREEMENT SHALL BE LIABLE FOR THE OTHER'S LOST PROFITS OR SPECIAL, INCIDENTAL OR CONSEQUENTIAL DAMAGES, WHETHER IN AN ACTION IN CONTRACT OR TORT, EVEN IF THE PARTY HAS BEEN ADVISED BY THE OTHER PARTY OF THE POSSIBILITY OF SUCH DAMAGES.

17. Notices

All notices and other communications in connection with this Agreement shall be in writing and shall be considered given as follows:

- when delivered personally to the recipient's address as stated on this Agreement
- three days after being deposited in the United States mail, with postage prepaid to the recipient's address as stated on this Agreement, or

- when sent by fax or electronic mail, such notice is effective upon receipt provided that a duplicate copy of the notice is promptly given by first class mail, or the recipient delivers a written confirmation of receipt.

18. No Partnership

This Agreement does not create a partnership relationship. Neither party has authority to enter into contracts on the other's behalf.

19. Applicable Law

This Agreement will be governed by the laws of the state of _____.

20. Assignment and Delegation (Optional)

(Check if applicable.)

☐ Either Consultant or Client may assign its rights or may delegate its duties under this Agreement.

Signatures

Client: _____
Name of Client

By: _____
Signature

_____
Typed or Printed Name

Title: _____

Date: _____

Consultant: _____
Name of Consultant

By: _____
Signature

_____
Typed or Printed Name

Title: _____

Taxpayer ID Number: _____

Date: _____

INDEPENDENT CONTRACTOR AGREEMENT FOR HOUSEHOLD WORKER

This Agreement is made between _____ ("Client")

whose address is _____

and _____ ("Contractor")

with a principal place of business at _____ .

1. Services to Be Performed

(Check and complete applicable provisions.)

Contractor agrees to perform the following services:

a. Cleaning Interior

☐ Contractor will clean the following rooms and areas: _____

_____

_____

b. Cleaning Exterior

Contractor will clean the following:

☐ Front porch or deck: _____

☐ Back porch or deck: _____

☐ Garage: _____

☐ Pool, hot tub, or sauna: _____

☐ Other exterior areas: _____

c. Gardening

☐ Contractor will perform the following gardening services:

_____

_____

d. Other Responsibilities

☐ Cooking: _____

☐ Laundry: _____

☐ Ironing: _____

☐ Shopping and errands: _____

☐ Other: _____

2. Payment

(Choose Alternative A or B.)

☐ ALTERNATIVE A

In consideration for the services to be performed by Contractor, Client agrees to pay Contractor $_____ .

☐ ALTERNATIVE B

In consideration for the services to be performed by Contractor, Client agrees to pay Contractor at the rate of
$_____ per ☐ hour, ☐ day, ☐ week, ☐ month according to the terms of payment set forth below.

3. Terms of Payment

(Choose Alternative A or B.)

☐ ALTERNATIVE A

Upon completion of Contractor's services under this Agreement, Contractor shall submit an invoice. Client shall
pay Contractor the compensation described within _____ [15, 30, 45, 60] days after receiving
Contractor's invoice.

☐ ALTERNATIVE B

Contractor shall invoice Client on a monthly basis for all hours worked pursuant to this Agreement during the
preceding month. Invoices shall specify an invoice number, the dates covered in the invoice, the hours
expended, and the work performed (in summary) during the invoice period. Client shall pay Contractor's fee
within _____ [15, 30, 45, 60] days after receiving Contractor's invoice.

4. Late Fees (Optional)

(Check and complete if applicable.)

☐ Late payments by Client shall be subject to late penalty fees of _____ % per month from the due date
until the amount is paid.

5. Expenses

(Choose Alternative A or B.)

☐ ALTERNATIVE A

Contractor shall be responsible for all expenses incurred while performing services under this Agreement. This
includes license fees, memberships, and dues; automobile and other travel expenses; meals and entertainment;
insurance premiums; and all salary, expenses, and other compensation paid to employees or contract personnel
the Contractor hires to complete the work under this Agreement.

☐ ALTERNATIVE B

Client shall reimburse Contractor for the following expenses that are attributable directly to work performed under this Agreement: _____

_____

_____

    Contractor shall submit an itemized statement of Contractor's expenses. Client shall pay Contractor within 30
days after receipt of each statement.

6. Materials (Optional)

(Check if applicable.)

(Choose Alternative A or B.)

☐ ALTERNATIVE A

Contractor will furnish all equipment, tools, materials, and supplies needed to perform the services required by
this Agreement.

☐ ALTERNATIVE B

Client shall make available to Contractor, at Client's expense, the following equipment and/or materials:

_____

_____. These items will be provided to Contractor by

_____ [date].

7. Work Schedule

Contractor shall perform the services at _____

during reasonable hours on a schedule to be mutually agreed upon by Client and Contractor based upon Client's needs and Contractor's availability to perform such services.

8. Term of Agreement

This agreement will become effective when signed by both parties and will terminate on the earlier of:

- the date Contractor completes the services required by this Agreement
- _____ [date], or
- the date a party terminates the Agreement as provided below.

9. Terminating the Agreement

(Choose Alternative A or B.)

☐ ALTERNATIVE A

With reasonable cause, either Client or Contractor may terminate this Agreement, effective immediately upon giving written notice.

Reasonable cause includes:

- a material violation of this Agreement, or
- nonpayment of contractor's compensation after 20 days' written demand for payment.

Contractor shall be entitled to full payment for services performed prior to the effective date of termination.

☐ ALTERNATIVE B

Either party may terminate this Agreement any time by giving _____ days' written notice to the other party of the intent to terminate.

10. Independent Contractor Status

Contractor is an independent contractor, not Client's employee. Contractor's employees or contract personnel are not Client's employees. Contractor and Client agree to the following rights consistent with an independent contractor relationship (check all that apply):

- ☐ Contractor has the right to perform services for others during the term of this Agreement.
- ☐ Contractor has the sole right to control and direct the means, manner, and method by which the services required by this Agreement will be performed.
- ☐ Contractor has the right to hire assistants as subcontractors or to use employees to provide the services required by this Agreement.
- ☐ The Contractor or Contractor's employees or contract personnel shall perform the services required by this Agreement; Client shall not hire, supervise, or pay any assistants to help Contractor.
- ☐ Neither Contractor nor Contractor's employees or contract personnel shall receive any training from Client in the skills necessary to perform the services required by this Agreement.

☐ Client shall not require Contractor or Contractor's employees or contract personnel to devote full time to performing the services required by this Agreement.

11. State and Federal Taxes

Client will not:

- withhold FICA (Social Security and Medicare taxes) from Contractor's payments or make FICA payments on Contractor's behalf
- make state or federal unemployment compensation contributions on Contractor's behalf, or
- withhold state or federal income tax from Contractor's payments.

The charges included here do not include taxes. If Contractor is required to pay any federal, state, or local sales, use, property, or value added taxes based on the services provided under this Agreement, the taxes shall be billed separately to Client. Client shall be responsible for paying any interest or penalties incurred due to late payment or nonpayment of any taxes by Client.

Signatures

Client: _____
Name of Client

By: _____
Signature

_____
Typed or Printed Name

Title: _____

Date: _____

Contractor: _____
Name of Contractor

By: _____
Signature

_____
Typed or Printed Name

Title: _____

Taxpayer ID Number: _____

Date: _____

INDEPENDENT CONTRACTOR AGREEMENT FOR DIRECT SELLER

This Agreement is made between _____ ("Client")

with a principal place of business at _____

and _____ ("Contractor")

with a principal place of business at _____ .

1. Services to be Performed

Contractor agrees to sell the following product or merchandise for client: _____

_____

(Optional: Check if applicable.)

☐ Contractor shall seek sales of the product in the homes of various individuals.

2. Compensation

In consideration for the services to be performed by Contractor, Client agrees to pay Contractor a commission

on completed sales as follows: _____

_____

_____

Contractor acknowledges that no other compensation is payable by Client and that all of Contractor's
compensation will depend on sales made by Contractor. None of Contractor's compensation shall be based on
the number of hours worked by Contractor.

3. Late Fees (Optional)

(Check and complete if applicable.)

☐ Late payments by Client shall be subject to late penalty fees of _____ % per month from the due date until
the amount is paid.

4. Expenses

(Choose Alternative A or B.)

☐ ALTERNATIVE A

Contractor shall be responsible for all expenses incurred while performing services under this Agreement. This
includes license fees, memberships and dues; automobile and other travel expenses; meals and entertainment;
insurance premiums; and all salary, expenses, and other compensation paid to employees or contract personnel
the Contractor hires to complete the work under this Agreement.

☐ ALTERNATIVE B

Client shall reimburse Contractor for the following expenses that are directly attributable to work performed under this Agreement: _____

_____

_____

Contractor shall submit an itemized statement of Contractor's expenses. Client shall pay Contractor within 30 days after receipt of each statement.

5. Materials (Optional)

(Check if applicable.)

☐ Contractor will furnish all materials and equipment used to provide the services required by this Agreement.

6. Term of Agreement

This agreement will become effective when signed by both parties and will terminate on the earlier of:

- the date Contractor completes the services required by this Agreement
- _____ [date], or
- the date a party terminates the Agreement as provided below.

7. Terminating the Agreement

(Choose Alternative A or B.)

☐ ALTERNATIVE A

With reasonable cause, either party may terminate this Agreement effective immediately by giving written notice of cause for termination. Reasonable cause includes:

- a material violation of this Agreement, or
- nonpayment of Contractor's compensation after 20 days' written demand for payment.

Contractor shall be entitled to full payment of all commissions earned on orders received by Client prior to the effective date of termination.

☐ ALTERNATIVE B

Either party may terminate this Agreement at any time by giving _____ days' written notice of termination. Contractor shall be entitled to full payment of all commissions earned on orders received by Client prior to the effective date of termination.

8. Independent Contractor Status

Contractor is an independent contractor, not Client's employee. Contractor's employees or contract personnel are not Client's employees. Contractor and Client agree to the following rights consistent with an independent contractor relationship (check all that apply):

☐ Contractor has the right to perform services for others during the term of this Agreement.

☐ Contractor shall have no obligation to perform any services other than the sale of the product described here.

☐ Contractor has the sole right to control and direct the means, manner, and method by which the services required by this Agreement will be performed. Consistent with this freedom from Client's control, Contractor:

  ☐ does not have to pursue or report on leads furnished by Client
  ☐ is not required to attend sales meetings organized by Client
  ☐ does not have to obtain Client's pre-approval for orders, and
  ☐ shall adopt and carry out Contractor's own sales strategy.

☐ Subject to any restrictions on Contractor's sales territory contained in this Agreement, Contractor has the right to perform the services required by this Agreement at any location or time.

☐ Contractor has the right to hire assistants as subcontractors or to use employees to provide the services required by this Agreement, except that Client may supply Contractor with sales forms.

☐ The Contractor or Contractor's employees or contract personnel shall perform the services required by this Agreement; Client shall not hire, supervise, or pay any assistants to help Contractor.

☐ Neither Contractor nor Contractor's employees or contract personnel shall receive any training from Client in the skills necessary to perform the services required by this Agreement.

☐ Client shall not require Contractor or Contractor's employees or contract personnel to devote full time to performing the services required by this Agreement.

9. Local, State, and Federal Taxes

Contractor shall pay all income taxes and FICA (Social Security and Medicare taxes) incurred while performing services under this Agreement. Client will not:

- withhold FICA from Contractor's payments or make FICA payments on Contractor's behalf
- make state or federal unemployment compensation contributions on Contractor's behalf, or
- withhold state or federal income tax from Contractor's payments.

The charges included here do not include taxes. If Contractor is required to pay any federal, state, or local sales, use, property, or value added taxes based on the services provided under this Agreement, the taxes shall be separately billed to Client. Client shall be responsible for paying any interest or penalties incurred due to late payment or nonpayment of any taxes by Client.

10. Exclusive Agreement

This is the entire Agreement between Contractor and Client.

11. Confidentiality (Optional)

(Check if applicable.)

☐ During the term of this Agreement and for _____ ☐ months ☐ years afterward, Contractor will use reasonable care to prevent the unauthorized use or dissemination of Client's confidential information. Reasonable care means at least the same degree of care Contractor uses to protect its own confidential information from unauthorized disclosure.

Confidential information is limited to information clearly marked as confidential, or disclosed orally and summarized and identified as confidential in a writing delivered to Contractor within 15 days of disclosure.

Confidential information does not include information that:

- the Contractor knew before Client disclosed it
- is or becomes public knowledge through no fault of Contractor
- Contractor obtains from sources other than Client who owe no duty of confidentiality to Client, or
- Contractor independently develops.

12. Resolving Disputes

(Choose Alternative A, B, or C and any desired optional clauses.)

☐ ALTERNATIVE A

If a dispute arises under this Agreement, any party may take the matter to court.

(Optional: Check if applicable.)

☐ If any court action is necessary to enforce this Agreement, the prevailing party shall be entitled to reasonable attorney fees, costs, and expenses in addition to any other relief to which the party may be entitled.

☐ ALTERNATIVE B

If a dispute arises under this Agreement, the parties agree to first try to resolve the dispute with the help of a mutually agreed-upon mediator in _____ [list city or county where mediation will occur]. Any costs and fees other than attorney fees associated with the mediation shall be shared equally by the parties. If the dispute is not resolved within 30 days after it is referred to the mediator, any party may take the matter to court.

(Optional: Check if applicable.)

☐ If any court action is necessary to enforce this Agreement, the prevailing party shall be entitled to reasonable attorney fees, costs, and expenses in addition to any other relief to which the party may be entitled.

☐ ALTERNATIVE C

If a dispute arises under this Agreement, the parties agree to first try to resolve the dispute with the help of a mutually agreed-upon mediator in _____ [list city or county where mediation will occur]. Any costs and fees other than attorney fees associated with the mediation shall be shared equally by the parties. If it proves impossible to arrive at a mutually satisfactory solution through mediation, the parties agree to submit the dispute to a mutually agreed-upon arbitrator in _____ [list city or county where arbitration will occur]. Judgment upon the award rendered by the arbitrator may be entered in any court having jurisdiction to do so. Costs of arbitration, including attorney fees, will be allocated by the arbitrator.

13. Notices

All notices and other communications in connection with this Agreement shall be in writing and shall be considered given as follows:

- when delivered personally to the recipient's address as stated on this Agreement
- three days after being deposited in the United States mail, with postage prepaid to the recipient's address as stated on this Agreement, or
- when sent by fax or electronic mail, such notice is effective upon receipt provided that a duplicate copy of the notice is promptly given by first class mail, or the recipient delivers a written confirmation of receipt.

14. No Partnership

This Agreement does not create a partnership relationship. Contractor does not have authority to enter into contracts on Client's behalf.

15. Applicable Law

This Agreement will be governed by the laws of the state of _____ .

16. Assignment and Delegation (Optional)

(Check if applicable.)

☐ Either Contractor or Client may assign its rights and may delegate its duties under this Agreement.

Signatures

Client: _____
Name of Client

By: _____
Signature

_____
Typed or Printed Name

Title: _____

Date: _____

Contractor: _____
Name of Contractor

By: _____
Signature

_____
Typed or Printed Name

Title: _____

Taxpayer ID Number: _____

Date: _____

INDEPENDENT CONTRACTOR AGREEMENT
FOR ACCOUNTANT & BOOKKEEPER

This Agreement is made between _____ ("Client")

with a principal place of business at _____

and _____ ("Contractor")

with a principal place of business at _____ .

1. Services to be Performed

(Choose Alternative A or B.)

☐ ALTERNATIVE A

Contractor agrees to perform the following services: _____

_____

_____

_____

_____

☐ ALTERNATIVE B

Contractor agrees to perform the services described in Exhibit A, which is attached to and made part of this Agreement.

2. Payment

(Choose Alternative A or B and optional provision, if desired.)

☐ ALTERNATIVE A

In consideration for the services to be performed by Contractor, Client agrees to pay Contractor $ _____ .

☐ ALTERNATIVE B

In consideration for the services to be performed by Contractor, Client agrees to pay Contractor at the rate of $ _____ per hour.

(Optional: Check and complete if applicable.)

☐ Contractor's total compensation shall not exceed $ _____ without Client's written consent.

3. Terms of Payment

(Choose Alternative A, B, or C.)

☐ ALTERNATIVE A

Upon completing Contractor's services under this Agreement, Contractor shall submit an invoice. Client shall pay Contractor within _____ days from the date of Contractor's invoice.

☐ ALTERNATIVE B

Contractor shall be paid $_____ upon signing this Agreement and the remaining amount due when Contractor completes the services and submits an invoice. Client shall pay Contractor within _____ days from the date of Contractor's invoice.

☐ ALTERNATIVE C

Contractor shall send Client an invoice monthly. Client shall pay Contractor within _____ days from the date of each invoice.

4. Late Fees (Optional)

(Check and complete if applicable.)

☐ Late payments by Client shall be subject to late penalty fees of _____ % per month from the due date until the amount is paid.

5. Expenses

Client shall reimburse Contractor for the following expenses that are attributable directly to work performed under this Agreement:

- travel expenses other than normal commuting, including airfares, rental vehicles, and highway mileage in company or personal vehicles at _____ cents per mile
- telephone, fax, online, and telegraph charges
- postage and courier services
- printing and reproduction
- computer services, and
- other expenses resulting from the work performed under this Agreement.

Contractor shall submit an itemized statement of Contractor's expenses. Client shall pay Contractor within 30 days from the date of each statement.

6. Materials

Client shall make available to Contractor, at Client's expense, the following materials: _____

_____

_____

These items will be provided to Contractor by _____ .

7. Term of Agreement

This Agreement will become effective when signed by both parties and will end no later than _____.

8. Terminating the Agreement

(Choose Alternative A or B.)

☐ ALTERNATIVE A

With reasonable cause, either party may terminate this Agreement effective immediately by giving written notice of cause for termination. Reasonable cause includes:

- a material violation of this Agreement, or
- nonpayment of Contractor's compensation after 20 days' written demand for payment.

Contractor shall be entitled to full payment for services performed prior to the effective date of termination.

☐ ALTERNATIVE B

Either party may terminate this Agreement at any time by giving _____ days' written notice of termination. Contractor shall be entitled to full payment for services performed prior to the date of termination.

9. Independent Contractor Status

Contractor is an independent contractor, not Client's employee. Contractor's employees or subcontractors are not Client's employees. Contractor and Client agree to the following rights consistent with an independent contractor relationship (check all that apply):

☐ Contractor has the right to perform services for others during the term of this Agreement.

☐ Contractor has the sole right to control and direct the means, manner, and method by which the services required by this Agreement will be performed.

☐ Contractor has the right to hire assistants as subcontractors or to use employees to provide the services required by this Agreement.

☐ The Contractor or Contractor's employees or subcontractors shall perform the services required by this Agreement; Client shall not hire, supervise, or pay any assistants to help Contractor.

☐ Neither Contractor nor Contractor's employees or subcontractors shall receive any training from Client in the skills necessary to perform the services required by this Agreement.

☐ Client shall not require Contractor or Contractor's employees or subcontractors to devote full time to performing the services required by this Agreement.

☐ Neither Contractor nor Contractor's employees or subcontractors are eligible to participate in any employee pension, health, vacation pay, sick pay, or other fringe benefit plan of Client.

10. Professional Obligations

Contractor shall perform all services under this Agreement in accordance with generally accepted accounting practices and principles. This Agreement is subject to the laws, rules, and regulations governing the accounting profession imposed by government authorities or professional associations of which Contractor is a member.

11. Local, State, and Federal Taxes

Contractor shall pay all income taxes and FICA (Social Security and Medicare taxes) incurred while performing services under this Agreement. Client will not:

• withhold FICA from Contractor's payments or make FICA payments on Contractor's behalf

• make state or federal unemployment compensation contributions on Contractor's behalf, or

• withhold state or federal income tax from Contractor's payments.

The charges included here do not include taxes. If Contractor is required to pay any federal, state, or local sales, use, property, or value added taxes based on the services provided under this Agreement, the taxes shall be separately billed to Client. Client shall be responsible for paying any interest or penalties incurred due to late payment or nonpayment of any taxes by Client.

12. Exclusive Agreement

This is the entire Agreement between Contractor and Client.

13. Modifying the Agreement (Optional)

(Check if applicable.)

☐ This Agreement may be modified only by a writing signed by both parties.

14. Resolving Disputes

(Choose Alternative A, B or C and any desired optional clauses.)

☐ ALTERNATIVE A

If a dispute arises under this Agreement, any party may take the matter to court.

(Optional: Check if applicable.)

☐ If any court action is necessary to enforce this Agreement, the prevailing party shall be entitled to reasonable attorney fees, costs, and expenses in addition to any other relief to which the party may be entitled.

☐ ALTERNATIVE B

If a dispute arises under this Agreement, the parties agree to first try to resolve the dispute with the help of a mutually agreed-upon mediator in _____ [list city or county where mediation will occur]. Any costs and fees other than attorney fees associated with the mediation shall be shared equally by the parties. If the dispute is not resolved within 30 days after it is referred to the mediator, any party may take the matter to court.

(Optional: Check if applicable.)

☐ If any court action is necessary to enforce this Agreement, the prevailing party shall be entitled to reasonable attorney fees, costs, and expenses in addition to any other relief to which the party may be entitled.

☐ ALTERNATIVE C

If a dispute arises under this Agreement, the parties agree to first try to resolve the dispute with the help of a mutually agreed-upon mediator in _____ [list city or county where mediation will occur]. Any costs and fees other than attorney fees associated with the mediation shall be shared equally by the parties. If it proves impossible to arrive at a mutually satisfactory solution through mediation, the parties agree to submit the dispute to a mutually agreed-upon arbitrator in _____ [list city or county where arbitration will occur]. Judgment upon the award rendered by the arbitrator may be entered in any court having jurisdiction to do so. Costs of arbitration, including attorney fees, will be allocated by the arbitrator.

15. Notices

All notices and other communications in connection with this Agreement shall be in writing and shall be considered given as follows:

- when delivered personally to the recipient's address as stated on this Agreement
- three days after being deposited in the United States mail, with postage prepaid to the recipient's address as stated on this Agreement, or
- when sent by fax or electronic mail, such notice is effective upon receipt provided that a duplicate copy of the notice is promptly given by first class mail, or the recipient delivers a written confirmation of receipt.

16. No Partnership

This Agreement does not create a partnership relationship. Neither party has authority to enter into contracts on the other's behalf.

17. Applicable Law

This Agreement will be governed by the laws of the state of _____ .

18. Assignment and Delegation

Either Contractor or Client may assign its rights or delegate any of its duties under this Agreement.

Signatures

Client: _____

Name of Client

By: _____

Signature

_____

Typed or Printed Name

Title: _____

Date: _____

Contractor: _____

Name of Contractor

By: _____

Signature

_____

Typed or Printed Name

Title: _____

Taxpayer ID Number: _____

Date: _____

INDEPENDENT CONTRACTOR AGREEMENT
FOR SOFTWARE CONSULTANT

This Agreement is made between _____ ("Client")

with a principal place of business at _____

and _____ ("Consultant")

with a principal place of business at _____ .

1. Services to Be Performed

(Choose Alternative A or B.)

☐ ALTERNATIVE A

Consultant agrees to perform the following services for Client: _____

_____

_____

_____

☐ ALTERNATIVE B

Consultant agrees to perform the services described in Exhibit A, which is attached to and made part of this Agreement.

2. Payment

(Choose Alternative A, B, or C and optional provision, if desired.)

☐ ALTERNATIVE A

Consultant shall be paid $_____ upon execution of this agreement and $_____ upon completion of the work as detailed in Clause 1.

☐ ALTERNATIVE B

Client shall pay Consultant a fixed fee of $_____, in _____ installments according to the payment schedule described in Exhibit _____ , which is attached to and made part of this Agreement.

☐ ALTERNATIVE C

Consultant shall be compensated at the rate of $_____ per _____ specify ☐ hour, ☐ day, ☐ week, ☐ month.

(Optional: Check if applicable.)

☐ Unless otherwise agreed upon in writing by Client, Client's maximum liability for all services performed during the term of this Agreement shall not exceed $_____ .

3. Invoices

Consultant shall submit invoices for all services rendered. Client shall pay the amounts due within _____ days of the date of each invoice.

4. Late Fees (Optional)

(Check if applicable.)

☐ Late payments by Client shall be subject to late penalty fees of _____% per month from the due date until the amount is paid.

5. Expenses

(Choose Alternative A or B and optional provision, if desired.)

☐ ALTERNATIVE A

Consultant shall be responsible for all expenses incurred while performing services under this Agreement.

(Optional: Check if applicable.)

☐ However, Client shall reimburse Consultant for all reasonable travel and living expenses necessarily incurred by Consultant while away from Consultant's regular place of business to perform services under this Agreement. Consultant shall submit an itemized statement of such expenses. Client shall pay Consultant within 30 days from the date of each statement.

☐ ALTERNATIVE B

Client shall reimburse Consultant for the following expenses that are attributable directly to work performed under this Agreement:

- travel expenses other than normal commuting, including airfares, rental vehicles, and highway mileage in company or personal vehicles at _____ cents per mile
- telephone, fax, online, and telegraph charges
- postage and courier services
- printing and reproduction
- computer services, and
- other expenses resulting from the work performed under this Agreement.

Consultant shall submit an itemized statement of Consultant's expenses. Client shall pay Consultant within 30 days from the date of each statement.

6. Materials

Consultant will furnish all materials, equipment, and supplies used to provide the services required by this Agreement.

7. Term of Agreement

This agreement will become effective when signed by both parties and will terminate on the earlier of:

- the date Consultant completes the services required by this Agreement
- _____ [date], or
- the date a party terminates the Agreement as provided below.

8. Terminating the Agreement

(Choose Alternative A or B.)

☐ ALTERNATIVE A

With reasonable cause, either party may terminate this Agreement effective immediately by giving written notice of termination for cause. Reasonable cause includes:

- a material violation of this agreement, or
- nonpayment of Consultant's compensation after 20 days' written demand for payment.

Consultant shall be entitled to full payment for services performed prior to the effective date of termination.

☐ ALTERNATIVE B

Either party may terminate this Agreement at any time by giving _____ days' written notice of termination without cause. Consultant shall be entitled to full payment for services performed prior to the effective date of termination.

9. Independent Contractor Status

Consultant is an independent contractor, and neither Consultant nor Consultant's staff is, or shall be deemed, Client's employees. In its capacity as an independent contractor, Consultant agrees and represents, and Client agrees, as follows (check all that apply):

☐ Consultant has the right to perform services for others during the term of this Agreement subject to noncompetition provisions set out in this Agreement, if any.

☐ Consultant has the sole right to control and direct the means, manner, and method by which the services required by this Agreement will be performed.

☐ Consultant has the right to perform the services required by this Agreement at any place or location and at such times as Consultant may determine.

☐ Consultant will furnish all equipment and materials used to provide the services required by this Agreement, except to the extent that Consultant's work must be performed on or with Client's computer or existing software.

☐ The services required by this Agreement shall be performed by Consultant, or Consultant's staff, and Client shall not be required to hire, supervise, or pay any assistants to help Consultant.

☐ Consultant is responsible for paying all ordinary and necessary expenses of its staff.

☐ Neither Consultant nor Consultant's staff shall receive any training from Client in the professional skills necessary to perform the services required by this Agreement.

☐ Neither Consultant nor Consultant's staff shall be required to devote full time to the performance of the services required by this Agreement.

☐ Client shall not provide insurance coverage of any kind for Consultant or Consultant's staff.

☐ Client shall not withhold from Consultant's compensation any amount that would normally be withheld from an employee's pay.

10. Intellectual Property Ownership

(Choose Alternative A or B and optional provision, if desired.)

☐ ALTERNATIVE A

Consultant assigns to Client its entire right, title, and interest in anything created or developed by Consultant for Client under this Agreement ("Work Product") including all patents, copyrights, trade secrets, and other proprietary rights. This assignment is conditioned upon full payment of the compensation due Consultant under this Agreement.

Consultant shall, at no charge to Client, execute and aid in the preparation of any papers that Client may consider necessary or helpful to obtain or maintain—at Client's expense—any patents, copyrights, trademarks, or other proprietary rights. Client shall reimburse Consultant for reasonable out-of-pocket expenses incurred under this provision.

(Optional: Check if applicable, then check applicable provision.)

☐ Client grants to Consultant a nonexclusive

    ☐ irrevocable license to use the Work Product

      OR:

    ☐ license for the term of _____ years to use the Work Product.

☐ ALTERNATIVE B

Consultant shall retain all copyright, patent, trade secret, and other intellectual property rights Consultant may have in anything created or developed by Consultant for Client under this Agreement ("Work Product"). Consultant grants Client a nonexclusive worldwide license to use and sublicense the use of the Work Product for the purpose of developing and marketing its products, but not for the purpose of marketing Work Product separate from its products. The license shall have a perpetual term and may not be transferred by Client. This license is conditioned upon full payment of the compensation due Consultant under this Agreement.

11. Consultant's Materials

Consultant owns or holds a license to use and sublicense various materials in existence before the start date of this Agreement ("Consultant's Materials"). Consultant may, at it's option, include Consultant's Materials in the work performed under this Agreement.

(Choose Alternative A or B.)

☐ ALTERNATIVE A

Consultant retains all right, title, and interest, including all copyright, patent rights, and trade secret rights, in Consultant's Materials. Subject to full payment of the consulting fees due under this Agreement, Consultant grants Client a nonexclusive worldwide license to use and sublicense the use of Consultant's Materials for the purpose of developing and marketing its products, but not for the purpose of marketing Consultant's Materials separate from its products. The license shall have a perpetual term and may not be transferred by Client. Client shall make no other commercial use of Consultant's Materials without Consultant's written consent.

(Optional: Check and complete if applicable.)

☐ This license is granted subject to the following terms: _____.

☐ Consultant's Materials include, but are not limited to, those items identified in Exhibit _____ , attached to and made part of this Agreement.

☐ ALTERNATIVE B

Consultant retains all right, title, and interest, including all copyright, patent rights, and trade secret rights, in Consultant's Materials. Subject to full payment of the consulting fees due under this Agreement, Consultant grants Client a nonexclusive worldwide license to use Consultant's Materials in the following product(s):

_____

The license shall have a perpetual term and may not be transferred by Client. Client shall make no other commercial use of Consultant's Materials without Consultant's written consent.

☐ Consultant's Materials include, but are not limited to, those items identified in Exhibit _____, attached to and made part of this Agreement.

12. Confidentiality

During the term of this Agreement and for _____ ☐ months ☐ years afterward, Consultant will use reasonable care to prevent the unauthorized use or dissemination of Client's confidential information. Reasonable care means at least the same degree of care Consultant uses to protect its own confidential information from unauthorized disclosure.

Confidential information is limited to information clearly marked as confidential, or disclosed orally and summarized and identified as confidential in a writing delivered to Consultant within 15 days of disclosure.

Confidential information does not include information that:

- the Consultant knew before Client disclosed it
- is or becomes public knowledge through no fault of Consultant
- Consultant obtains from sources other than Client who owe no duty of confidentiality to Client, or
- Consultant independently develops.

13. Warranties

(Choose Alternative A or B.)

☐ ALTERNATIVE A

THE GOODS OR SERVICES FURNISHED UNDER THIS AGREEMENT ARE PROVIDED AS IS WITHOUT ANY EX-PRESS OR IMPLIED WARRANTIES OR REPRESENTATIONS; INCLUDING, WITHOUT LIMITATION, ANY IMPLIED WARRANTIES OF MERCHANTABILITY OR FITNESS FOR A PARTICULAR PURPOSE.

☐ ALTERNATIVE B

Consultant warrants that all services performed under this Agreement shall be performed consistent with gener-ally prevailing professional or industry standards. Client must report any deficiencies in Consultant's services to Consultant in writing within _____ days of performance to receive warranty remedies.

Client's exclusive remedy for any breach of the above warranty shall be the reperformance of Consultant's services. If Consultant is unable to reperform the services, Client shall be entitled to recover the fees paid to Consultant for the deficient services.

THIS WARRANTY IS EXCLUSIVE AND IN LIEU OF ALL OTHER WARRANTIES, WHETHER EXPRESS OR IMPLIED, INCLUDING ANY IMPLIED WARRANTIES OF MERCHANTABILITY OR FITNESS FOR A PARTICULAR PURPOSE AND ANY ORAL OR WRITTEN REPRESENTATIONS, PROPOSALS, OR STATEMENTS MADE PRIOR TO THIS AGREEMENT.

14. Limitation on Consultant's Liability to Client

- In no event shall Consultant be liable to Client for lost profits of Client or special, incidental, or consequen-tial damages (even if Consultant has been advised of the possibility of such damages).
- Consultant's total liability under this Agreement for damages, costs, and expenses, regardless of cause, shall not exceed the total amount of fees paid to Consultant by Client under this Agreement.
- Client shall indemnify Consultant against all claims, liabilities, and costs, including reasonable attorney fees, of defending any third-party claim or suit, other than for infringement of intellectual property rights,

arising out of or in connection with Consultant's performance under this Agreement. Consultant shall promptly notify Client in writing of such claim or suit, and Client shall have the right to fully control the defense and any settlement of the claim or suit.

15. Taxes

The charges included here do not include taxes. If Consultant is required to pay any federal, state, or local sales, use, property, or value added taxes based on the services provided under this Agreement, the taxes shall be billed separately to Client. Client shall be responsible for paying any interest or penalties incurred due to late payment or nonpayment of such taxes by Client.

16. Contract Changes

Client and Consultant recognize that:

- Consultant's original cost and time estimates may be too low due to unforeseen events, or to factors unknown to Consultant when this Agreement was made
- Client may desire a mid-project change in Consultant's services that would add time and cost to the project and possibly inconvenience Consultant, or
- other provisions of this Agreement may be difficult to carry out due to unforeseen circumstances.

If any intended changes or any other events beyond the parties' control require adjustments to this Agreement, the parties shall make a good faith effort to agree on all necessary particulars. Such agreements shall be put in writing, signed by the parties, and added to this Agreement.

17. Resolving Disputes

(Choose Alternative A, B or C and any desired optional clauses.)

☐ ALTERNATIVE A

If a dispute arises under this Agreement, any party may take the matter to court.

(Optional: Check if applicable.)

☐ If any court action is necessary to enforce this Agreement, the prevailing party shall be entitled to reasonable attorney fees, costs, and expenses in addition to any other relief to which the party may be entitled.

☐ ALTERNATIVE B

If a dispute arises under this Agreement, the parties agree to first try to resolve the dispute with the help of a mutually agreed-upon mediator in _____ [list city or county where mediation will occur]. Any costs and fees other than attorney fees associated with the mediation shall be shared equally by the parties. If the dispute is not resolved within 30 days after it is referred to the mediator, any party may take the matter to court.

(Optional: Check if applicable.)

☐ If any court action is necessary to enforce this Agreement, the prevailing party shall be entitled to reasonable attorney fees, costs, and expenses in addition to any other relief to which the party may be entitled.

☐ ALTERNATIVE C

If a dispute arises under this Agreement, the parties agree to first try to resolve the dispute with the help of a mutually agreed-upon mediator in _____ [list city or county where mediation will occur]. Any costs and fees other than attorney fees associated with the mediation shall be shared equally by the parties. If it proves impossible to arrive at a mutually satisfactory solution through mediation, the parties agree to submit the dispute to a mutually agreed-upon arbitrator in _____

[list city or county where arbitration will occur]. Judgment upon the award rendered by the arbitrator may be entered in any court having jurisdiction to do so. Costs of arbitration, including attorney fees, will be allocated by the arbitrator.

18. Exclusive Agreement

This is the entire Agreement between Consultant and Client.

19. Applicable Law

This Agreement will be governed by the laws of the state of _____ .

20. Notices

All notices and other communications in connection with this Agreement shall be in writing and shall be considered given as follows:

- when delivered personally to the recipient's address as stated on this Agreement
- three days after being deposited in the United States mail, with postage prepaid to the recipient's address as stated on this Agreement, or
- when sent by fax or electronic mail, such notice is effective upon receipt provided that a duplicate copy of the notice is promptly given by first class mail, or the recipient delivers a written confirmation of receipt.

21. No Partnership

This Agreement does not create a partnership relationship. Neither party has authority to enter into contracts on the other's behalf.

Signatures

Client: _____
Name of Client

By: _____
Signature

_____
Typed or Printed Name

Title:_____

Date: _____

Consultant: _____
Name of Consultant

By: _____
Signature

_____
Typed or Printed Name

Title:_____

Taxpayer ID Number: _____

Date: _____

INDEPENDENT CONTRACTOR AGREEMENT
FOR CREATIVE CONTRACTOR

This Agreement is made between _____ ("Client")

with a principal place of business at _____

and _____ ("Contractor"), with a principal place

of business at _____ .

1. Services to be Performed

(Choose Alternative A or B.)

☐ ALTERNATIVE A

Contractor agrees to perform the following services on Client's behalf:_____

_____

_____

_____

☐ ALTERNATIVE B

Contractor agrees to perform the services described in Exhibit A, which is attached to this Agreement.

2. Payment

(Choose Alternative A or B and optional provision, if desired.)

☐ ALTERNATIVE A

In consideration for the services to be performed by Contractor, Client agrees to pay Contractor $_____ .

☐ ALTERNATIVE B

In consideration for the services to be performed by Contractor, Client agrees to pay Contractor at the rate of

$_____ per _____ ☐ hour, ☐ day, ☐ week, ☐ other_____.

(Optional: Check and complete if applicable.)

☐ Contractor's total compensation shall not exceed $_____ without Client's written consent.

3. Terms of Payment

(Choose Alternative A, B, C, or D.)

☐ ALTERNATIVE A

Upon completing Contractor's services under this Agreement, Contractor shall submit an invoice. Client shall pay Contractor within _____ days from the date of Contractor's invoice.

☐ ALTERNATIVE B

Contractor shall be paid $ _____ upon signing this Agreement and the remaining amount due when Contractor completes the services and submits an invoice. Client shall pay Contractor within _____ days from the date of Contractor's invoice.

☐ ALTERNATIVE C

Contractor shall be paid according to the Schedule of Payments set forth in Exhibit _____ attached to and made part of this Agreement.

☐ ALTERNATIVE D

Contractor shall send Client an invoice monthly. Client shall pay Contractor within _____ days from the date of each invoice.

4. Late Fees (Optional)

(Check and complete if applicable.)

☐ Late payments by Client shall be subject to late penalty fees of _____% per month from the due date until the amount is paid.

5. Expenses

(Choose Alternative A or B and optional provision, if desired.)

☐ ALTERNATIVE A

Contractor shall be responsible for all expenses incurred while performing services under this Agreement.

(Optional: Check if applicable.)

☐ However, Client shall reimburse Contractor for all reasonable travel and living expenses necessarily incurred by Contractor while away from Contractor's regular place of business to perform services under this Agreement. Contractor shall submit an itemized statement of such expenses. Client shall pay Contractor within 30 days from the date of each statement.

☐ ALTERNATIVE B

Client shall reimburse Contractor for the following expenses that are attributable directly to work performed under this Agreement:

- travel expenses other than normal commuting, including airfares, rental vehicles, and highway mileage in company or personal vehicles at _____ cents per mile
- telephone, fax, online, and telegraph charges
- postage and courier services
- printing and reproduction
- computer services, and
- other expenses resulting from the work performed under this Agreement.

Contractor shall submit an itemized statement of Contractor's expenses. Client shall pay Contractor within 30 days from the date of each statement.

6. Materials

Contractor will furnish all materials and equipment used to provide the services required by this Agreement.

7. Intellectual Property Ownership

(Choose Alternative A or B.)

☐ ALTERNATIVE A

Contractor hereby licenses to Client the following intellectual property rights in the work created or developed by Contractor under this Agreement: _____.

This license is conditioned upon full payment of the compensation due Contractor under this Agreement. Contractor reserves all rights not expressly granted to Client by this Agreement.

(Check applicable provision.)

☐ The rights granted above are exclusive to Client.

OR:

☐ The rights granted above are nonexclusive.

☐ ALTERNATIVE B

Contractor assigns to Client all patent, copyright, and trade secret rights in anything created or developed by Contractor for Client under this Agreement. This assignment is conditioned upon full payment of the compensation due Contractor under this Agreement.

Contractor shall help prepare any documents Client considers necessary to secure any copyright, patent, or other intellectual property rights at no charge to Client. However, Client shall reimburse Contractor for reasonable out-of-pocket expenses.

8. Reusable Materials (Optional)

(Check if applicable.)

☐ Contractor owns or holds a license to use and sublicense various materials in existence before the start date of this Agreement ("Contractor's Materials"). Contractor's Materials include, but are not limited to, those items identified in Exhibit _____, attached to and made part of this Agreement. Contractor may, at its option, include Contractor's Materials in the work performed under this Agreement. Contractor retains all right, title, and interest, including all copyrights, patent rights, and trade secret rights, in Contractor's Materials. Contractor grants Client a royalty-free nonexclusive license to use any Contractor's Materials incorporated into the work performed by Contractor under this Agreement. The license shall have a perpetual term and may not be transferred by Client.

9. Releases

(Choose Alternative A or B.)

☐ ALTERNATIVE A

Client shall obtain all necessary copyright permissions and privacy releases for materials included in the Designs at Client's request. Client shall indemnify Contractor against all claims and expenses, including reasonable attorney fees, due to Client's failure to obtain such permissions or releases.

☐ ALTERNATIVE B

Contractor shall obtain all necessary copyright permissions and privacy releases for materials included in the Designs at Client's request.

10. Copyright Notice and Credit Line

A copyright notice and credit line in Contractor's name shall accompany any reproduction of the Designs in the following form:

_____

_____

11. Term of Agreement

This agreement will become effective when signed by both parties and will terminate on the earlier of:

- the date Contractor completes the services required by this Agreement
- _____ [date], or
- the date a party terminates the Agreement as provided below.

12. Terminating the Agreement

(Choose Alternative A or B.)

☐ ALTERNATIVE A

With reasonable cause, either party may terminate this Agreement effective immediately by giving written notice of termination for cause. Reasonable cause includes:

- a material violation of this Agreement, or
- nonpayment of Contractor's compensation after 20 days' written demand for payment.

Contractor shall be entitled to full payment for services performed prior to the effective date of termination.

☐ ALTERNATIVE B

Either party may terminate this Agreement at any time by giving ____ days' written notice of termination. Contractor shall be entitled to full payment for services performed prior to the date of termination.

13. Independent Contractor Status

Contractor is an independent contractor, not Client's employee. Contractor's employees or subcontractors are not Client's employees. Contractor and Client agree to the following rights consistent with an independent contractor relationship (check all that apply):

☐ Contractor has the right to perform services for others during the term of this Agreement.

☐ Contractor has the sole right to control and direct the means, manner, and method by which the services required by this Agreement will be performed.

☐ Contractor has the right to hire assistants as subcontractors or to use employees to provide the services required by this Agreement.

☐ Contractor or Contractor's employees or subcontractors shall perform the services required by this Agreement; Client shall not hire, supervise, or pay any assistants to help Contractor.

☐ Neither Contractor nor Contractor's employees or subcontractors shall receive any training from Client in the skills necessary to perform the services required by this Agreement.

☐ Client shall not require Contractor or Contractor's employees or subcontractors to devote full time to performing the services required by this Agreement.

☐ Neither Contractor nor Contractor's employees or subcontractors are eligible to participate in any employee pension, health, vacation pay, sick pay, or other fringe benefit plan of Client.

14. Local, State, and Federal Taxes

Contractor shall pay all income taxes and FICA (Social Security and Medicare taxes) incurred while performing services under this Agreement. Client will not:

- withhold FICA from Contractor's payments or make FICA payments on Contractor's behalf
- make state or federal unemployment compensation contributions on Contractor's behalf, or
- withhold state or federal income tax from Contractor's payments.

The charges included here do not include taxes. If Contractor is required to pay any federal, state, or local sales, use, property, or value added taxes based on the services provided under this Agreement, the taxes shall be separately billed to Client. Client shall be responsible for paying any interest or penalties incurred due to late payment or nonpayment of any taxes by Client.

15. Exclusive Agreement

This is the entire Agreement between Contractor and Client.

16. Modifying the Agreement (Optional)

(Check if applicable.)

☐ Client and Contractor recognize that:

- Contractor's original cost and time estimates may be too low due to unforeseen events or to factors unknown to Contractor when this Agreement was made
- Client may desire a mid-project change in Contractor's services that would add time and cost to the project and possibly inconvenience Contractor, or
- other provisions of this Agreement may be difficult to carry out due to unforeseen circumstances.

If any intended changes or any other events beyond the parties' control require adjustments to this Agreement, the parties shall make a good faith effort to agree on all necessary particulars. Such agreements shall be put in writing, signed by the parties, and added to this Agreement.

17. Resolving Disputes

(Choose Alternative A, B, or C and any desired optional clauses.)

☐ ALTERNATIVE A

If a dispute arises under this Agreement, any party may take the matter to court.

(Optional: Check if applicable.)

☐ If any court action is necessary to enforce this Agreement, the prevailing party shall be entitled to reasonable attorney fees, costs, and expenses in addition to any other relief to which the party may be entitled.

☐ ALTERNATIVE B

If a dispute arises under this Agreement, the parties agree to first try to resolve the dispute with the help of a mutually agreed-upon mediator in _____ [list city or county where mediation will occur]. Any costs and fees other than attorney fees associated with the mediation shall be shared equally by the parties. If the dispute is not resolved within 30 days after it is referred to the mediator, any party may take the matter to court.

(Optional: Check if applicable.)

☐ If any court action is necessary to enforce this Agreement, the prevailing party shall be entitled to reasonable attorney fees, costs, and expenses in addition to any other relief to which the party may be entitled.

☐ ALTERNATIVE C

If a dispute arises under this Agreement, the parties agree to first try to resolve the dispute with the help of a mutually agreed-upon mediator in _____ [list city or county where mediation will occur]. Any costs and fees other than attorney fees associated with the mediation shall be shared equally by the parties. If it proves impossible to arrive at a mutually satisfactory solution through mediation, the parties agree to submit the dispute to a mutually agreed-upon arbitrator in _____ [list city or county where arbitration will occur]. Judgment upon the award rendered by the arbitrator may be entered in any court having jurisdiction to do so. Costs of arbitration, including attorney fees, will be allocated by the arbitrator.

18. Limited Liability (Optional)

(Check and complete if applicable.)

☐ This provision allocates the risks under this Agreement between Contractor and Client. Contractor's pricing reflects the allocation of risk and limitation of liability specified below.

Contractor's total liability to Client under this Agreement for damages, costs, and expenses shall not exceed $_____ or the compensation received by Contractor under this Agreement, whichever is less.

However, contractor shall remain liable for bodily injury or personal property damage resulting from grossly negligent or willful actions of Contractor or Contractor's employees or agents while on Client's premises to the extent such actions or omissions were not caused by Client.

NEITHER PARTY TO THIS AGREEMENT SHALL BE LIABLE FOR THE OTHER'S LOST PROFITS, OR SPECIAL, INCIDENTAL, OR CONSEQUENTIAL DAMAGES, WHETHER IN AN ACTION IN CONTRACT OR TORT, EVEN IF THE PARTY HAS BEEN ADVISED BY THE OTHER PARTY OF THE POSSIBILITY OF SUCH DAMAGES.

19. Notices

All notices and other communications in connection with this Agreement shall be in writing and shall be considered given as follows:

- when delivered personally to the recipient's address as stated on this Agreement
- three days after being deposited in the United States mail with postage prepaid to the recipient's address as stated on this Agreement, or
- when sent by fax or electronic mail, such notice is effective upon receipt provided that a duplicate copy of the notice is promptly given by first class mail or the recipient delivers a written confirmation of receipt.

20. No Partnership

This Agreement does not create a partnership relationship. Neither party has authority to enter into contracts on the other's behalf.

21. Applicable Law

This Agreement will be governed by the laws of the state of _____.

22. Assignment and Delegation (Optional)

(Check if applicable.)

☐ Either Contractor or Client may assign rights or may delegate duties under this Agreement.

Signatures

Client: _____
Name of Client

By: _____
Signature

_____
Typed or Printed Name

Title: _____

Date: _____

Contractor: _____
Name of Contractor

By: _____
Signature

_____
Typed or Printed Name

Title: _____

Taxpayer ID Number: _____

Date: _____

INDEPENDENT CONTRACTOR AGREEMENT FOR CONSTRUCTION CONTRACTOR

This Agreement is made between _____ ("Owner")
with a principal place of business at _____
and _____ ("Contractor"), with a principal place of
business at _____ .

1. Services to Be Performed

Contractor shall furnish all labor and materials to construct and complete the project shown on the contract documents contained in Exhibit A, which is attached to and made part of this Agreement.

2. Payment

(Choose Alternative A or B.)

☐ ALTERNATIVE A

Owner shall pay Contractor for all labor and materials the sum of $ _____ .

☐ ALTERNATIVE B

Owner shall pay Contractor $_____ for labor. Materials shall be paid for by Owner upon delivery to the worksite or as follows: _____

_____

_____

3. Terms of Payment

(Choose Alternative A, B, or C.)

☐ ALTERNATIVE A

Upon completing Contractor's services under this Agreement, Contractor shall submit an invoice. Owner shall pay Contractor within _____ days from the date of Contractor's invoice.

☐ ALTERNATIVE B

Contractor shall be paid $_____ upon signing this Agreement and the remaining amount due when Contractor completes the services and submits an invoice. Owner shall pay Contractor within

_____ days from the date of Contractor's invoice.

☐ ALTERNATIVE C

Contractor shall be paid according to the Schedule of Payments set forth in Exhibit _____ , attached to and made part of this agreement.

4. Late Fees (Optional)

(Check and complete if applicable.)

☐ Late payments by Owner shall be subject to late penalty fees of _____ % per month from the due date until the amount is paid.

5. Time of Completion

The work to be performed under this Agreement shall commence on _____ and be substantially completed on or before _____ .

6. Permits and Approvals

(Choose Alternative A or B.)

☐ ALTERNATIVE A

Owner shall be responsible for determining which state and local permits are necessary for performing the specified work and for obtaining and paying for the permits.

☐ ALTERNATIVE B

Contractor shall be responsible for determining which state and local permits are necessary for performing the specified work and for obtaining and paying for the permits.

7. Warranty

Contractor warrants that all work shall be completed in a good workmanlike manner and in compliance with all building codes and other applicable laws.

8. Site Maintenance

Contractor agrees to be bound by the following conditions when performing the specified work:

- Contractor shall remove all debris and leave the premises in a broom-clean condition.
- Contractor shall perform the specified work during the following hours:

  _____ .

- Contractor agrees that disruptively loud activities shall be performed only at the following times:

  _____ .

- At the end of each day's work, Contractor's equipment shall be stored in the following location:

  _____ .

9. Subcontractors

Contractor may at its discretion engage subcontractors to perform services under this Agreement, but Contractor shall remain responsible for proper completion of this Agreement.

10. Independent Contractor Status

Contractor is an independent contractor, not Owner's employee. Contractor's employees or subcontractors are not Owner's employees. Contractor and Owner agree to the following rights consistent with an independent contractor relationship (check all that apply):

- ☐ Contractor has the right to perform services for others during the term of this Agreement.
- ☐ Contractor has the sole right to control and direct the means, manner, and method by which the services required by this Agreement will be performed.
- ☐ The Contractor or Contractor's employees or subcontractors shall perform the services required by this Agreement; Owner shall not hire, supervise, or pay any assistants to help Contractor.
- ☐ Owner shall not require Contractor or Contractor's employees or subcontractors to devote full time to performing the services required by this Agreement.
- ☐ Neither Contractor nor Contractor's employees or subcontractors are eligible to participate in any employee pension, health, vacation pay, sick pay, or other fringe benefit plan of Owner.

11. Local, State, and Federal Taxes

Contractor shall pay all income taxes and FICA (Social Security and Medicare taxes) incurred while performing services under this Agreement. Owner will not:

- withhold FICA from Contractor's payments or make FICA payments on Contractor's behalf
- make state or federal unemployment compensation contributions on Contractor's behalf, or
- withhold state or federal income tax from Contractor's payments.

The charges included here do not include taxes. If Contractor is required to pay any federal, state, or local sales, use, property, or value added taxes based on the services provided under this Agreement, the taxes shall be billed separately to Owner. Owner shall be responsible for paying any interest or penalties incurred due to late payment or nonpayment of any taxes by Owner.

12. Insurance

Contractor agrees to obtain adequate business liability insurance for injuries to its employees and others incurring loss or injury as a result of the acts of Contractor or its employees or subcontractors.

13. Terminating the Agreement

(Choose Alternative A or B.)

☐ ALTERNATIVE A

With reasonable cause, either Owner or Contractor may terminate this Agreement effective immediately by giving written notice of cause for termination. Reasonable cause includes:

- a material violation of this Agreement, or
- nonpayment of Contractor's compensation after 20 days' written demand for payment.

Contractor shall be entitled to full payment for services performed prior to the effective date of termination.

☐ ALTERNATIVE B

Either Owner or Contractor may terminate this Agreement at any time by giving _____ days' written notice of termination. Contractor shall be entitled to full payment for services performed prior to the date of termination.

14. Exclusive Agreement

This is the entire Agreement between Contractor and Owner.

15. Modifying the Agreement (Optional)

(Check if applicable.)

☐ Owner and Contractor recognize that:

- Contractor's original cost and time estimates may be too low due to unforeseen events or to factors unknown to Contractor when this Agreement was made
- Owner may desire a mid-project change in Contractor's services that would add time and cost to the project and possibly inconvenience Contractor, or
- other provisions of this Agreement may be difficult to carry out due to unforeseen circumstances.

If any intended changes or any other events beyond the parties' control require adjustments to this Agreement, the parties shall make a good faith effort to agree on all necessary particulars. Such agreements shall be put in writing, signed by the parties, and added to this Agreement.

16. Resolving Disputes

(Choose Alternative A, B, or C and any desired optional clauses.)

☐ ALTERNATIVE A

If a dispute arises under this Agreement, any party may take the matter to court.

(Optional: Check if applicable.)

☐ If any court action is necessary to enforce this Agreement, the prevailing party shall be entitled to reasonable attorney fees, costs, and expenses in addition to any other relief to which the party may be entitled.

☐ ALTERNATIVE B

If a dispute arises under this Agreement, the parties agree to first try to resolve the dispute with the help of a mutually agreed-upon mediator in _____ [list city or county where mediation will occur]. Any costs and fees other than attorney fees associated with the mediation shall be shared equally by the parties. If the dispute is not resolved within 30 days after it is referred to the mediator, any party may take the matter to court.

(Optional: Check if applicable.)

☐ If any court action is necessary to enforce this Agreement, the prevailing party shall be entitled to reasonable attorney fees, costs, and expenses in addition to any other relief to which the party may be entitled.

☐ ALTERNATIVE C

If a dispute arises under this Agreement, the parties agree to first try to resolve the dispute with the help of a mutually agreed-upon mediator in _____ [list city or county where mediation will occur]. Any costs and fees other than attorney fees associated with the mediation shall be shared equally by the parties. If it proves impossible to arrive at a mutually satisfactory solution through mediation, the parties agree to submit the dispute to a mutually agreed-upon arbitrator in

_____ [list city or county where arbitration will occur]. Judgment upon the award rendered by the arbitrator may be entered in any court having jurisdiction to do so. Costs of arbitration, including attorney fees, will be allocated by the arbitrator.

17. Notices

All notices and other communications in connection with this Agreement shall be in writing and shall be considered given as follows:

- when delivered personally to the recipient's address as stated on this Agreement
- three days after being deposited in the United States mail, with postage prepaid to the recipient's address as stated on this Agreement, or
- when sent by fax or electronic mail, such notice is effective upon receipt provided that a duplicate copy of the notice is promptly given by first class mail, or the recipient delivers a written confirmation of receipt.

18. No Partnership

This Agreement does not create a partnership relationship. Neither party has authority to enter into contracts on the other's behalf.

19. Applicable Law

This Agreement will be governed by the laws of the state of _____ .

Signatures

Owner: _____

Name of Owner

By: _____

Signature

_____

Typed or Printed Name

Title: _____

Date: _____

Contractor: _____

Name of Contractor

By: _____

Signature

_____

Typed or Printed Name

Title: _____

Taxpayer ID Number: _____

Date: _____

INDEPENDENT CONTRACTOR AGREEMENT FOR COURIERS, MESSENGERS, AND DELIVERY PEOPLE

This Agreement is made between _____ ("Client")

with a principal place of business at _____

and _____ ("Contractor"),

with a principal place of business at _____ .

1. Services to Be Performed

(Check and complete applicable provision.)

☐ ALTERNATIVE A

Contractor agrees to perform the following services: _____

_____

☐ ALTERNATIVE B

Contractor agrees to perform the services described in Exhibit A, which is attached to this Agreement.

2. Payment

In consideration for the services to be performed by Contractor, Client agrees to pay Contractor at the following rates:_____ .

    Contractor shall be paid within a reasonable time after Contractor submits an invoice to Client. The invoice should include the following: an invoice number, the dates covered by the invoice, and a summary of the work performed.

3. Terms of Payment

Contractor shall send Client an invoice

☐ Upon completion of the services required by this Agreement

OR

☐ _____ (choose time period—for example, monthly].

Client shall pay Contractor within _____ days from the date of each invoice.

4. Late Fees

(Check if applicable.)

☐ Late payments by Client shall be subject to late penalty fees of _____% per month from the
   due date until the amount is paid.

5. Expenses

(Choose Alternative A or B.)

☐ ALTERNATIVE A

Contractor shall be responsible for all expenses incurred while performing services under this Agreement.

☐ ALTERNATIVE B

Client shall reimburse Contractor for the following expenses that are attributable directly to work performed under this Agreement: _____ .

6. Equipment and Materials (Optional)

(Check if applicable.)

☐ Contractor will furnish all equipment and materials used to provide the services required by this Agreement.

7. Term of Agreement

This agreement will become effective when signed by both parties and will terminate on the earlier of:

- the date Contractor completes the services required by this Agreement
- _____ [date], or
- the date a party terminates the Agreement as provided below.

8. Terminating the Agreement

(Choose Alternative A or B.)

☐ ALTERNATIVE A

With reasonable cause, either Client or Contractor may terminate this Agreement, effective immediately upon giving written notice.

Reasonable cause includes:

- a material violation of this Agreement, or
- nonpayment of Contractor's compensation after 20 days written demand for payment.

☐ ALTERNATIVE B

Either party may terminate this Agreement at any time by giving _____ days' written notice to the other party of the intent to terminate. Contractor shall be entitled to full payment for services performed prior to the date of termination.

9. Independent Contractor Status

Contractor is an independent contractor, and neither Contractor nor Contractor's employees or contract personnel are, or shall be deemed, Client's employees. In its capacity as an independent contractor, Contractor agrees and represents, and Client agrees, as follows (check all that apply):

- Contractor has the right to perform services for others during the term of this Agreement.
- Contractor has the sole right to control and direct the means, manner, and method by which the services required by this Agreement will be performed. Contractor shall select the routes taken, starting and quitting times, days of work, and order the work is performed.
- Contractor has the right to hire assistants as subcontractors, or to use employees to provide the services required by this Agreement.
- Neither Contractor nor Contractor's employees or contract personnel shall be required to wear any uniforms provided by Client.
- The services required by this Agreement shall be performed by Contractor, Contractor's employees or contract personnel. Client shall not hire, supervise, or pay any assistants to help Contractor.
- Neither Contractor nor Contractor's employees or contract personnel shall receive any training from Client in the professional skills necessary to perform the services required by this Agreement.
- Neither Contractor nor Contractor's employees or contract personnel shall be required by Client to devote full time to the performance of the services required by this Agreement.

10. Local, State, and Federal Taxes

Contractor shall pay all income taxes and FICA (Social Security and Medicare taxes) incurred while performing services under this Agreement. Client will not:

- withhold FICA from Contractor's payments or make FICA payments on Contractor's behalf
- make state or federal unemployment compensation contributions on Contractor's behalf, or
- withhold state or federal income tax from Contractor's payments.

Contractor shall not be required to pay any federal, state, or local sales taxes based on Contractor's compensation for the services provided under this Agreement. If client is charged for such sales taxes, the taxes shall be separately billed to Client. Client shall be responsible for any interest or penalties incurred due to late payment or nonpayment of any taxes by Client.

11. Exclusive Agreement

This is the entire Agreement between Contractor and Client.

12. Modifying the Agreement (Optional)

(Check if applicable.)

This Agreement may be modified only by a writing signed by both parties.

13. Resolving Disputes

(Choose Alternative A, B, or C and any desired optional clauses.)

☐ ALTERNATIVE A

If a dispute arises under this Agreement, any party may take the matter to court.

(Optional: Check if applicable.)

If any court action is necessary to enforce this Agreement, the prevailing party shall be entitled to reasonable attorney fees, costs, and expenses in addition to any other relief to which the party may be entitled.

☐ ALTERNATIVE B

If a dispute arises under this Agreement, the parties agree to first try to resolve the dispute with the help of a mutually agreed-upon mediator in _____ [list city or county where mediation will occur]. Any costs and fees other than attorney fees associated with the mediation shall be shared equally by the parties. If the dispute is not resolved within 30 days after it is referred to the mediator, any party may take the matter to court.

(Optional: Check if applicable.)

If any court action is necessary to enforce this Agreement, the prevailing party shall be entitled to reasonable attorney fees, costs, and expenses in addition to any other relief to which the party may be entitled.

☐ ALTERNATIVE C

If a dispute arises under this Agreement, the parties agree to first try to resolve the dispute with the help of a mutually agreed-upon mediator in _____ [list city or county where mediation will occur]. Any costs and fees other than attorney fees associated with the mediation shall be shared equally by the parties. If it proves impossible to arrive at a mutually satisfactory solution through mediation, the parties agree to submit the dispute to a mutually agreed-upon arbitrator in _____
[list city or county where arbitration will occur]. Judgment upon the award rendered by the arbitrator may be

entered in any court having jurisdiction to do so. Costs of arbitration, including attorney fees, will be allocated by the arbitrator.

14. Limited Liability (Optional)

(Check if applicable.)

☐ This provision allocates the risks under this Agreement between Contractor and Client. Contractor's pricing reflects the allocation of risk and limitation of liability specified below.

Contractor's total liability to Client under this Agreement for damages, costs, and expenses, shall not exceed $_____ or the compensation received by Contractor under this Agreement, whichever is less. However, contractor shall remain liable for bodily injury or personal property damage resulting from grossly negligent or willful actions of Contractor or Contractor's employees or agents while on Client's premises to the extent such actions or omissions were not caused by Client.

NEITHER PARTY TO THIS AGREEMENT SHALL BE LIABLE FOR THE OTHER'S LOST PROFITS, OR SPECIAL, INCIDENTAL, OR CONSEQUENTIAL DAMAGES, WHETHER IN AN ACTION IN CONTRACT OR TORT, EVEN IF THE PARTY HAS BEEN HAS BEEN ADVISED BY THE OTHER PARTY OF THE POSSIBILITY OF SUCH DAMAGES.

15. Notices

All notices and other communications in connection with this Agreement shall be in writing and shall be considered given as follows:

- when delivered personally to the recipient's address as stated on this Agreement
- three days after being deposited in the United States mail, with postage prepaid to the recipient's address as stated on this Agreement, or
- when sent by fax or electronic mail, such notice is effective upon receipt provided that a duplicate copy of the notice is promptly given by first class mail, or the recipient delivers a written confirmation of receipt.

16. No Partnership

This Agreement does not create a partnership relationship. Neither party has authority to enter into contracts on the other's behalf.

17. Applicable Law

This Agreement will be governed by the laws of the state of _____ .

18. Assignment and Delegation

(Choose Alternative A or B.)

☐ ALTERNATIVE A

Either Contractor or Client may assign rights and may delegate duties under this Agreement.

☐ ALTERNATIVE B

Contractor may not assign or subcontract any rights or delegate any of its duties under this Agreement without Client's prior written approval.

19. Signatures

Owner:

_____
Name of Owner

By: _____
Signature

_____
Typed or Printed Name

Title: _____

Date: _____

Contractor: _____
Name of Contractor

By: _____
Signature

_____
Typed or Printed Name

Title: _____

Taxpayer ID Number: _____

Date: _____

CONTRACT AMENDMENT FORM

This Amendment is made between _____ and

_____ to amend the Original Agreement titled

_____ , signed by them on _____ .

The Original Agreement is amended as follows:

_____

_____

_____

_____

_____

_____

_____

_____

_____

_____

All provisions of the Original Agreement, except as modified by this Amendment, remain in full force and effect and are reaffirmed. If there is any conflict between this Amendment and any provision of the Original Agreement, the provisions of this Amendment shall control.

Client: _____
Name of Client

By: _____
Signature

_____
Typed or Printed Name

Title: _____

Date: _____

Consultant/Contractor: _____
Name of Consultant/Contractor

By: _____
Signature

_____
Typed or Printed Name

Title: _____

Taxpayer ID Number: _____

Date: _____

PUBLICITY/PRIVACY RELEASE

For good and valuable consideration, the receipt and sufficiency of which is hereby acknowledged, I hereby grant _____ permission to use, adapt, modify, reproduce, distribute, publicly perform and display, in any form now known or later developed, the Materials specified in this release (as indicated by my initials) throughout the world, by incorporating them into one or more Works and/or advertising and promotional materials relating thereto.

This release is for the following Materials [initial appropriate lines]:

_____ Name

_____ Voice

_____ Visual likeness (on photographs, video, film, etc.)

_____ Photographs, graphics, or other artwork as specified: _____

_____

_____ Film, videotape, or other audiovisual materials as specified: _____

_____

_____ Music or sound recordings as specified: _____

_____

_____ Other:

_____

I warrant and represent that the Materials identified above are either owned by me and/or are original to me and/or that I have full authority from the owner of the Materials to grant this release.

I release _____, its agents, employees, licensees, and assigns from any and all claims I may have now or in the future for invasion of privacy, right of publicity, copyright infringement, defamation, or any other cause of action arising out of the use, reproduction, adaptation, distribution, broadcast, performance, or display of the Works.

I waive any right to inspect or approve any Works that may be created containing the Materials.

I understand and agree that _____ is and shall be the exclusive owner of all right, title, and interest, including copyright, in the Works, and any advertising or promotional materials containing the Materials, except as to preexisting rights in any of the Materials released hereunder.

I am of full legal age and have read this release and am fully familiar with its contents.

_____
Signature

_____
Typed or Printed Name

_____
Date

INVOICE

_____

_____

_____

_____

Date: _____

Invoice number: _____   Order number: _____

Terms: _____   Time period of: _____

To: _____

Services: _____   _____

_____   _____

_____   _____

_____   _____

_____   _____

Material costs: _____   _____

_____   _____

_____   _____

_____   _____

_____   _____

_____   _____

Expenses: _____   _____

_____   _____

_____   _____

_____   _____

TOTAL AMOUNT OF THIS INVOICE: _____

Signed by: _____

Index

CATALOG

...more from Nolo

BUSINESS

	PRICE	CODE
Becoming a Mediator: Your Guide to Career Opportunities	$29.99	BECM
Business Buyout Agreements (Book w/CD-ROM)	$49.99	BSAG
The CA Nonprofit Corporation Kit (Binder w/CD-ROM)	$69.99	CNP
California Workers' Comp: How to Take Charge When You're Injured on the Job	$34.99	WORK
The Complete Guide to Buying a Business	$24.99	BUYBU
The Complete Guide to Selling Your Business	$24.99	SELBU
Consultant & Independent Contractor Agreements (Book w/CD-ROM)	$29.99	CICA
The Corporate Records Handbook (Book w/CD-ROM)	$69.99	CORMI
Create Your Own Employee Handbook (Book w/CD-ROM)	$49.99	EMHA
Dealing With Problem Employees	$44.99	PROBM
Deduct It! Lower Your Small Business Taxes	$34.99	DEDU
Effective Fundraising for Nonprofits	$24.99	EFFN
The Employer's Legal Handbook	$39.99	EMPL
Federal Employment Laws	$49.99	FELW
Form Your Own Limited Liability Company (Book w/CD-ROM)	$44.99	LIAB
Home Business Tax Deductions: Keep What You Earn	$34.99	DEHB
How to Create a Noncompete Agreement (Book w/CD-ROM)	$44.95	NOCMP
How to Form a California Professional Corporation (Book w/CD-ROM)	$59.99	PROF
How to Form a Nonprofit Corporation (Book w/CD-ROM)—National Edition	$49.99	NNP
How to Form a Nonprofit Corporation in California (Book w/CD-ROM)	$49.99	NON
How to Form Your Own California Corporation (Binder w/CD-ROM)	$59.99	CACI
How to Form Your Own California Corporation (Book w/CD-ROM)	$34.99	CCOR
How to Get Your Business on the Web	$29.99	WEBS
How to Run a Thriving Business: Strategies for Success & Satisfaction	$19.99	THRV

Prices subject to change.

	PRICE	CODE
How to Write a Business Plan	$34.99	SBS
Incorporate Your Business	$49.99	NIBS
The Independent Paralegal's Handbook	$34.99	PARA
Legal Guide for Starting & Running a Small Business	$34.99	RUNS
Legal Forms for Starting & Running a Small Business (Book w/CD-ROM)	$29.99	RUNSF
LLC or Corporation?	$24.99	CHENT
The Manager's Legal Handbook	$39.99	ELBA
Marketing Without Advertising	$20.00	MWAD
Mediate, Don't Litigate	$24.99	MEDL
Music Law (Book w/CD-ROM)	$34.99	ML
Negotiate the Best Lease for Your Business	$24.99	LESP
Nolo's Guide to Social Security Disability	$29.99	QSS
Nolo's Quick LLC	$29.99	LLCQ
Nondisclosure Agreements (Book w/CD-ROM)	$39.95	NAG
The Partnership Book: How to Write a Partnership Agreement (Book w/CD-ROM)	$39.99	PART
The Performance Appraisal Handbook	$29.99	PERF
The Small Business Start-up Kit (Book w/CD-ROM)	$29.99	SMBU
The Small Business Start-up Kit for California (Book w/CD-ROM)	$24.99	OPEN
Starting & Running a Successful Newsletter or Magazine	$29.99	MAG
Tax Savvy for Small Business	$36.99	SAVVY
Workplace Investigations: A Step by Step Guide	$39.99	CMPLN
Working for Yourself: Law & Taxes for Independent Contractors, Freelancers & Consultants	$39.99	WAGE
Working With Independent Contractors (Book w/CD-ROM)	$29.99	HICI
Your Crafts Business: A Legal Guide (Book w/CD-ROM)	$26.99	VART
Your Limited Liability Company: An Operating Manual (Book w/CD-ROM)	$49.99	LOP
Your Rights in the Workplace	$29.99	YRW

CONSUMER

How to Win Your Personal Injury Claim	$29.99	PICL
Nolo's Encyclopedia of Everyday Law	$29.99	EVL

	PRICE	CODE
Nolo's Guide to California Law	$24.99	CLAW

ESTATE PLANNING & PROBATE

	PRICE	CODE
8 Ways to Avoid Probate	$19.99	PRAV
Estate Planning Basics	$21.99	ESPN
The Executor's Guide: Settling a Loved One's Estate or Trust	$34.99	EXEC
How to Probate an Estate in California	$49.99	PAE
Make Your Own Living Trust (Book w/CD-ROM)	$39.99	LITR
Nolo's Simple Will Book (Book w/CD-ROM)	$36.99	SWIL
Plan Your Estate	$44.99	NEST
Quick & Legal Will Book	$16.99	QUIC
Special Needs Trust: Protect Your Child's Financial Future	$34.99	SPNT

FAMILY MATTERS

	PRICE	CODE
Building a Parenting Agreement That Works	$24.99	CUST
The Complete IEP Guide	$34.99	IEP
Divorce & Money: How to Make the Best Financial Decisions During Divorce	$34.99	DIMO
Do Your Own California Adoption: Nolo's Guide for Stepparents and Domestic Partners (Book w/CD-ROM)	$34.99	ADOP
Every Dog's Legal Guide: A Must-Have for Your Owner	$19.99	DOG
Get a Life: You Don't Need a Million to Retire Well	$24.99	LIFE
The Guardianship Book for California	$39.99	GB
A Legal Guide for Lesbian and Gay Couples	$34.99	LG
Living Together: A Legal Guide (Book w/CD-ROM)	$34.99	LTK
Living Wills and Powers of Attorney in California (Book w/CD-ROM)	$21.99	CPOA
Nolo's IEP Guide: Learning Disabilities	$29.99	IELD
Prenuptial Agreements: How to Write a Fair & Lasting Contract (Book w/CD-ROM)	$34.99	PNUP
Using Divorce Mediation: Save Your Money & Your Sanity	$29.99	UDMD

	PRICE	CODE

GOING TO COURT

Beat Your Ticket: Go To Court & Win! (National Edition)	$21.99	BEYT
The Criminal Law Handbook: Know Your Rights, Survive the System	$34.99	KYR
Everybody's Guide to Small Claims Court (National Edition)	$26.99	NSCC
Everybody's Guide to Small Claims Court in California	$29.99	CSCC
Fight Your Ticket & Win in California	$29.99	FYT
How to Change Your Name in California	$34.99	NAME
How to Collect When You Win a Lawsuit (California Edition)	$29.99	JUDG
The Lawsuit Survival Guide	$29.99	UNCL
Nolo's Deposition Handbook	$29.99	DEP
Represent Yourself in Court: How to Prepare & Try a Winning Case	$34.99	RYC
Win Your Lawsuit: A Judge's Guide to Representing Yourself in California Superior Court	$29.99	SLWY

HOMEOWNERS, LANDLORDS & TENANTS

California Tenants' Rights	$27.99	CTEN
Deeds for California Real Estate	$24.99	DEED
Every Landlord's Legal Guide (National Edition, Book w/CD-ROM)	$44.99	ELLI
Every Landlord's Tax Deduction Guide	$34.99	DELL
Every Tenant's Legal Guide	$29.99	EVTEN
For Sale by Owner in California	$29.99	FSBO
How to Buy a House in California	$34.99	BHCA
The California Landlord's Law Book: Rights & Responsibilities (Book w/CD-ROM)	$44.99	LBRT
The California Landlord's Law Book: Evictions (Book w/CD-ROM)	$44.99	LBEV
Leases & Rental Agreements	$29.99	LEAR
Neighbor Law: Fences, Trees, Boundaries & Noise	$26.99	NEI
The New York Landlord's Law Book (Book w/CD-ROM)	$39.99	NYLL
New York Tenants' Rights	$27.99	NYTEN
Renters' Rights (National Edition)	$24.99	RENT

	PRICE	CODE

IMMIGRATION

Becoming A U.S. Citizen: A Guide to the Law, Exam and Interview	$24.99	USCIT
Fiancé & Marriage Visas (Book w/CD-ROM)	$44.99	IMAR
How to Get a Green Card	$29.99	GRN
Student & Tourist Visas	$29.99	ISTU
U.S. Immigration Made Easy	$44.99	IMEZ

MONEY MATTERS

101 Law Forms for Personal Use (Book w/CD-ROM)	$29.99	SPOT
Bankruptcy: Is It the Right Solution to Your Debt Problems?	$21.99	BRS
Chapter 13 Bankruptcy: Repay Your Debts	$36.99	CHB
Credit Repair (Book w/CD-ROM)	$24.99	CREP
Getting Paid: How to Collect From Bankrupt Debtors	$29.99	CRBNK
How to File for Chapter 7 Bankruptcy	$29.99	HFB
IRAs, 401(k)s & Other Retirement Plans: Taking Your Money Out	$34.99	RET
Solve Your Money Troubles	$19.99	MT
Stand Up to the IRS	$29.99	SIRS
Surviving an IRS Tax Audit	$24.95	SAUD
Take Control of Your Student Loan Debt	$26.95	SLOAN

PATENTS AND COPYRIGHTS

All I Need is Money: How to Finance Your Invention	$19.99	FINA
The Copyright Handbook: How to Protect and Use Written Works (Book w/CD-ROM)	$39.99	COHA
Copyright Your Software (Book w/CD-ROM)	$34.95	CYS
Getting Permission: How to License and Clear Copyrighted Materials Online and Off (Book w/CD-ROM)	$34.99	RIPER
How to Make Patent Drawings	$29.99	DRAW

	PRICE	CODE
The Inventor's Notebook	$24.99	INOT
License Your Invention (Book w/CD-ROM)	$39.99	LICE
Nolo's Patents for Beginners	$29.99	QPAT
Patent, Copyright & Trademark	$39.99	PCTM
Patent It Yourself...	$49.99	PAT
Patent Pending in 24 Hours	$29.99	PEND
Patenting Art & Entertainment: New Strategies for Protecting Creative Ideas	$39.99	PATAE
The Public Domain	$34.99	PUBL
Trademark: Legal Care for Your Business and Product Name	$39.99	TRD
Web and Software Development: A Legal Guide (Book w/ CD-ROM)	$44.99	SFT
What Every Inventor Needs to Know About Business & Taxes (Book w/CD-ROM)	$21.99	ILAX

RESEARCH & REFERENCE

Legal Research: How to Find & Understand the Law	$39.99	LRES

SENIORS

Long-Term Care: How to Plan & Pay for It	$21.99	ELD
Social Security, Medicare & Goverment Pensions	$29.99	SOA

SOFTWARE

**Call or check our website at www.nolo.com
for special discounts on Software!**

	PRICE	CODE
Incorporator Pro	89.99	STNC1
LLC Maker—Windows	$89.95	LLP1
Patent Pending Now!	$199.99	PP1
PatentEase—Windows	$349.00	PEAS
Personal RecordKeeper 5.0 CD—Windows	$59.95	RKD5
Quicken Legal Business Pro 2006—Windows	$109.99	SBQB6
Quicken WillMaker Plus 2006—Windows	$79.99	WQP6

Special Upgrade Offer

Save 35% on the latest edition of your Nolo book

Because laws and legal procedures change often, we update our books regularly. To help keep you up-to-date, we are extending this special upgrade offer. Cut out and mail the title portion of the cover of your old Nolo book and we'll give you **35% off** the retail price of the NEW EDITION of that book when you purchase directly from Nolo. This offer is to individuals only.

Call us today at 1-800-728-3555

Prices and offer subject to change without notice.

Order Form

Name _____

Address _____

City _____

State, Zip _____

Daytime Phone _____

E-mail _____

Item Code	Quantity	Item	Unit Price	Total Price

Subtotal	
Add your local sales tax (California only)	
Shipping: RUSH $9, Basic $5 (See below)	
"I bought 3, ship it to me FREE!"(Ground shipping only)	
TOTAL	

Method of payment

☐ Check ☐ VISA ☐ MasterCard
☐ Discover Card ☐ American Express

Account Number _____

Expiration Date _____

Signature _____

Shipping and Handling

Rush Delivery—Only $9

We'll ship any order to any street address in the U.S. by UPS 2nd Day Air* for only $9!

* Order by noon Pacific Time and get your order in 2 business days. Orders placed after noon Pacific Time will arrive in 3 business days. P.O. boxes and S.F. Bay Area use basic shipping. Alaska and Hawaii use 2nd Day Air or Priority Mail.

Basic Shipping—$5

Use for P.O. Boxes, Northern California and Ground Service.

Allow 1-2 weeks for delivery. U.S. addresses only.

For faster service, use your credit card and our toll-free numbers

**Call our customer service group
Monday thru Friday 7am to 7pm PST**

Phone	1-800-728-3555
Fax	1-800-645-0895
Mail	Nolo 950 Parker St. Berkeley, CA 94710

Order 24 hours a day @
www.nolo.com

Remember:

Little publishers have big ears.
We really listen to you.

Take 2 Minutes & Give Us Your 2 cents

Your comments make a big difference in the development and revision of Nolo books and software. Please take a few minutes and register your Nolo product—and your comments—with us. Not only will your input make a difference, you'll receive special offers available only to registered owners of Nolo products on our newest books and software. Register now by:

PHONE
1-800-728-3555

FAX
1-800-645-0895

EMAIL
cs@nolo.com

or **MAIL** us
this registration card

fold here

Registration Card

NAME _____ DATE _____

ADDRESS _____

_____

CITY _____ STATE _____ ZIP _____

PHONE _____ E-MAIL _____

WHERE DID YOU HEAR ABOUT THIS PRODUCT? _____

WHERE DID YOU PURCHASE THIS PRODUCT? _____

DID YOU CONSULT A LAWYER? (PLEASE CIRCLE ONE)   YES   NO   NOT APPLICABLE

DID YOU FIND THIS BOOK HELPFUL?   (VERY)   5   4   3   2   1   (NOT AT ALL)

COMMENTS _____

_____

_____

WAS IT EASY TO USE?   (VERY EASY)   5   4   3   2   1   (VERY DIFFICULT)

We occasionally make our mailing list available to carefully selected companies whose products may be of interest to you.

❑   If you do not wish to receive mailings from these companies, please check this box.

❑   You can quote me in future Nolo promotional materials.
     Daytime phone number _____.

CICA 5.0

Nolo in the NEWS

"Nolo helps lay people perform legal tasks without the aid—or fees—of lawyers."

—USA TODAY

Nolo books are ..."written in plain language, free of legal mumbo jumbo, and spiced with witty personal observations."

—ASSOCIATED PRESS

"...Nolo publications...guide people simply through the how, when, where and why of law."

—WASHINGTON POST

"Increasingly, people who are not lawyers are performing tasks usually regarded as legal work... And consumers, using books like Nolo's, do routine legal work themselves."

—NEW YORK TIMES

"...All of [Nolo's] books are easy-to-understand, are updated regularly, provide pull-out forms...and are often quite moving in their sense of compassion for the struggles of the lay reader."

—SAN FRANCISCO CHRONICLE

- - - - - - - - - - - - - - - - - - fold here - - - - - - - - - - - - - - - - - - -

Place
stamp here

Nolo
950 Parker Street
Berkeley, CA 94710-9867

Attn: CICA 5.0